QUANTITATIVE SEISMOLOGY

A Series of Books in Geology

EDITOR: Allan Cox

QUANTITATIVE SEISMOLOGY
Theory and Methods

VOLUME II

Keiiti Aki

MASSACHUSETTS INSTITUTE OF TECHNOLOGY

Paul G. Richards

COLUMBIA UNIVERSITY

W. H. FREEMAN AND COMPANY
SAN FRANCISCO

Sponsoring Editor: John H. Staples
Manuscript Editor: Dick Johnson
Designer: Marie Carluccio
Production Coordinator: Linda Jupiter
Illustration Coordinator: Cheryl Nufer
Artist: Catherine Brandel
Compositor: Syntax International
Printer: The Maple-Vail Book Manufacturing Group

Library of Congress Cataloging in Publication Data

Aki, Keiiti, 1930–
    Quantitative seismology.
    (A Series of books in geology)

    Includes bibliographies and index.
    1. Seismology—Mathematics. I. Richards, Paul G.,
1943–   joint author. II. Title.
QE539.A64   551.2′2   79-17434
ISBN 0-7167-1059-5(v. 2)

Printed in the United States of America

1 2 3 4 5 6 7 8 9

Contents

VOLUME I

Preface

In the past decade, seismology has matured as a quantitative science through an extensive interplay between theoretical and experimental workers. Several specialized journals recorded this progress in thousands of pages of research papers, yet such a forum does not bring out key concepts systematically. Because many graduate students have expressed their need for a textbook on this subject and because many methods of seismogram analysis now used almost routinely by small groups of seismologists have never been adequately explained to the wider audience of scientists and engineers who work in the peripheral areas of seismology, we have here attempted to give a unified treatment of those methods of seismology which are currently used in interpreting actual data.

We develop the theory of seismic wave propagation in realistic Earth models. We study specialized theories of fracture and rupture propagation as models of an earthquake, and we supplement these theoretical subjects with practical descriptions of how seismographs work and how data are analyzed and inverted.

Our text is arranged in two volumes. Volume I gives a systematic development of the theory of seismic-wave propagation in classical Earth models, in which material properties vary only with depth. It concludes with a chapter on seismometry. This volume is intended to be used as a textbook in basic courses for advanced students of seismology. Volume II summarizes progress made in the major frontiers of seismology during the past decade. It covers a wide range of special subjects, including chapters on data analysis and inversion, on successful methods for quantifying wave propagation in media varying laterally (as well as with depth), and on the kinematic and dynamic aspects of motions near a fault

plane undergoing rupture. The second volume may be used as a textbook in graduate courses on tectonophysics, earthquake mechanics, inverse problems in geophysics, and geophysical data processing.

Many people have helped us. Armando Cisternas worked on the original plan for the book, suggesting part of the sequence of subjects we eventually adopted. Frank Press's encouragement was a major factor in getting this project started. Chapter 12, on inverse problems, grew out of a course given at M.I.T. by one of the authors and Theodore R. Madden, to whom we are grateful for many helpful discussions. Our students' Ph.D. theses have taught us much of what we know, and have been freely raided. We drew upon the explicit ideas and results of several hundred people, many of them colleagues, and hope their contributions are correctly acknowledged in the text. Here, we express our sincere thanks.

Critical readings of all or part of the manuscript were undertaken by Roger Bilham, Jack Boatwright, David Boore, Roger Borcherdt, Michel Bouchon, Arthur Cheng, Tom Chen, Wang-Ping Chen, Bernard Chouet, George Choy, Vernon Cormier, Allan Cox, Shamita Das, Bill Ellsworth, Mike Fehler, Neil Frazer, Freeman Gilbert, Neal Goins, Anton Hales, David Harkrider, Lane Johnson, Bruce Julian, Gerry LaTorraca, Wook Lee, Dale Morgan, Bill Menke, Gerhard Müller, Albert Ng, Howard Patton, Steve Roecker, Tony Shakal, Euan Smith, Teng-Fong Wong, Mai Yang, and George Zandt. We appreciate their attention, their advice, and their encouragement.

About fifteen different secretaries typed for us over the four years during which we prepared this text. Linda Murphy at Lamont-Doherty carried the major burden, helping us to salvage some self-respect in the way we handled deadlines. We also thank our manuscript editor, Dick Johnson, for his sustained efforts and skill in clarifying the original typescript.

We acknowledge support from the Alfred P. Sloan Foundation and the John Simon Guggenheim Memorial Foundation (P.G.R.). This book could not have been written without the support given to our research projects over the years by several funding agencies: the U.S. Geological Survey and the Department of Energy (K.A.); the Advanced Research Projects Agency, monitored through the Air Force Office of Scientific Research (K.A. and P.G.R.); and the National Science Foundation (K.A. and P.G.R.).

Keiiti Aki
Paul G. Richards

June 1979

Introduction

The practicing seismologist needs a thorough grasp of wave-propagation theory for media in which properties vary quite strongly with depth. He can often model sources adequately by imagining them to act solely at a point within the medium. Our first volume covers these basic subjects and describes how seismic data are acquired with a variety of instruments. Building upon this material, the present volume describes certain special subjects that have been responsible for the great progress made in seismology during the past decade.

We begin here with an account of how seismic data are analyzed and inverted to obtain model parameters. Thus, in Chapter 11 (numbers continue on from the 10 chapters of Volume I), seismic data are considered to be composed of signal and noise. A signal is the part of the data that can be used to gain some information about the Earth and seismic source. After a historical sketch of how discoveries were made with each introduction of a new class of data, we describe various methods for improving the quality of data, i.e., the signal-to-noise ratio.

In Chapter 12, we come to one of the ultimate tasks of seismology: inversion of data to determine the structure of the Earth's interior. We first consider the travel-time data and describe the classic Herglotz-Wiechert method that, by 1910, had established the gross structure of the Earth, with a solid mantle and liquid core. We then describe the extension of the method by Gerver and Markushevitch (1966, 1967) to cases in which low-velocity layers exist.

Another inverse problem that has been completely solved uses as data the reflection seismogram for vertically incident waves in a vertically heterogeneous medium. We find

that we can recover the impedance as a function of vertical travel time from the reflection seismogram, in contrast to the seismic velocity as a function of depth recovered from the travel-time data.

Exact solutions are unknown for any inverse problem in seismology other than the above two simplified problems. More generally, the inverse solution may be obtained by a linearization-iteration process: starting with an initial model; solving for the forward problem for the initial model; setting up linear equations for the residual between observable and calculated values for the initial model, in terms of a perturbation in model parameters; solving the linear equations for the perturbation; revising the initial model; and then repeating the whole process. This method is applicable to almost any seismological data, including surface-wave phase velocities and free-oscillation periods, and has been widely used by seismologists since the stimulating work of Backus and Gilbert (1967, 1968, 1970), who introduced, among other things, the concept of the resolution kernel as a measure of nonuniqueness of solution. Chapter 12 concludes with the application of the linear-inversion method to the determination of three-dimensional seismic structure of the lithosphere under a large-aperture seismic array, using teleseismic travel-time data.

The methods of calculating seismograms for two- and three-dimensionally heterogeneous Earth models are in the developmental stage. These methods are clearly in demand, not only for such practical purposes as exploration of natural resources, but also for purely scientific motivations, such as studying the splitting of spectral peaks of free oscillations due to lateral heterogeneity in the Earth. Even in the deep Earth, the departure from lateral homogeneity is becoming a crucial problem. Consider, for example, the GH-branch of the body wave PKP, a branch that Bolt (1962, 1964) proposed as being due to a discontinuity in the Earth's fluid core, about 500 km above the inner-core/outer-core boundary. Several other seismologists developed core models of this type to explain the data, but such discontinuities were questioned by Doornbos and Husebye (1972), who studied the speed with which these waves were observed to sweep across large-aperture arrays of seismometers. King et al. (1974) give convincing evidence in support of Haddon's (1972) theory that lateral heterogeneity near the core-mantle boundary is responsible for the GH-branch. We see here a clear demonstration of danger in forcing a simple one-dimensional framework on the real Earth, since this may result in a false vertical structure.

In Chapter 13, we describe several attempts to allow two- and three-dimensional flexibility in our Earth models. A natural extension of the preceding chapters is to start with an initial model of a vertically heterogeneous Earth model and introduce small perturbations in the medium property or in the depth of interface. The problems involving seismic scattering in three-dimensionally heterogeneous media are classified according to the scale of inhomogeneity and the travel distance relative to the wavelength. Problem areas are identified with applicable methods, such as ray-theory, perturbation-theory, finite-difference, finite-element, and other numerical methods (see Fig. 13.11). Complicated problems defying a deterministic approach are treated by the use of a random medium.

The models in seismology are essentially mathematical. The physics involved in seismology is rather simple, mostly contained in the equation of motion, Hooke's law, and a few other constitutive relations. The challenge to a seismologist is in reducing the observed complex vector-wave phenomenon in three space dimensions with wiggly temporal variation in an orderly manner to a description of wave source and propagation medium. It is therefore very important to have an adequate model of the source of seismic waves. Chapters 14 and 15 are devoted, respectively, to the kinematic and dynamic models of

an earthquake fault. In the kinematic model, we study the relation between the fault-slip function and seismic radiation in far field and near field. We find that sources of finite spatial extent can in practice have seismic radiation differing from that emanating from a point source, even for receivers at great distances from the source. In the dynamic model, the slip function is derived from the initial condition of tectonic stress and from frictional and cohesive properties of the fault zone. These models are important for the study of earthquake source mechanisms and current tectonic activities in the Earth. They will also be useful for the practical purpose of predicting earthquake strong motions for an active fault.

QUANTITATIVE SEISMOLOGY

CHAPTER 11

Analysis of Seismological Data

The record of Earth motion obtained by a seismograph contains information about the nature of the seismic source that generated the motion as well as about properties of the media through which the seismic disturbances were propagated. Our problem now is to extract such information from the observed record. The methods of inverting data to find the properties of source and media will be the subject of Chapter 12. We shall be concerned, in the present chapter, with the preparation of data to be used in the inversion. Our task is to improve the quality of data as much as possible in order to learn about the Earth's interior with accuracy and in great detail.

11.1   Seismological Data and Associated Discoveries

The data collected by seismologists are quite diversified because of (i) the variety of seismic sources; (ii) the great range in source size, travel distance, and receiver-array dimension; and (iii) the different types of measurements made with seismic waves.

Natural sources of seismic waves include earthquakes, volcanic eruptions, winds, waves in the ocean, and meteorite impacts. The simplest and most

common artificial sources are buried explosives, but other energy sources, such as air guns, gas exploders, electric sparkers, and hydraulic shakers, are becoming increasingly more popular. These sources provide a remarkable range of intensity. The equivalent point force of the largest earthquake is about 10^{18} times larger than point forces of the smallest earthquakes investigated in microseismicity studies. The amount of explosives used in seismic-wave generation ranges from a few grams to more than a megaton (a factor of 10^{12}). The size of a seismograph network ranges from less than 100 meters for probing the basement rock for an engineering structure to more than 10,000 km for studying earthquake source mechanisms and the Earth's deep interior (a factor of 10^6). Seismic signals range in frequency from 0.0001 to 100 Hz (a factor of 10^6).

With these great ranges in signal frequency, receiver network dimension, and source size, we have a tremendous variety of data sets to be analyzed.

We shall describe the contents of seismological data in more detail, separately for the natural source and for the artificial source, by reviewing historically how new data were created and how they led to major discoveries in seismology.

11.1.1 Natural sources

Anyone who looks, for the first time, at a seismogram marked by seismologists with symbols such as P, PP, PPP, PcP, S, SS, ScS, etc., wonders how they are identified. The founders of instrumental seismology also puzzled over wiggles on the seismogram. (See Dewey and Byerly, 1969, for a detailed survey on the inventors of early seismographs.) In March 1881, when the first seismogram of a local earthquake was obtained by means of a horizontal pendulum at the University of Tokyo, Ewing (1881) and Milne (1881) disagreed about whether the initial wave group was a compressional or a shear wave. Within 20 years, however, most people (with an important exception, as mentioned in the first chapter of volume I) agreed about the identification of P, S, and surface waves. The first recording of a distant earthquake was made at Potsdam by Rebeur-Paschwitz on April 17, 1889, for an earthquake in Japan, and the first travel-time curve was published by Oldham in 1900.

The initial global network of seismograph stations was set up through the efforts of Milne by the Committee on Seismology, which was formed in 1896 by the British Association for the Advancement of Science. The International Seismological Association was founded in 1903, and recordings and station bulletins were improved. Some stations reported not only the arrival times but also amplitudes of P, S, etc., which were later used by Gutenberg and Richter (1954) in the determination of magnitude. For example, they were able to use amplitude data from about 30 station bulletins for earthquakes dating back as far as 1910.

The year 1918 marked the start of the International Seismological Summary (ISS), which contained the data sent from world-wide stations, including the

origin time and epicenter determined from those data. The number of stations reporting to the ISS in 1951 was 602, of which 108 were in Japan. In 1975, about 850 stations were reporting. The ISS was succeeded by publications of the International Seismological Center (ISC), situated in Edinburgh until 1975, and subsequently in Newbury, England. The bulletin of the ISC has been published since 1964. A magnetic tape image of the bulletin is available upon request. The data from 1964 to 1970 are stored in 15 tapes, each 2400 feet long with a density of 1600 bits per inch (total 700 megabits). At present, data composed of 50,000 to 100,000 observations are processed each month. A 2-year wait is necessary before data from some remote key stations, such as those in the South Pacific, can be processed. The ISC also publishes a semiannual bibliography of seismology.

An efficient epicenter location program has been operated for many years by the U.S. Coast and Geodetic Survey (later Environmental Science Services Administration, NOAA, now U.S. Geological Survey), which publishes the preliminary determination of epicenters (PDE) card. They receive about 40,000 readings per month from more than 700 stations around the world and distribute the PDE cards to 800 stations and research centers.

The quality of travel-time data in these bulletins has been improved with time by an iterative process. In the beginning, the location of an epicenter was determined by macroseismic effects observed by human eyes, and the origin time by the starting time of shaking in the epicentral area. The travel-time curve must initially be constructed using travel times measured from the known origin time and distances to the stations from the known epicenter. For each new earthquake, there are four unknown source parameters (longitude, latitude, depth of focus, and origin time). If the travel-time curve is known, a single P-wave identified at four stations can, in principle, determine these parameters. Therefore, the number of stations in excess of four supply additional information for revising the travel-time curve.

For example, Jeffreys and Bullen started with the Zöppritz-Turner table and calculated source parameters of each earthquake using data reported in the ISS. The residual of the observed travel time relative to the calculated one, based on the calculated source parameters and the initial table, is examined as a function of epicentral distance. Any systematic error found is then used as a correction to the initial table. The Jeffreys-Bullen table, published in 1940, was one of the greatest accomplishments in classic seismology based on arrival-time data. An attempt to revise the Jeffreys-Bullen table by a group headed by E. Herrin (1968) resulted in a new table for P-waves.

When a travel-time curve is found, it is natural to seek an Earth structure to explain the observations. The first quantitative Earth model was constructed by Wiechert and his colleagues around 1910. Some of their results are still valid. For example, the value of depth to the core-mantle boundary (2900 km) given by Gutenberg in 1913 is very close to the latest estimates (2885–2890 km).

Once an Earth model is specified, theoretical travel times can be calculated for a number of ray paths that have not been considered previously and can

then be compared with observations. By this process, new phases are identified, and the residual of observed time relative to the predicted time is used to revise the Earth model further.

Thus we have described an iterative process, starting with the identification of a wave type, followed by measurement of its time, epicenter location, residual calculation, correction to the travel-time table, and revision of the Earth model, leading to the identification of new wave types. With each cycle of iteration, our knowledge of the Earth's interior and the earthquake source is improved.

A dramatic event occurred in the study of earthquake sources in the late 1920's: the discovery of deep-focus earthquakes. Earlier, Turner (1922) had published a result that demonstrated great variability of focal depth. This result, based on data obtained during the preparation of the ISS, failed to convince many people, because he had included a so-called high-focus earthquake (in the air). Eventually, however, convincing evidence was obtained by Wadati (1928), who took full advantage of the dense network of stations spread over the Japanese islands. Using data from about 50 of these stations, he was able to draw a picture of the wavefronts of both *P*- and *S*-waves spreading from the epicenter. Comparing two events sharing a common epicenter, he found that the wavefront spreading was more than twice as fast for one event than for the other. The only explanation for this observation is a great difference in focal depth for the two events.

The temporal and spatial distribution of numerous small earthquakes in a tectonically active area have been recognized as basic data for understanding current tectonic processes ever since Asada (1957) reported more than 200 shocks per day for a normally active period at Tsukuba, Japan. His microearthquake records were obtained with a moving-coil transducer coupled to an electronic amplifier (which was not commonly used in earthquake seismology at that time). Approximately 50 permanent networks of seismograph stations that now exist throughout the world produce a large amount of data on the time and amplitude of waves from microearthquakes in each area. The most extensive network of this type has been operated by the U.S. Geological Survey in Central California since 1966. In this network, seismic signals from about 260 stations (in 1979) are telemetered via commercial telephone line to the headquarters at Menlo Park. For most shocks, the accuracies of epicenter location and focal depth are better than 2.5 and 5 km, respectively. The network is intended to monitor any phenomena that might precede a large earthquake. Travel-time data from local and teleseismic events are also used in the study of velocity structure in crust and upper mantle under the network.

Another important discovery was made in the early days of seismology using data from a dense network of stations. The first convincing quadrantal pattern of the sense of first motion was observed by Shida for a Japanese earthquake in 1917 May 18. Both theoretical and experimental studies of earthquake source mechanism were vigorously pursued by a number of Japanese seismologists around 1930. Honda (1962) made a complete summary of earlier work as well

as his own, including the result that the axes of the greatest or least principal stress for deep-focus earthquakes were parallel to the plane of the seismic zone, now a familiar picture of stress in the downgoing slab.

Global data on first-motion patterns were studied extensively by Hodgson and his colleagues in the 1950's (Hodgson, 1957), using Byerly's method of projecting the focal sphere on the map of seismograph stations. Since the sense of first motion is normally not reported in station bulletins, they collected data by questionnaires addressed to the station seismologists. The data collected in this way were apparently inadequate. Typically, 15 to 20 percent of the data were inconsistent with the final solution. A remarkable reduction in the inconsistency (to less than 1%) was accomplished by the introduction of new data in the 1960's. The new data were obtained by having the same observer pick the first motion in a consistent manner directly from the records of standardized long-period seismographs of the WWSSN (World-Wide Standard Seismograph Network). This method led Sykes (1967) to a confirmation of Wilson's idea of the transform fault, which revolutionized our thinking on earthquake tectonics by uniting three distinct types of earthquake zones (ridge, arc, and block tectonic (Richter, 1958)) into one global system. The result also remarkably increased the credibility of fault-plane solutions.

The WWSSN was created during the early 1960's. This network consists of some 120 continuously recording stations distributed over the world as shown in Figure 11.1. It is by far the finest general-purpose global system of seismograph stations ever operated. Each of these stations is equipped with three short-period and three long-period seismographs, allowing the measurement of vertical and two horizontal components of ground motion. The instruments consist of moving-coil pendulums coupled with recording galvanometers. The free periods of the pendulum and galvanometer are 1 and 0.75 sec, respectively, for the short-period seismograph, and 15 or 30 and 100 sec for the long-period seismograph. The peak magnification of the short-period system can be varied from 3125 to 400,000 in 6-db steps. The long-period magnification ranges from 750 to 6000. Each station produces six sheets of record (29.2 by 91.4 cm) per day. A data center (National Geophysical and Solar-Terrestrial Data Center, NOAA, at Boulder, Colorado) collects the records and distributes copies to anyone requesting them; they are available in the full-size form, in 35-mm film rolls, in 70-mm film chips, and in a microfiche (10.5 × 15 cm) containing 24 seismograms.

The real impact of WWSSN data is only beginning to show up. The data we have so far discussed in this chapter are time, amplitude, and sense of onset of a particular wave packet, the rest of the seismogram being neglected. The easy availability of well-calibrated seismograms on a global scale has opened up the wide possibility of using the waveform or spectrum of a wave group or even the entire record in studies of the Earth's structure and seismic source mechanisms. This is making seismology a truly quantitative science. We shall refer to this as the wave-theoretical approach as opposed to the ray-theoretical

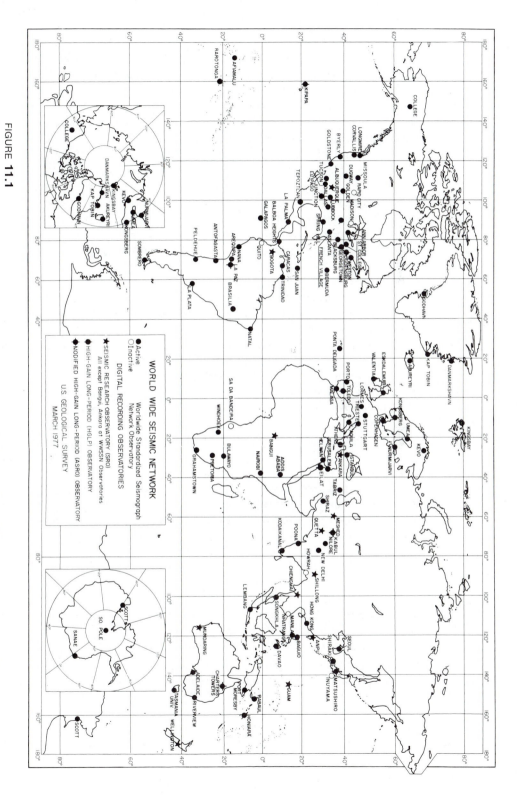

FIGURE **11.1**

Stations of the World Wide Standard Seismograph Network (WWSSN). [From U.S. Geological Survey; 1977.]

approach, in which only the time and amplitude of a wave packet are used as data.

The wave-theoretical approach in earthquake seismology has a long history, dating back to Lamb's (1904) classic paper on seismic wave generation by a point force in a half-space. Ewing, Jardetzky, and Press (1957) summarized the theoretical results and discussed their application to various experiments. The successful application described in their book, together with the advent of the digital computer, encouraged the wave-theoretical approach in the 1960's. This approach has been productive mostly for long waves, because they are insensitive to fine details of the Earth's structure, and hence a simple model is applicable.

The simplest form of data in the wave-theoretical approach can be the unprocessed record itself. Synthetic seismograms calculated for a model that includes instrument response may then be directly compared with the observed record. The first successful case of such a comparison is probably Pekeris' (1948) work on acoustic waves in the ocean. The more common form of the data is a selected portion of the seismogram that is believed to consist of a small number of known types of waves, such as Rayleigh waves, Love waves, and their higher modes.

If the portion is primarily made up from a single mode, the observed wave $f(x, t)$ may be approximated by the simple expression

$$f(x, t) = \frac{1}{2\pi} \int_{-\infty}^{\infty} |f(x, \omega)| \exp\left[-i\omega\left(t - \frac{x}{c(\omega)} \right) + i\phi(\omega) \right] d\omega,$$

where t is time, x is the distance from the source to receiver, $\phi(\omega)$ is the source phase delay, $c(\omega)$ is the phase velocity, and $|f(x, \omega)|$ is the amplitude spectral density. This approximation was successfully applied by Satô (1955), and since has been used extensively in the measurement of attenuation, phase velocity, and source phase. Thus the amplitude and phase spectrum of a selected portion of a seismogram are now important data in seismology.

Press (1956) made the first phase-velocity measurement using records of the Benioff long-period seismographs (pendulum period 1 sec, galvanometer period 90 sec) at the southern California network operated by the California Institute of Technology. He was able to achieve the accuracy needed for a study of crustal structure under the network.

Satô (1958) made the first measurement of phase velocity and attenuation of surface waves making multiple round-trips along a complete great circle. He used a single station record obtained by another Benioff seismograph designed for measuring long-period waves.

Around 1960, when Aki (1960), Brune (1960), and Ben-Menahem (1961) initiated the study of source mechanisms using the amplitude and phase spectra of long-period surface waves, the main data were either the records of Benioff seismographs operated by the Seismological Laboratory, California Institute

of Technology, or those of a global network of long-period seismographs installed and operated during the International Geophysical Year (1957–1958) and thereafter by the Lamont-Doherty Geological Observatory of Columbia University, in cooperation with local institutions. The latter network consisted of about a dozen stations distributed over the world and may be considered as the prototype of the WWSSN.

In the early 1960's, the Fourier transform of the entire seismogram became an important set of data after Benioff et al. (1961), Ness et al. (1961), and Alsop et al. (1961a) discovered in the power spectra of the records of a 1960 Chilean earthquake spectral peaks that correspond to various modes of free oscillation of the whole Earth. Although most data used for the measurement of spectral peaks were from earthquakes with magnitudes around 8, modern low-noise seismographs, such as the Block-Moore accelerometer, can now produce data useful for free-oscillation analysis from medium-sized earthquakes.

Helmberger (1968) initiated a major advance in working with pulse shapes of body waves in the time domain, and Luh and Dziewonski (1975) showed that the whole wavetrain of long-period waves in the time domain could be regarded as the basic data set.

The WWSSN is an excellent source of the data required in the wave-theoretical approach (e.g., Müller, 1973). However, there are a few deficiencies that are being improved. Perhaps the most significant of these is the analogue recording adopted for the system. A comprehensive analysis of free-oscillation data for world-wide stations requires a few million bits of digitization, which means many months of manual labor for one earthquake. Digital recording is now being implemented to upgrade some of the WWSSN stations. About 20 digital recording stations are operated around the world (in 1979), including SRO and IDA stations described in Section 10.3.5. These digital data are also available from the same data center as the WWSSN data.

The superiority and effectiveness of digital recording were demonstrated by the Large Aperture Seismic Array (LASA), designed primarily for the purpose of detection and identification of distant underground nuclear explosions. The one in Montana, U.S., once included 525 short-period vertical seismographs (grouped into 21 subarrays) and 21 3-component long-period seismographs spread in an area about 200 km in diameter (Fig. 11.2). The signal was sampled at the rate of 20 per sec and digitized to a 15-bit word. The flow of information amounted to more than 150 kilobits per sec. (This data rate has been reduced in several stages until the array ceased operation.) The array data have been used to enhance teleseismic signals for measurement of phase slowness ($dT/d\Delta$), for estimation of wavenumber spectra, and for the study of structure beneath the array. We shall discuss the array data further, later in this chapter.

Other deficiencies of the WWSSN are its poor linearity and limited dynamic range. Large earthquakes make the galvanometer deflection go off scale or make the record trace invisible. Nearby small earthquakes may introduce a

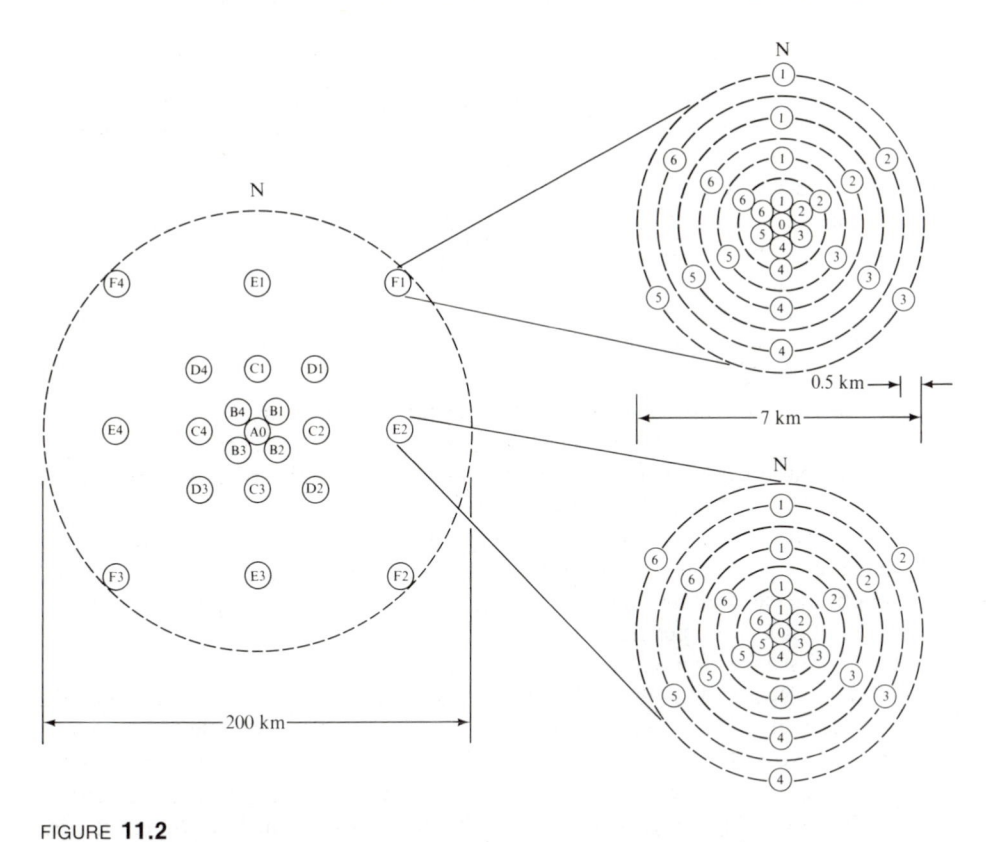

FIGURE **11.2**

The Large Aperture Seismic Array (LASA) in eastern Montana consisted of 21
subarrays, each with 25 short-period vertical seismographs and one three-component
long-period seismograph.

spurious long-period signal by nonlinear cross-coupling of the higher modes
of the instrument, which are excited by short-period strong motion.

Accelerographs specifically designed to register earthquake strong motion
have been extensively operated in several seismically active regions. A large
number of digitized accelerograms have been published, and the available data
are rapidly expanding.

Since the primary interest in accelerograms was in providing the basis for
designing earthquake-resistant structures, most accelerographs were installed
in buildings or in soil, and very rarely on hard rock. For the purpose of studying
the earthquake mechanism, accelerograms obtained from equipment installed
on harder rocks at shorter epicentral distances are most desirable because they
are less contaminated by the complex effects of wave path, such as scattering
and attenuation. An accelerogram obtained by the U.S. Coast and Geodetic

Survey for the Parkfield, California, earthquake of 1966 was the first to be analyzed wave-theoretically, by Aki (1968) and Haskell (1969), using the elastic dislocation model. Not only do the data from the near field offer information on the source parameters unattainable from the far-field data alone, but the wave-theoretical study of the near field will eventually provide us with the capability—much requested from earthquake engineers—of predicting the seismic motion due to an expected earthquake in a nearby fault zone.

11.1.2   Artificial sources

An earthquake source is the most powerful natural source of seismic waves, but it is not readily available. Because earthquakes are rare events, the operation of a continuously recording seismograph waiting for earthquakes is economically a very inefficient way of collecting useful data.

An artificial source allows the efficient collection of high-quality data, because one can often afford a high-accuracy measurement both in time and amplitude, as well as dense coverage over space, if the experiment can be finished in a short period of time.

The first measurement of seismic waves by the use of explosive sources was made by Robert Mallet (1848), who had a clear conception of the value of the seismic method as a geologic tool. In a paper entitled "On the dynamics of earthquakes: Being an attempt to reduce their observed phenomena to the known laws of wave motion in solids and fluids," he suggested the seismic method of making a geologic map of the ocean floor, which, although it constitutes three-fourths of the total surface of the Earth, was then almost totally unexplored. He stressed the necessity for experiments to measure seismic velocity accurately in order to establish a relation between velocity and rock formation.

He actually carried out such an experiment on a sand beach in Ireland and in granitic rock of a nearby island. He buried a few pounds of gun powder four or five feet under the ground and extended a wire circuit with a battery over a measured mile to the recording station. His seismograph was an eleven-power magnifier focused at cross-hairs through a reflecting surface of mercury in a container. The powder was fired by the experimenter, who detected the arrival of seismic waves when the image of cross-hairs was blurred or disappeared. The results of these experiments gave seismic velocities through sand of 824 feet/sec and through granite of 1306 and 1664 feet/sec. These velocities are low, and probably associated with surface waves. With the crude seismoscope, the experimenter must have missed the first motion.

There are two fundamentally different sets of seismic data acquired using artificial sources. One is obtained by refraction and the other by reflection. In the refraction method, we measure the time of arrival of waves as a function of distance from the source. In the reflection method, we measure the reflected

signal as a function of vertical travel time spent in a round trip from source to reflector to receiver. The former method gives information on the velocities of layers, the latter on the impedance contrasts at the layer boundaries (see Chapter 12).

Clearly, Robert Mallet was the discoverer of the refraction method. In an excellent review on the early history of applied geophysics, DeGolyer (1935) gives credit to Reginald Fessenden for discovering the reflection method. In 1914, Fessenden filed a patent application for determining ocean depths by acoustic methods. In 1917, he filed another application on methods and apparatus for locating ore bodies, which covers clearly both refraction and reflection as used to locate geologic formations. The first discovery of an oil field by the seismic method was made in 1924 in Fort Bend County, Texas, by a crew sent from a company headed by L. Mintrop of Germany.

The first refraction measurements at sea were made in October 1935 by Ewing, Crary, and Rutherford (1937) on the ship *Atlantis* of the Woods Hole Oceanographic Institution, at four stations on a line extending south from Woods Hole, Massachusetts. In these early measurements, a small whale boat was used to fire explosives. The ship carrying the recording apparatus was anchored, and geophones were lowered to the ocean bottom by means of cables that connected them with a photographically recording oscillograph. The geophones were vertical pendulums with natural frequency 35 Hz. Their output was fed to the 90-Hz galvanometers of the oscillograph. Time marks were placed on the record by a 50-Hz tuning fork.

The first deep-water refraction profile was obtained by Ewing et al. (1950) in February, 1949, using two ships, *Atlantis* and *Caryn*, both from Woods Hole. The detector was a hydrophone (pressure detector) floated at a depth of 50 feet (below the depth of disturbance by wind action) at a distance of 100 feet from the receiver ship *Caryn*. Shots were set off from the *Atlantis* as it moved along the line of refraction measurements. This and subsequent work by Ewing and his colleagues revealed one of the most striking features of the Earth's structure: a very thin crust (about 7 km) under the ocean as compared to the average continental crust (about 35 km).

A sono-radio buoy system was developed at the University of Cambridge, England (Hill, 1963), that required only one ship. The seismic signal was transmitted by radio from hydrophones suspended by a buoy to the ship from which the charges were fired. This technique was extended to the distance range 700–2000 km by a group from the University of Hawaii (Odegard and Sutton, 1972), who used an aircraft as the receiver during the Cannikin nuclear test conducted in the Aleutians.

By 1960, well over 100 refraction profiles had been recorded in various parts of deep basins in the Atlantic ocean. A comparable number of measurements were made on the continental shelves and in the Mediterranean Sea. The Pacific and Indian oceans have been studied mainly by the Scripps Institution of Oceanography, which collected data at 229 stations according to the summary

by Shor and Raitt (1969). A typical station consists of two reversed profiles, each about 40 miles long. One station requires about 1600 lb of TNT in about 80 shots. A shot point farthest from the receiving station usually requires 50–100 lb of TNT, but in bad weather 300 lb is not uncommon.

A dense areal coverage by these shooting and recording stations revealed anisotropy of P-wave velocity in the upper mantle under the ocean. In analyzing these data, Raitt et al. (1969) used the extended time-term method (Scheidegger and Willmore, 1957), in which the effects of lateral heterogeneity and of anisotropy were separated by expressing the arrival time as the sum of the source term, the receiver term, and the term that depends only on the distance and azimuth from source to station.

Let us now turn to the refraction data obtained on land, and see what discoveries have been made from that source. In 1909, a Serbian seismologist, A. Mohorovičić, studied arrival times of first motions from an earthquake in the Balkans, and discovered that the travel time as a function of distance consisted of two straight lines, one corresponding to the direct waves through the crust, and the other to the waves refracted by what has come to be known as the Mohorovičić discontinuity, often abbreviated as the M discontinuity, or Moho.

Great impetus was given to refraction work on both sea and land when, in 1946, large quantities of surplus high explosives from World War II became available in many countries. Several European seismologists joined an experiment in 1947 when a large explosion was detonated at Heligoland. The velocity in the upper mantle found by this experiment was significantly greater than that implied in the Jeffreys-Bullen table.

In the U.S.S.R., an extensive program of refraction work called "deep seismic sounding" (DSS) has been undertaken using the phase-correlation technique inspired by G. A. Gamburtsev. Seismometers are spaced at intervals of 100–200 meters along lines 30–40 (sometimes 70–80) km long, with gaps between the lines of 10–30 km, covering a total length of nearly 500 km. Two or three tons of explosives was used at several shot points along the line. According to the summary report by Kosminskaya and Riznichenko (1964), detailed profiles totaling 10,000 km were covered by DSS in the Soviet Union; in seas and oceans, more than 15,000 km.

The data obtained by DSS enable seismologists to trace a particular wave group from seismometer to seismometer over a certain distance, facilitating the identification of refracted and reflected waves from a particular discontinuity in the crust and upper mantle. The relative amplitudes of these waves are also used in the study of the detailed characteristics of these boundaries. The refraction work has been done in many parts of the world, and the results have been tabulated by McConnell et al. (1966).

Efforts to extend refraction profiles to more than 1000 km have also been made in North America. In 1963, the Carnegie Institution of Washington coordinated a large, cooperative, international seismic experiment centered in

Lake Superior (Steinhart, 1964). Besides the surprisingly complicated crustal structure under Lake Superior discovered by the experiment, it was found that seismic waves from 1-ton shots were detected at distances of more than 2000 km. Encouraged by this, the U.S. Geological Survey made a long-offset refraction measurement from Lake Superior to central Arizona in 1964 (Roller and Jackson, 1966). This led to another experiment, called Project Early Rise, in which thirty-eight 5-ton shots were detonated in Lake Superior. Several groups from North American universities and government agencies participated in a recording program along the profiles shown in Figure 11.3. The seismograms obtained over the distance range 400–1350 km from Lake Superior to Colorado are shown in Figure 11.4 (Hales, 1972).

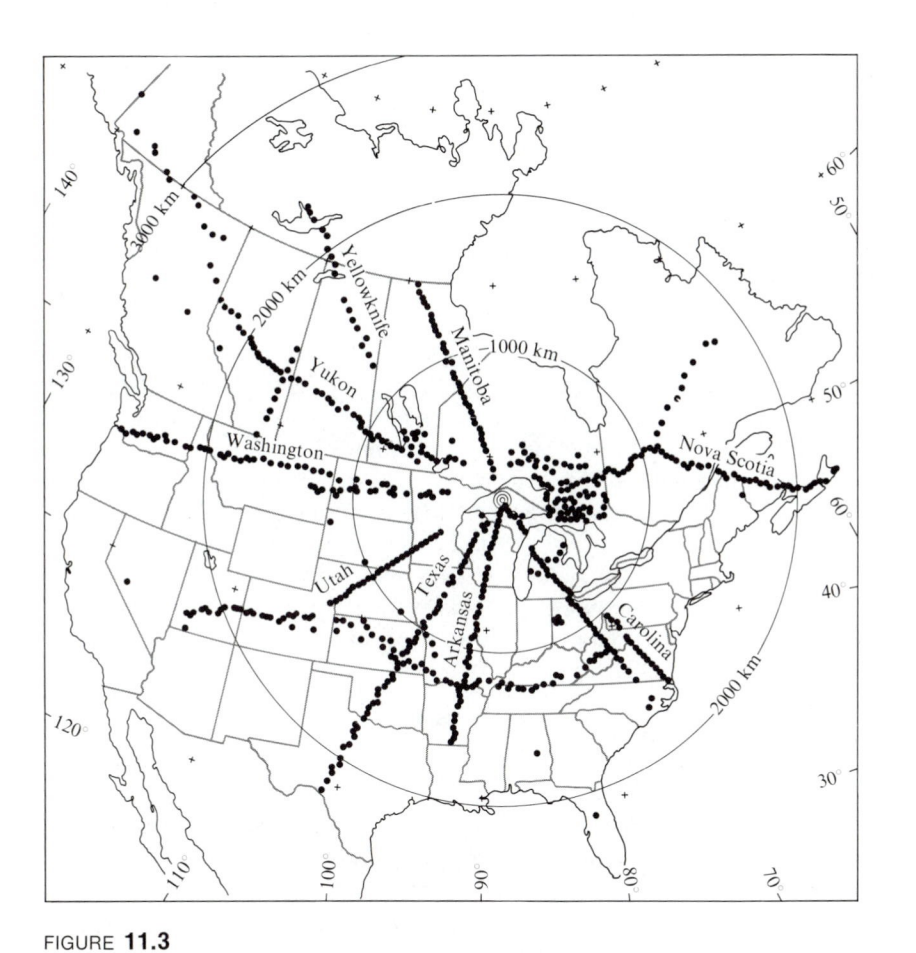

FIGURE **11.3**

Array of seismographs of Project Early Rise. Shot location is shown by concentric circles in Lake Superior. [Reproduced from Iyer et al. (1969); copyrighted by American Geophysical Union.]

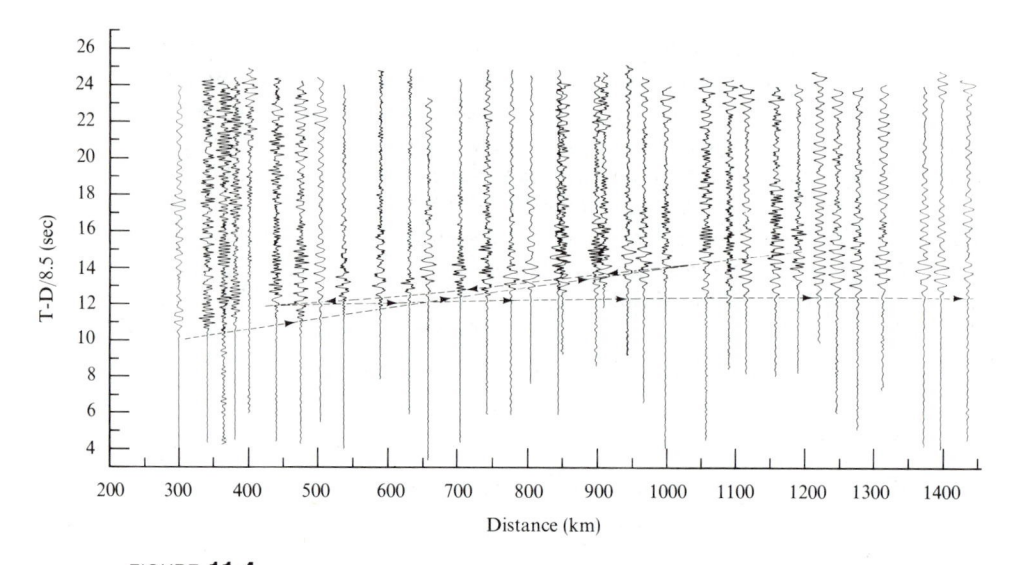

FIGURE **11.4**

Record section of Early Rise Arkansas Profile. Seismograms are arranged according to the epicentral distance and the time relative to the arrival of waves with surface velocity 8.5 km/sec. [From Hales, 1972.]

The underground nuclear explosions at the Nevada Test Site and other areas also offer a large amount of seismic data. Up to December 31, 1976, the United States had conducted 614 nuclear explosions, and the presumed ones in the Soviet Union amounted to 354. Lists of all unclassified explosions are given in Griggs and Press (1961) and Springer and Kinnaman (1971). About 500 shots are listed with origin time, location, depth, yield, shot medium, and information on water tables and lithology of the site. They include five shots with yield about 1 megaton or greater. The magnitude (m_b) of a 1-megaton shot is about 6.3, large enough to be recorded by standard observatory seismographs at almost any place in the world. The data are of course useful for checking the results obtained from earthquakes.

About 40 mobile seismic stations equipped with short- and long-period seismographs, both 3-component, were deployed for the observation of seismic waves within the United States from the nuclear tests. Shots were recorded on 35-mm film and on one-inch 14-channel magnetic tape. These mobile stations are called **LRSM** (long-range seismic measurement) vans; together with other permanent stations, they have produced high-quality data used for both amplitude and travel-time studies. The explosion Gnome in New Mexico, detonated in December 1961, conclusively demonstrated the lateral heterogeneity of the upper mantle; the P_n velocity on top of the mantle varies from 7.5 to 8.5 km/sec within the United States, and the travel-time residual relative

to a standard table is different between the eastern and the western U.S. by as much as 11 seconds (Herrin and Taggert, 1962). It was shown that errors of epicenter location can be reduced by an order of magnitude if this lateral heterogeneity effect is taken into account. Regional variations in the amplitude-distance relation were also discovered using LRSM data (Evernden, 1967).

In a summary report on travel times of P-waves for distances of $3-30°$, Hales (1972) compares upper mantle models obtained from the Early Rise experiment (Lake Superior Source) and those based on observations for the NTS shots and earthquakes in the western U.S. The latter models have well-developed low-velocity zones in the upper mantle, but the former do not, at least for P-waves. Since the low-velocity zone plays a crucial role in plate tectonics, whether the zone exists globally or not is a fundamentally important question with regard to the dynamics of the lithosphere.

In the long-range refraction measurements, the profile will occupy different geophysical provinces. It becomes difficult to decide whether an observed feature of a seismogram should be due to the vertical heterogeneity or to the lateral one. We need three-dimensional Earth models for a unique interpretation, and such models require amounts of data that are many orders of magnitude greater.

Since the wave paths for refracted waves are nearly horizontal, the travel-time data represent the seismic velocity averaged in some way over a horizontal distance. Therefore, the refraction method has poor spatial resolution in the horizontal direction. The near-vertical reflection method has potentially much greater resolving power in the horizontal direction, and becomes increasingly important when an efficient seismic method is demanded for a detailed study of lateral heterogeneity of the Earth.

The collection, processing, and analysis of reflection seismic data has been done extensively by exploration geophysicists to find natural deposits of gas and oil. A revolutionary advance in this field, called the "digital revolution," was accomplished in 1960's by switching from analogue recording, processing, and analysis to digital form. This improved the quality of data, increased the flexibility of processing by the digital computer, and facilitated better man-computer interaction, so that the analyst may eventually be able to interrogate the data to review, modify, and test various hypotheses in real time. Advances in the U.S. are summarized by Schneider (1971).

A typical set of reflection data in exploration seismology is obtained in the following way. Twenty-four geophone groups are spread over a distance of 8000–9000 feet. Seismic signals from the twenty-four groups are fed to amplifiers with binary gain control. The gain of this amplifier is doubled when the input signal level falls below a certain threshold. It is again doubled when the signal level falls below half the threshold. The doubling occurs between the sampling times, and the gain change is separately recorded. The dynamic range referred to input signal can be as high as 170 db (3×10^8 to 1). Twenty-four channel data may be digitized to 15-bit words at 2-msec intervals. Since

the length of a typical reflection record is 6 sec, one shot will produce about 1 megabit of information.

For a 24-channel system, the shot point is advanced one group interval for each shot. The spread is also advanced one group per shot simply by removing a group at one end and adding it to the other end of the spread. This procedure will produce seismogram traces needed for 12-fold common-depth-point (CDP) stacking, which is described in Section 11.5.1. The data produced for one profile 10 km long would be almost 100 megabits.

Marine reflection seismic data are collected typically for 24-fold stacking with 8000-foot streamers, which are marine cables incorporating pressure-sensing hydrophones that are towed through the water; the ship does not stop while data are being collected.

For continuous reflection profiling in the sea, nonexplosive sources such as electro-acoustic transducers, electric sparkers, gas exploders, and air guns have been used for many years. On the land, additional nonexplosive sources such as VIBROSEIS* (hydraulic vibrator) are increasingly used in place of high explosives. This device saves much time, because holes do not have to be drilled, and it also allows data to be acquired in places (e.g., towns) where explosions would not be tolerated.

The data for deeper near-vertical reflections, such as those expected from the Mohorovičić discontinuity, have not yet been collected extensively. In order to identify an arrival of energy unambiguously as near-vertical reflection, several criteria must be satisfied. The wavefront must be nearly parallel to the horizontal to eliminate the possibility of slow surface waves or back-scattering from lateral heterogeneity. An acceptable value of average velocity must be obtained from the slope of X^2 versus T^2 (see Chapter 12), where X is the horizontal distance from the shot point and T is the travel time. The multiple reflections at shallower discontinuities must not be mistaken for a deeper primary reflection. The variation of amplitudes caused by near-surface irregularities must be eliminated. The effect of lateral heterogeneity becomes stronger for deeper reflections than for the shallower ones because of the longer travel distances. Again, as in the case of refraction measurements, we face the need for three-dimensional interpretation, which will demand greater amounts of data.

In a few favorable cases, however, the above criteria are met, and deep reflections are unambiguously identified. For example, the Riel discontinuity was discovered by Kanasewich and Cumming (1965) in western Canada from strong reflections arriving about 12 sec after the shot time. This discontinuity is at a depth of about 35 km, and separates the upper crustal layer, with a velocity 6.5 km/sec, from the lower one, with a velocity of 7.2 km/sec. In an extended reflection study (Clowes et al., 1968), this discontinuity was found to have structural relief of 8 km over a distance of 25 km, thus demonstrating high resolving power of the reflection method.

---

* Registered trade mark of the Continental Oil Company.

A major effort to use reflection methods for studying deep layers within the crust, and for uppermost layers within the mantle, has been reported by the COCORP group (Consortium for Continental Reflection Profiling: see Oliver et al., 1976). Using the VIBROSEIS technique, with several large vibrators and waves of relatively low frequency (10–32 Hz), reflections were obtained in Hardeman County, Texas, from depths as great as 45 km. In the lower part of the section, it was found that reflections are not, in general, continuous laterally over more than a few kilometers.

BOX **11.1**

Recording media

We mentioned in Section 11.1.2 the "digital revolution" that took place in the recording and processing of reflection seismic data used in exploring for natural deposits of gas and oil. Digital recording is superior to analogue recording in dynamic range and in ease of processing by computer. However, its drawback is relatively low density of information per unit area of recording medium. This aspect is important for observation of natural seismic sources, for which we must wait a long time before an event is recorded.

Analogue recording on magnetic tape is superior to digital recording with regard to information density. For example, a dozen channels of seismic signal with frequency response up to 100 Hz can be frequency-modulated, mixed, and recorded on a single channel of a tape recorder capable of recording up to 10 kHz. With correction for tape speed, a dynamic range of more than 50 db is possible for each seismic channel. Since 50 db corresponds to about 8 bits of information, and since 100 Hz signal requires a sampling rate of 200 per sec (see Section 11.2.2), the information stored by the above analogue recording amounts to 20,000 bits/sec. For a tape speed of $7\frac{1}{2}$ inches/sec (which is appropriate for recording 10 kHz), the information density per unit length of tape is about 3000 bits/inch. This is considerably higher than the density obtainable by digital recording (typically 800 bits/inch for $\frac{1}{2}$-inch tape width).

Recently, digital event recorders with digital memories and decision capability have been developed (e.g., Ambuter and Solomon, 1974). Such a device will trigger the tape recorder at the arrival of a signal and record it completely. The early part of the signal is not lost while the tape recorder comes up to speed, because at *any* instant the memory contains the signal for the previous few seconds. Once triggered, the tape recorder records the contents of the memory, as well as subsequent signal.

The smoked paper recording, in which a needle point scratches a thin sheet covered with soot, is still one of the highest-density forms of recording in seismology. The recording is made helically on a sheet wrapped around a drum that moves slowly along its axis as it rotates. This device has been in use for many years, and is still used in expeditions for microseismicity study.

At standard seismograph stations, such as those of the WWSSN, the recording is made on photographic papers using helical drums. More details of the recording at WWSSN are described in Section 11.1.1.

A very deep reflection in the Earth was observed by Engdahl et al. (1970), using the data obtained by the Montana LASA for 1-megaton underground nuclear explosions in Nevada. These are near-vertical reflections (wave slowness $dT/d\Delta$ across LASA; 0.3 sec-deg^{-1}) from the boundary between the liquid outer core and the solid inner core ($PKiKP$) that arrived 16 min, 37 sec after shot time.

11.2   Quality of Seismological Data

In Section 11.1, we described a variety of seismological data and gave a brief historical review of how the creation of new data of improved quality led to exciting discoveries about the Earth's interior. In this section, we shall consider what determines the quality of seismological data and how to improve them for a more detailed and accurate study of the Earth's interior.

For this purpose, we shall introduce the concepts of signal and noise. We define the signal as the desired part of the data and the noise as the unwanted part. Our definition of signal and noise is subjective in the sense that a given part of the data is "signal" for those who know how to analyze and interpret the data, but it is "noise" for those who do not. For example, for many years the times of the first arrivals of P- and S-waves were the only signals conveyed by an earthquake, and the rest of the seismogram, such as surface waves and coda waves, had to be considered as useless until appropriate methods of interpretation were found.

Thus, through the application of a new technique to old data, an analyst can experience a moment of discovery as joyful as a data gatherer does using a new observational device. With the advance of analysis techniques, the signal-to-noise ratio of given data increases. This ratio determines the quality of information contained in given data.

11.2.1   Sampling theorems

In order to discuss the signal-noise problem in seismological data more quantitatively, let us digitize a seismogram at discrete time points. Considering a transient seismogram $f(t)$, for which the Fourier transform $F(\omega)$ exists (cf. (10.20) and (10.21)), we find

$$f(t) = \frac{1}{2\pi} \int_{-\infty}^{\infty} F(\omega) \exp(-i\omega t) \, d\omega. \tag{11.1}$$

We shall assume that $f(t)$ does not contain frequencies higher than W Hz. Then we can rewrite (11.1) as

$$f(t) = \frac{1}{2\pi} \int_{-2\pi W}^{2\pi W} F(\omega) \exp(-i\omega t) \, d\omega. \tag{11.2}$$

Since $F(\omega)$ is defined for $-2\pi W \leq \omega \leq 2\pi W$, it can be expressed by a Fourier series with the fundamental wavelength $4\pi W$:

$$F(\omega) = \sum_{n=-\infty}^{\infty} F_n e^{i\omega n/2W}. \tag{11.3}$$

The coefficients F_n are determined by

$$F_n = \frac{1}{4\pi W} \int_{-2\pi W}^{2\pi W} F(\omega) e^{-i\omega n/2W} \, d\omega. \tag{11.4}$$

Comparing the above equation with (11.2), we find

$$F_n = \frac{f\left(\dfrac{n}{2W}\right)}{2W}. \tag{11.5}$$

Putting (11.5) into (11.3) and then into (11.2), we find that $f(t)$ is completely determined by its values sampled at discrete points with the interval $1/2W$, as shown below.

$$f(t) = \frac{1}{4\pi W} \sum_{n=-\infty}^{\infty} f\left(\frac{n}{2W}\right) \int_{-2\pi W}^{2\pi W} \exp\left(\frac{i\omega n}{2W} - i\omega t\right) d\omega \tag{11.6}$$

$$= \sum_{n=-\infty}^{\infty} f\left(\frac{n}{2W}\right) \frac{\sin\{\pi(2Wt - n)\}}{\pi(2Wt - n)}. \tag{11.7}$$

For $t = m/2W$, where m is an integer, both sides become $f(m/2W)$.

Thus the seismogram $f(t)$, which does not contain frequencies higher than W Hz, can be completely described by discrete samples taken at an interval $1/2W$. If the time length of $f(t)$ is T sec., the total number of samples required

---

BOX **11.2**

Notation

Until this stage, we have avoided special notation such as $F(\omega)$ for the Fourier transform of $f(t)$, and have used $f(\omega)$. This is convenient when the transform is an abstraction. But in data processing, both the original time series and its transform may be an array of numbers, with each array defined in some computer memory. We therefore introduce $F(\omega)$ as the Fourier transform of $f(t)$, and F_n for the discrete spectrum.

We use an explicit summation symbol in this chapter, rather than an implied summation over repeated subscripts.

to describe $f(t)$ is $2WT$. The amount of information in the seismogram $f(t)$ may be measured by the size of the computer memory needed to store it. If the data at each sample point are digitized to an integer in the range from 0 to $2^m - 1$, then a memory size of $2WTm$ bits is required to store $f(t)$. If we increase m, the amount of information will increase proportionally. However, since data always contain some noise, a limit exists for the useful information carried by the signal.

According to Shannon (1949), a simple formula exists for the limiting amount of useful information when the signal and noise are both band-limited white noise. Representing the powers of signal and noise by S^2 and N^2, respectively, the limiting amount of useful information is given by

$$WT \log_2 \frac{S^2 + N^2}{N^2} \tag{11.8}$$

This corresponds to the memory size required when the data are digitized to an integer in the range from 0 to $\sqrt{S^2 + N^2}/\sqrt{N^2}$.

11.2.2   Aliasing

As shown in the preceding section, when the seismogram contains frequencies up to W Hz, the discrete samples at an interval of $1/2W$ sec completely determine the seismogram. If we sample at an interval shorter than $1/2W$ sec, the data will be redundant, but no error is introduced. However, if the sampling interval is longer than $1/2W$ sec, a very serious thing will happen. Figure 11.5 shows a sinusoid with frequency W Hz, sampled at discrete time points equally spaced at an interval longer than $1/2W$ sec. If the sampled points are smoothly connected, they show periodicity that did not exist in the original data. This undesirable effect was called *aliasing* by Blackman and Tukey (1958).

The aliasing occurs because the sampling of $f(t)$ is essentially a multiplication in the time domain by a Dirac comb of delta functions, $\sum_i \delta(t - i \Delta t)$. This becomes convolution in the frequency domain, but the transform of the Dirac comb becomes a comb in frequency, $\sum_i \delta(\omega/2\pi - i/\Delta t)/\Delta t$. Convolution with such a comb will repeat the spectrum of f at intervals spaced by $(1/\Delta t)$ Hz. If

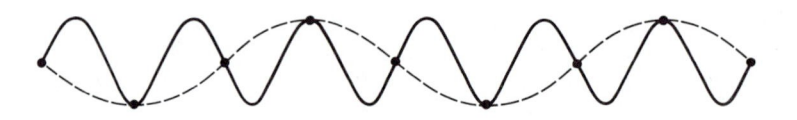

FIGURE **11.5**

Aliasing: spurious signal (broken line) is sampled because the sample interval is longer than half the shortest period contained in the signal.

the bandwidth of f, say $2W$ Hz, is greater than $(1/\Delta t)$, then some undesirable overlap will occur, giving a false spectrum for the sampled function.

11.2.3   Measurement of the first motion

The arrival time and the sense of the first motion of a wave group are the most widely collected seismological data. They are the basic data for (i) the construction of travel-time curves, (ii) the determination of epicenter and focal depth of an earthquake, and (iii) the fault-plane solution. A major part of our knowledge of the internal structure and seismicity of the Earth, of earthquake mechanisms, and of current tectonics was obtained through measuring the first motion. This is because the simple and powerful ray-theoretical interpretation is most applicable to the first motion.

To identify the first motion, one must be able to say whether the signal is present or not. In other words, the seismologist needs one bit of information. In Section 11.2.1, we determined how many bits of information are contained in a time-unit length of data characterized by the signal-to-noise ratio and the frequency range. Using equation (11.8), Pakiser and Steinhart (1964) suggested that, since one bit of information is contained in the time length Δt given by

$$\Delta t = \frac{1}{W \log_2\left(1 + \dfrac{S^2}{N^2}\right)}, \tag{11.9}$$

Δt may be considered as the measure of error in the measurement of arrival time of the first motion.

The above equation shows that the error in the measurement of arrival time of first motion decreases inversely in proportion to the frequency bandwidth W. However, the error depends logarithmically on the signal to noise ratio. For a large signal-to-noise ratio, the increase of signal amplitude improves the accuracy of time measurement only slightly, because the error is inversely proportional to the logarithm of the ratio.

There is a problem in applying equation (11.9) to the actual seismogram. It is implicit in (11.9) that the average signal amplitude S is stationary in time. On the other hand, the first motion has zero amplitude by definition. Obviously it is meaningless to put $S = 0$ in equation (11.9).

To resolve this problem, Aki (1976) selected noise-free seismograms of local earthquakes and mixed them with computer-generated noise having various mean square amplitudes. The outputs were measured by several seismologists, and the errors in the arrival time and sense of first motion were studied as functions of signal-to-noise ratio.

Modeling first motion by a stationary time series is obviously naive, and the dependence of observed arrival-time error on noise power does not agree well

with that calculated using (11.9). In the absence of a better model, Aki forced the matching between observed and calculated errors, with the result that the best matching was obtained when the RMS signal amplitude S of the first motion is set to $1/20$ of the maximum amplitude of the P-wave group. In other words, roughly speaking, the signal level of "first motion" is about $1/20$ of the maximum amplitude of the P-wave group.

An interesting result was found concerning the error in reading the first motion. When the noise level is high, the probability of reading the sense of first motion incorrectly can exceed 0.5. In such a case, the opposite of what a seismologist picks as the sense of first motion is more likely to be correct. The probable reason is that, in many cases, the amplitude of the second half-cycle of P-waves is greater than that of the first half-cycle.

11.2.4   *Measurement of phase velocity*

The measurement of phase velocity by the conventional method presupposes that the wave group under consideration is purely a single kind of wave possessing a propagation velocity determined only by frequency. In other words, we assume that the wave group propagating along the x-axis can be expressed as

$$f(x, t) = \frac{1}{2\pi} \int_{-\infty}^{\infty} |f(x, \omega)| \exp\left[-i\omega\left(t - \frac{x}{c(\omega)}\right) + i\phi(\omega)\right] d\omega, \quad (11.10)$$

where $|f(x, \omega)|$ is the amplitude spectral density, $\phi(\omega)$ is the phase term due to factors other than propagation, and $c(\omega)$ is the phase velocity. The phase delay $\omega x/c(\omega)$ due to propagation can be obtained by the Fourier analysis of the seismogram. The Fourier transform of (11.10) can be written as

$$\int_{-\infty}^{\infty} f(x, t) \exp(i\omega t)\, dt = |f(x, \omega)| \exp\left[i\phi(\omega) + i\frac{\omega x}{c(\omega)}\right] \quad (11.11)$$

By measuring the difference in phase spectra between two stations at distances x_1 and x_2, which is $[\omega/c(\omega)](x_1 - x_2) \pm 2n\pi$, we can obtain the phase velocity $c(\omega)$. The uncertain integer n can be determined either by an approximate *a priori* knowledge of $c(\omega)$ or by measurement at more than two stations.

Let us consider the effect of noise on the accuracy of phase-velocity measurement. Suppose that the seismogram $d(t)$ consists of signal $s(t)$ and noise $n(t)$. Then

$$d(t) = s(t) + n(t). \quad (11.12)$$

Let us write the Fourier transform of $s(t)$ as $|S(\omega)| \exp[i\phi_s(\omega)]$ and that of $n(t)$ (for the duration $2T$ of signal) as $|N_T(\omega)| \exp[i\phi_n(\omega)]$. The power spectral

density $P(\omega)$ of $n(t)$ is related to $N_T(\omega)$ by

$$P(\omega) = \lim_{T \to \infty} \frac{|N_T(\omega)|^2}{2T}. \tag{11.13}$$

This can be obtained from (10.25) as follows:

$$
\begin{aligned}
P(\omega) &= \int_{-\infty}^{\infty} \langle n(t)n(t + \tau) \rangle \exp(i\omega\tau) \, d\tau \\
&= \lim_{T \to \infty} \int_{-\infty}^{\infty} \frac{1}{2T} \int_{-T}^{T} n(t)n(t + \tau) \exp(i\omega t) \, dt \, d\tau \\
&= \lim_{T \to \infty} \frac{1}{2T} \int_{-T}^{T} n(t) \exp(-i\omega t) \, dt \int_{-\infty}^{\infty} n(t + \tau) \exp[i\omega(t + \tau)] \, d\tau \\
&= \lim_{T \to \infty} \frac{N_T^*(\omega)N_T(\omega)}{2T}. \tag{11.14}
\end{aligned}
$$

The Fourier transform $D(\omega)$ of the seismogram can be written as

$$|D(\omega)| \exp[i\phi_d(\omega)] = |S(\omega)| \exp[i\phi_s(\omega)] + |N_T(\omega)| \exp[i\phi_n(\omega)]. \tag{11.15}$$

For small $|N_T(\omega)|/|S(\omega)|$, we can write

$$\phi_d(\omega) \sim \phi_s(\omega) + \frac{|N_T(\omega)|}{|S(\omega)|} \sin[\phi_n(\omega) - \phi_s(\omega)]. \tag{11.16}$$

The second term is the fluctuation of phase due to the noise. If $n(t)$ is uncorrelated with $s(t)$, and $\phi_n - \phi_s$ is randomly and uniformly distributed from 0 to 2π, the RMS fluctuation of $\phi_d(\omega)$ will be equal to

$$\Delta\phi_{RMS} = \frac{1}{\sqrt{2}} \frac{|N_T(\omega)|}{|S(\omega)|}. \tag{11.17}$$

The RMS error of phase difference between two stations that share the same signal and noise level will be $|N_T(\omega)|/|S(\omega)|$. Therefore, the RMS error of phase-velocity measurement will be

$$\frac{\Delta c_{RMS}}{c} = \frac{1}{2\pi} \frac{|N_T(\omega)|}{|S(\omega)|} \frac{\lambda}{\Delta x}, \tag{11.18}$$

where λ is the wavelength and Δx is the distance over which the phase difference is measured. As suggested in the preceding section, the error in the measurement of first motion may be only logarithmically dependent on the signal amplitude.

The error in measurement of phase velocity, however, decreases inversely in proportion to the signal amplitude, because the entire waveform is used in the latter measurement whereas only one bit of information is utilized in the former.

11.2.5 Accuracy of phase-velocity measurement: A case history of improving the quality of data

Equation (11.18) points to the following two basic ways of improving the measurement of seismic phase velocity: either the signal to noise ratio must be increased or the wavelength relative to the distance between stations must be decreased.

The most straightforward way of improving the accuracy may seem to be to increase the signal amplitude by increasing the power of the seismic source. The error would decrease inversely in proportion to the intensity of the source. However, the noise—the unwanted part of the data according to our definition—may increase with the power of the source. As mentioned before, in the phase-velocity measurement we presume that the signal is a single kind of wave. Therefore, if other waves coexist with the data, they must be considered as noise. If our signal is the fundamental mode of Rayleigh waves, then all the higher modes of Rayleigh waves, body waves, and even the fundamental-mode Rayleigh waves, which approach with different propagation directions because of lateral heterogeneity, must be considered as noise. Obviously such noise increases with the power of the source. Therefore, once the main part of the noise becomes such signal-generated noise, no further improvement in accuracy is possible by increasing the power of the seismic source.

Further improvement of the signal-to-noise ratio is possible by increasing the number of measurements for an appropriate arrangement of sources and receivers. If the signal-generated noise can be made uncorrelated among these measurements, the error of velocity measurement will decrease inversely in proportion to the square root of the number of measurements.

When the phase-velocity method was first applied to the California area by Press (1956), special care was taken in selecting noise-free Rayleigh-wave seismograms. He used a large earthquake in Samoa with the path to California lying entirely in the deep ocean and showing little evidence of the multi-path interference effect. He used a tripartite method, in which phase-delay times measured at three stations are combined to give the phase velocity and azimuth of wave approach. Although the ratio of wavelengths to the travel distance was small, around 1, a good signal-to-noise ratio made possible an accuracy of about 1%.

Taking advantage of the dense network of seismograph stations operated by the Japan Meteorological Agency, Aki (1961) found a significant variation of phase velocity within Japan. About 50 stations were divided into several

geographical groups. A large number of stations made possible an accuracy of about 1.5% for each group, although the wavelength was comparable to or greater than the station distance.

The accuracy of velocity measurement may be improved by increasing the travel distance Δx at the expense of the spatial resolution. For a relatively uniform area, such as shield and deep ocean floor, a long travel distance has been effectively used to attain an accurate average phase-velocity measurement. For example, Brune and Dorman (1963) obtained a good result for the Canadian Shield using the two-station method, in which the phase velocity was obtained from the records of earthquakes whose epicenters lie within 4° of the great-circle path connecting two receivers. The wavelength λ ranged from 80 to 300 km, and the station distance, Δx, was about 3000 km. Thus the ratio $\lambda/\Delta x$ ranged from 1/40 to 1/10. Equation (11.18) gives the corresponding range of velocity error:

$$\frac{\Delta c}{c} \sim \frac{|N_T(\omega)|}{|S(\omega)|} (0.004\text{–}0.015). \tag{11.19}$$

From the scatter of measured phase velocities from a mean smooth curve, they estimated an RMS error of about 0.5%, roughly independent of wavelength. From (11.19), we find for this phase-velocity error that the signal-to-noise ratio must have been about 1 for the shortest waves and about 3 for the longest waves. In fact, they stated that surface waves with periods less than about 20 sec become very low in amplitude or else rather incoherent, and the correlation of such short periods over long distances is difficult.

In the tripartite measurement of the phase velocity, if the propagation direction is parallel to one leg of the tripartite net, the magnitude of phase velocity is determined by the time difference between the two stations of the leg, and is unaffected by a small fluctuation in the measured time at the third station. Recognizing this fact, Knopoff et al. (1967) showed experimentally that the two-station method is superior to triangulation in the matter of minimizing errors in phase shifts in the presence of lateral inhomogeneity. The result of an extensive application of the method to the WWSSN data was summarized by Knopoff (1972).

Further support for the superiority of the two-station method over the three-station method may be found in wave-scattering theory. According to Chernov (1960), the phase fluctuation of waves propagating in a random medium has much greater correlation distance (smoother variation) in the direction parallel to the propagation path than perpendicular to it.

In some cases, the single station method gives even better results than the two-station method. The single-station method is applicable to earthquakes with known source parameters needed for calculating the term $\phi(\omega)$ in equation (11.10), and has been successfully applied to many earthquakes in various regions

of the Earth. For example, Weidner (1974) showed that the phase velocities across the Atlantic Ocean determined by the two-station method using a Greek earthquake and a Nicaraguan earthquake agree well with those by the single-station method using a mid-Atlantic ridge earthquake recorded at both sides of the ocean, but the former showed a somewhat greater interference effect than the latter.

The distance Δx appearing in (11.18) is (in Weidner's analysis) the width of the whole Atlantic for the two-station method, but only half of this width for the one-station method. We therefore expect from Equation (11.18) a better accuracy for the former than for the latter. However, the ratio $\lambda/\Delta x$ is also a crucial factor in determining the phase fluctuation due to scattering. For a random medium characterized by correlation distance a, the RMS fluctuation of phase angle is proportional to $\sqrt{a \, \Delta x/\lambda}$, according to Chernov. Thus the part of the velocity measurement error due to lateral heterogeneity may increase with the ratio $\Delta x/\lambda$. The limiting value of $\Delta x/\lambda = 40$ for the Canadian shield observed by Brune and Dorman may also be attributed to the effect of lateral heterogeneity.

Although the scattering due to lateral heterogeneity is the important factor in limiting the accuracy of velocity measurement for long waves, the attenuation also becomes an important factor for very short waves. For purposes such as monitoring the change of tectonic stress by measuring seismic velocity *in situ*, the precision required may be one part in 10^4 or more. Referring to (11.18), one may want to attain the precision by making $\lambda/\Delta x$ small. Since the distance in a typical experiment may be $\Delta x \sim 10$ km, very high frequency is required to make $\lambda/\Delta x$ small. Then, the signal amplitude would decay very quickly with distance in the shallow part of the Earth's crust, reducing the signal-to-noise ratio and increasing the error of velocity measurement. The only remedy for this case would be either to increase the power of the seismic source or to repeat the experiment many times.

11.3   Time-series Analysis for Noise Reduction

As shown in the preceding section for the example of phase-velocity measurement, the most important factor determining the quality of given seismological data is the signal-to-noise ratio. The signal is defined as the desired part of data and the noise as the unwanted part. In this section, we shall consider the case in which the data are the output of a single seismograph, i.e., a single function of time. Let us first consider the technique of deconvolution.

As discussed in this section, under certain assumptions and simplifications, a given seismogram may be considered as an output of a series of boxes, each representing a separate effect characterized by an impulse-response function. An obvious example of such a box is the seismograph. This instrumental effect can be eliminated from the record by using any of the following four methods.

i) The record and ground motion are related by a differential equation such as (10.2) or (10.63). Direct numerical integration of the equation may be used to find ground motion from the record. (See Bogdanov and Graizer, 1976, for a simple, stable method.)

ii) Take the Fourier transform of the record, divide it by the frequency response of the seismograph, such as given by (10.5), and compute the Fourier inverse transform of the ratio.

iii) Convolve the record with the impulse response of an inverse filter, i.e., the Fourier inverse transform of the reciprocal of the seismograph frequency response. (In both (2) and (3), some smooth tapering of the Fourier transform is needed because the reciprocal of the frequency response of a seismograph may become infinite at zero or infinite frequencies, as shown in (10.65) and (10.66) for a standard seismograph.)

iv) Apply a recursive filter as described in Section 11.3.4.

In general, the boxes into which we separate various effects on a seismogram do not have simple governing differential equations. Furthermore, in some cases, a box may be applicable to a small portion of the seismogram, and one may have to worry about which portion should be Fourier-transformed if method (ii) is used. The problem of selecting an appropriate portion of seismogram may be less if (iii) is used. On the other hand, with the advent of the Fast Fourier Transform (FFT) method, the computer time is much less for method (ii) than for method (iii).

Let us define the impulse response of a box as $g(t)$ and its Fourier transform as $G(\omega)$. The output $d(t)$ of the box is then expressed as the convolution of input $u(t)$ with the impulse response:

$$d(t) = u(t)*g(t)$$
$$= \int_0^\infty u(t - \tau)g(\tau)\, d\tau. \tag{11.20}$$

Writing the Fourier transforms of $u(t)$ and $d(t)$ as $U(\omega)$ and $D(\omega)$, respectively, we get

$$D(\omega) = U(\omega)G(\omega), \tag{11.21}$$

where

$$D(\omega) = \int_{-\infty}^\infty d(t)\exp(i\omega t)\, dt,$$

$$U(\omega) = \int_{-\infty}^\infty u(t)\exp(i\omega t)\, dt,$$

$$G(\omega) = \int_{-\infty}^\infty g(t)\exp(i\omega t)\, dt.$$

Method (ii) above recovers the input from output by

$$u(t) = \frac{1}{2\pi} \int_{-\infty}^{\infty} \frac{D(\omega)}{G(\omega)} \exp(-i\omega t)\, d\omega. \tag{11.22}$$

Method (iii) introduces the *inverse filter*, with impulse response given by

$$g^{-1}(t) = \frac{1}{2\pi} \int_{-\infty}^{\infty} \frac{\exp(-i\omega t)}{G(\omega)}\, d\omega$$

and recovers the input by the deconvolution

$$u(t) = \int_{-\infty}^{\infty} d(t - \tau)g^{-1}(\tau)\, d\tau. \tag{11.23}$$

11.3.1 Surface waves

Within the framework of linear elasticity, we have shown in Chapter 2 that the seismogram obtained at a point on the Earth due to an arbitrary seismic source can be represented as a space-time convolution of appropriate Green functions with the source function. When the observed wavelength is much longer than the source-space dimension—i.e., when the point-source approximation is valid—the space convolution becomes trivial. In that case, one can express the seismogram as a convolution of two time functions, one representing the source function and the other representing the Green function. Thus one can consider the source function as an input and the seismogram as the corresponding output of a box characterized only by the property of the wave path. We have further shown in Chapter 7 that the Green function for a layered Earth model may be decomposed into normal modes and leaky modes, each of which has an expression composed of factors separately depending on focal depth, travel distance, receiver depth, etc. Therefore, if we are able to pick up a portion of a seismogram that consists entirely of one single mode, we can express the seismogram as the output of a serial sequence of boxes, each of which is representing the effect of different factors separately. In the case of a single mode, even if the source size is greater than a wavelength, as long as the distance between the source and receiver is much greater than the source dimension, the effect of finite source size can also be separated, as discussed in Chapter 14.

Thus it is common practice in surface-wave seismology to express the spectrum of the observed record as the product of the spectra of the source-time function, the effects of finiteness, propagation, attenuation, and recording. When the spectra are measured from the seismograms at two stations along the same propagation direction from a common source, it is also common to attribute the spectral amplitude ratio to attenuation and the phase difference to phase

delay due to propagation. It must be cautioned that both of these practices are applicable only to a signal composed of a single mode. Coexisting other modes must be considered as noise, which limits the accuracy of the result of such analyses.

Noise of this kind may be eliminated if we can separate different modes from the observed seismogram. This "mode separation" can be achieved to the degree to which we know *a priori* the characteristics of individual modes composing the seismogram. If we know their group velocities, we may suppress the seismogram over the time range outside the interval of expected arrival of a given mode. This is called *group-velocity windowing*, in which the data are multiplied by a window function centered at the time of expected arrival of the mode. The group velocities to be used for mode separation may be directly measured by a process called *moving-window analysis* (Landisman et al., 1969), in which spectral amplitudes are estimated at various frequencies for successive seismogram portions multiplied by a window function whose width changes with the frequency while keeping constant resolution.

If the polarization characteristic (amplitude-phase relation among three components of displacement) is different between different modes, this information can also be used in designing the shape of a window. This windowing also distorts the waveform of the signal. Since the way in which it distorts the waveform depends not only on the shape of the window but also on the unknown signal form, its effect on the amplitude and phase spectra of the signal is generally unpredictable. We shall describe in Section 11.5 a method to minimize such undesirable distortion by the maximum use of *a priori* knowledge about the signal.

Another method of separating different modes is to combine the records of an array of stations, thereby eliminating the unwanted modes by using the differences in phase velocity. It can be done by a straightforward frequency-wavenumber spectral analysis or by multi-channel filtering on the basis of auto- and cross-correlation functions of the signal and noise. This subject will be discussed in Sections 11.4.3 and 11.5.3.

11.3.2 Body waves

A similar source-path separation may be justified for body waves observed at distances much greater than the source size. In the analysis of teleseismic body waves, the seismogram is usually considered as the output of a series of boxes representing the source, near-source shallow structure, deep interior of the Earth, near-receiver shallow structure, and recording instrument.

In that case, the effect of near-source shallow structure can be approximated by evaluating the far-field body waves transmitted into an imagined homogeneous half-space beneath the shallow structure. Likewise, the effect of near-receiver structure can be approximated by the response to the incident plane

P-, SV-, or SH-waves. The incidence angle for these plane waves can be calculated from the deep structure by using the geometric ray theory. If the deep structure itself is capable of changing the waveform, perhaps by the diffraction effects described in Section 9.5, then this effect too may be separated as an additional box.

The first deconvolution of a seismogram with respect to the near-receiver shallow structure was done by Takahasi and Hirano (1941) for the simplest case of a single layer and normally incident plane SH-wave source. The incident waveform $u(t)$ was obtained from the seismogram $d(t)$ at the surface simply by using

$$u(t) = \frac{1}{1 - r}\left[d(t) - r\, d\left(t - \frac{2H}{\beta_1}\right)\right], \tag{11.24}$$

where $r = (\beta_1\rho_1 - \beta_2\rho_2)/(\beta_1\rho_1 + \beta_2\rho_2)$ is the reflection coefficient, (β_1, ρ_1) and (β_2, ρ_2) are the shear velocity and density in the surface layer and substratum, respectively, and H is the layer thickness. In Chapter 12, we shall solve the

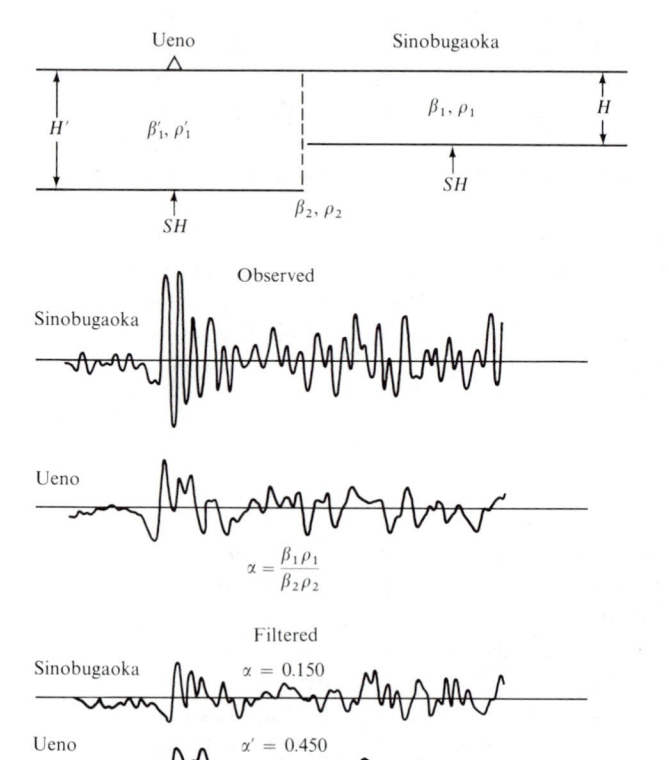

FIGURE **11.6**

The effect of a surface layer is removed from the seismogram by a simple two-point filter (11.24). The seismograms of an earthquake recorded at two locations with different layering show similarity after the filtering. [From Takahasi and Hirano, 1941.]

problem for a more general case and derive the above equation as a special case. Takahasi and Hirano applied the equation (11.24) to seismograms obtained at two stations in Tokyo, one on very soft alluvium and the other on a relatively harder formation. Figure 11.6 shows that the deconvolution process successfully reduced apparently different records at two locations to a nearly common incident waveform.

The waveform of incident plane body waves and the surface ground motion are related as the input and output of a linear system. Following Takahasi's successful work, Kanai and his colleagues made an extensive study of the response characteristics of the shallow-layered structure with special reference to the earthquake-hazard problem (Kanai and Tanaka, 1961).

11.3.3   *Reflection seismograms*

Deconvolution is also an important data-processing technique in reflection seismology for oil prospecting. For example, when an explosive source is buried at a certain depth, the waves initially propagated upward will be reflected downward at the free surface or at other high-contrast shallow interfaces. These reflected waves follow the primary downgoing waves and later show up as *ghost reflections*, degrading the resolution of the reflection seismogram. This effect may be approximated as a delayed superposition of reflection seismograms due to imaginary sources. The resultant seismogram may be expressed as the convolution of the primary reflection seismogram and the sequence of real and imaginary sources. The effect may then be removed by deconvolution.

Another example is the reverberation in a low-velocity layer, such as the water layer, in a submarine reflection survey. When plane waves are incident on such a layer from below, the incident waveform and the surface motion are related as the input-output of a linear filter, as shown by equation (11.24).

If we assume that the reflections coming back from the deeper layers are plane waves, then the water layer will act as a linear filter. In addition to this, however, there are waves that are multiply reflected in the water layer before penetrating into greater depths, as well as reflections from the deeper layers back to the water layer, which are then again reflected downward. A complete analysis of this process is given in Section 12.2. It suffices here to state that simple deconvolution can eliminate the effect of a water layer only approximately. The reflection seismogram expected when the water layer is stripped away is not related to the observed reflection seismogram as a simple input-output of a linear filter whose property is completely determined by the water layer.

For eliminating the ghost reflections or reverberations, accurate information on the layer parameters at shallow depths is needed for calculating the deconvolution filter. This information may not be readily available in practice. A method that can be applied without knowing the parameters was developed by

Robinson (1957, 1966). He assumes that the sequence of reflections in the absence of undesirable effects, such as ghosting and reverberations, are a random sequence of spikes with the δ-function autocorrelation, and that the observed seismogram is an output of a linear filter when this sequence of spikes is the input. The autocorrelation function of the seismogram is then the Fourier inverse transform of the square of the absolute value of the filter-frequency response. The phase characteristics of the filter response can be obtained from its absolute value by means of equation (15) in Box 5.8 if the frequency-response function has no zeros in the upper ω-plane. As discussed in Box 5.8, this corresponds to the minimum-delay system. Assuming that the filter has minimum delay, its response function and then that of the deconvolution filter can be determined. In practice, the filter coefficients are determined in the time domain by solving the linear equation (11.56), discussed in Section 11.3.6. This method, although based on bold assumptions and approximations, apparently works well in practical cases.

11.3.4   *Digital filtering*

In practice, filtering is now usually done on digitized data. Let us digitize the seismogram $d(t)$ at equal intervals Δt, as discussed in Section 11.2. The interval Δt is chosen to be less than $1/2W$ to avoid aliasing, where W is the maximum frequency (in Hz) contained in the data.

As an approximation to the Fourier transform $D(k\,\Delta\omega)$ in (11.21), we shall define the discrete Fourier transform D_k as

$$D_k = \sum_{n=0}^{N-1} d_n \exp(ik\,\Delta\omega n\,\Delta t)\,\Delta t \qquad (k = 0, 1, \ldots, N-1), \qquad (11.25)$$

where $\Delta\omega = 2\pi/(N\Delta t)$. For real d_n, D_k is complex and D_{N-k} is the complex conjugate of D_k. The inverse discrete Fourier transform is given by

$$d_m = \sum_{k=0}^{N-1} D_k \exp(-ik\,\Delta\omega m\,\Delta t)\,\frac{\Delta\omega}{2\pi} \qquad (m = 0, 1, \ldots, N-1), \qquad (11.26)$$

where $\Delta\omega = 2\pi/(N\,\Delta t)$.

Although d_n and D_k are approximations to the continuous Fourier-transform pairs, the inverse relation (11.26) holds exactly when (11.25) is valid, because the expression

$$\sum_{k=0}^{N-1} \exp(ik\,\Delta\omega \cdot n\,\Delta t - ik\,\Delta\omega \cdot m\,\Delta t) = \sum_{k=0}^{N-1} \exp\left\{i\left[\frac{2\pi(n-m)k}{N}\right]\right\}$$

is equal to N for $m = n$, and vanishes otherwise.

Let us consider the product of two discrete Fourier transforms D_k and F_k, which are derived by

$$D_k = \sum_{n=0}^{N-1} d_n \exp(ik\,\Delta\omega \cdot n\,\Delta t)\,\Delta t$$

$$F_k = \sum_{n=0}^{N-1} f_n \exp(ik\,\Delta\omega \cdot n\,\Delta t)\,\Delta t \qquad (k = 0, 1, \ldots, N-1). \tag{11.27}$$

In a manner analogous to convolution for continuous functions, we shall define the convolution of d_n and f_n as the inverse Fourier transform of the product $D_k \cdot F_k$:

$$u_l = \sum_{k=0}^{N-1} D_k \cdot F_k \exp(-ik\,\Delta\omega \cdot l\,\Delta t)\,\frac{\Delta\omega}{2\pi}. \tag{11.28}$$

Inserting the inverse transforms of (11.27) into the above equation, we get

$$u_l = \sum_{k=0}^{N-1} \left[\sum_{n=0}^{N-1} d_n \exp(ik\,\Delta\omega \cdot n\,\Delta t)\,\Delta t \right]$$

$$\times \left[\sum_{m=0}^{N-1} f_m \exp(ik\,\Delta\omega \cdot m\,\Delta t)\,\Delta t \exp(-ik\,\Delta\omega l\,\Delta t)\,\frac{\Delta\omega}{2\pi} \right]$$

$$= \frac{1}{N} \sum_{n=0}^{N-1} \sum_{m=0}^{N-1} d_n f_m \left\{ \sum_{k=0}^{N-1} \exp[ik\,\Delta\omega(m + n - l)\,\Delta t] \right\} \Delta t, \tag{11.29}$$

where

$$\sum_{k=0}^{N-1} \exp[ik\,\Delta\omega(m + n - l)\,\Delta t] = \sum_{k=0}^{N-1} \exp\left(ik2\pi\,\frac{m + n - l}{N}\right)$$

is equal to N when $m + n - l$ is zero or integral multiples $(\pm)$ of N, and vanishes otherwise. Therefore, we have

$$u_l = \sum_{m=0}^{N-1} d_{(l-m)} f_m\,\Delta t, \tag{11.30}$$

which is called a *circular convolution* (Gold and Rader, 1969).

In the above equation, the symbol $(l - m)$ means $l - m$ if $0 \le l - m < N$, and $l - m + N$ if $-N \le l - m < 0$. Because of the periodicity implied in the definition (11.26), the tail of the function d_n given in the range $0 \le n < N$ shows up ahead of the signal that begins with $n = 0$. This problem can be easily eliminated by increasing N, thereby adding as many zeros as may be needed in the tail of u_n and f_n.

With these preparations, let us now consider three basic ways of filtering the digital data d_m $(m = 1, 2, \ldots, N - 1)$. They are (i) convolution, (ii) Fast Fourier Transform (FFT), and (iii) recursion.

Suppose that the impulse response of the filter is known as f_i $(i = 0, 1, \ldots, M - 1)$; in that case, the simplest method is, of course, convolution:

$$u_l = \sum_{m=0}^{M-1} d_{l-m} f_m \, \Delta t \qquad (l = 1, \ldots, N - 1). \tag{11.31}$$

This operation, however, requires $M \times N$ multiplications and additions. If we can replace the above operation by the circular convolution (11.30), taking appropriate precautions for the periodicity problem, then the number of multiplications and additions can be drastically reduced by the use of the FFT. As we shall describe in the next section, the discrete Fourier transform (11.25) may be calculated by about $N \log_2 N$ additions and less than $(N/2) \log_2 N$ multiplications. As shown in (11.29), the circular convolution can be obtained by (i) calculating two FFT's, (ii) making N products of the two FFT's, and (iii) calculating one inverse FFT. The total number of additions is about $3N \log_2 N$ and of multiplications is less than $N(1 + \frac{3}{2} \log N)$. Therefore, if the length M of the filter impulse response is greater than $1 + \frac{3}{2} \log_2 N$, it saves time to use the FFT method rather than direct convolution in the time domain. For example, if $N = 1024$, then the FFT method is recommended for $M > 16$.

When the frequency response of the filter is known as F_i $(i = 0, 1, \ldots, N - 1)$, naturally the FFT method is preferred. In this case, only one FFT, N multiplications, and one inverse FFT are required for the operation.

In some cases, the recursive method gives a good result quite effectively. For example, the method of eliminating the surface-layer effect mentioned in Section 11.3.2 was done only by a two-point filter $(M = 2)$ as shown in (11.24). In general, one can write the filter as

$$u_n + \sum_{k=1}^{M_1} a_k u_{n-k} = \sum_{l=0}^{M_2} b_l d_{n-l}. \tag{11.32}$$

The output u_n of the filter is expressed as a linear combination of the past values of output at M_1 time points and the present and past values of input at M_2 time points. Because of this formulation, the output will be zero before any nonzero input is applied. Such a filter is called *physically realizable* or *causal*, and the frequency-response function will have no poles in the upper half ω-plane. The frequency response $F(\omega)$ of the above filter can be obtained by putting $d_n = D(\omega) \exp(-i\omega n \, \Delta t)$ and $u_n = U(\omega) \exp(-i\omega n \, \Delta t)$, which gives

$$F(\omega) = \frac{U(\omega)}{D(\omega)} = \frac{\displaystyle\sum_{l=0}^{M_2} b_l \exp(i\omega l \, \Delta t)}{1 + \displaystyle\sum_{k=1}^{M_1} a_k \exp(i\omega k \, \Delta t)}. \tag{11.33}$$

It is convenient to change the variable ω to z by $z = \exp(i\omega \, \Delta t)$, so that

$$F(z) = \frac{\displaystyle\sum_{l=0}^{M_2} b_l z^l}{1 + \displaystyle\sum_{k=1}^{M_1} a_k z^k}. \qquad (11.34)$$

If we extend z to complex values, we find that the real axis of the complex ω-plane corresponds to the unit circle in the z-plane, because $|\exp(i\omega \, \Delta t)| = 1$ for real ω. A given ω corresponds to a point on the unit circle at the phase angle $\omega \, \Delta t$. The absolute value of $F(z)$ may be visualized as the product of the distances from that point to all the roots of polynomial $\Sigma b_l z^l$ divided by the product of the distances from the same point to all the roots of polynomial $1 + \Sigma a_k z^k$. The phase angle of $F(z)$ may be visualized as the sum of the phase angles measured from the point on the unit circle to all the roots of $\Sigma b_l z^l$ minus the sum of the phase angles measured from the same point to all the roots of $1 + \Sigma a_k z^k$. With the aid of this geometrical interpretation, one can construct a filter of desired characteristics. (Try Problem 11.3.) If only a small number, $M_1 + M_2$, of coefficients is required for a given purpose, this method can be more efficient than the FFT method.

11.3.5   Fast Fourier transform

The discrete Fourier transform of a time series d_k $(k = 0, 1, 2, \ldots, N-1)$ was defined by (11.25) as

$$D_j = \sum_{k=0}^{N-1} d_k \exp(ij \, \Delta\omega k \, \Delta t) \, \Delta t \qquad (j = 0, 1, \ldots, N-1),$$

where $\Delta\omega = 2\pi/(N \, \Delta t)$. Putting $E = \exp(2\pi i/N)$, we can write

$$D_j = \sum_{k=0}^{N-1} d_k E^{jk}, \qquad (11.35)$$

where we have for convenience dropped the scaling factor Δt.

In the FFT computation introduced by Cooley and Tukey (1965), we assume that the number N of the data points is an integral power of 2:

$$N = 2^m. \qquad (11.36)$$

This is not a restrictive condition for a transient signal frequently encountered in seismology, because one can add an arbitrary number of zeros to the head

or tail of the record. A review of several variations of FFT algorithms, including those applicable to the number of data points other than 2^m, is given in a textbook on digital processing by Gold and Rader (1969).

Any integer k, $0 \le k < N = 2^m$, may be expressed as a binary number,

$$k = k_0 + k_1 \cdot 2 + \cdots + k_{m-1} \cdot 2^{m-1},$$

where k_i takes the value 0 or 1.

Similarly, an integer j, $0 \le j < 2^m$, may be expressed as

$$j = j_0 + j_1 \cdot 2 + \cdots + j_{m-1} \cdot 2^{m-1}.$$

Then equation (11.35) can be rewritten as

$$
\begin{aligned}
D_j &= D(j_{m-1}, j_{m-2}, \ldots, j_0) \\
&= \sum_{k_0=0}^{1} \sum_{k_1=0}^{1} \cdots \sum_{k_{m-1}=0}^{1} d_k E^{(\Sigma_{r=0}^{m-1} k_r \cdot 2^r)(\Sigma_{s=0}^{m-1} j_s \cdot 2^s)}.
\end{aligned}
\tag{11.37}
$$

The sum is taken for each k_r over 0 and 1. Since $2^m = N$ and $E^N = 1$, we have

$$E^{(\Sigma_{r=0}^{m-1} k_r 2^r)(\Sigma_{s=0}^{m-1} j_s 2^s)} = E^{(\Sigma_{r=0}^{m-1} k_r 2^r \Sigma_{s=0}^{m-r-1} j_s 2^s)}.
\tag{11.38}$$

Inserting (11.38) into (11.37) and putting $d_k = d(k_{m-1}, k_{m-2}, \ldots, k_0)$ we can rewrite (11.37) as

$$
\begin{aligned}
&D(j_{m-1}, j_{m-2}, \ldots, j_0) \\
&= \sum_{k_0} E^{(k_0 \Sigma_0^{(m-1)} j_s 2^s)} \sum_{k_1} E^{(k_1 \cdot 2 \cdot \Sigma_0^{(m-2)} j_s 2^s)} \cdots \sum_{k_r} E^{(k_r \cdot 2^r \cdot \Sigma_0^{(m-r-1)} j_s 2^s)} \\
&\quad \cdots \sum_{k_{m-1}} E^{(k_{m-1} \cdot 2^{m-1} \cdot j_0)} d(k_{m-1}, k_{m-2}, \ldots, k_0).
\end{aligned}
\tag{11.39}
$$

The above operation can be accomplished by the following recursive relation:

$$
\begin{aligned}
&d(j_0, j_1, \ldots, j_{m-r-1}, k_{r-1}, \ldots, k_0) \\
&= \sum_{k_r} E^{(k_r 2^r \Sigma_{s=0}^{m-r-1} j_s 2^s)} d(j_0, j_1, \ldots, j_{m-r-2}, k_r, \ldots, k_0).
\end{aligned}
\tag{11.40}
$$

Starting with $d(k_{m-1}, \ldots, k_0)$, we make the summation over k_{m-1} using (11.40) to get $d(j_0, k_{m-2}, \ldots, k_0)$. Then, by summing $d(j_0, k_{m-2}, \ldots, k_0)$ over k_{m-2} by (11.40), we obtain $d(j_0, j_1, k_{m-3}, \ldots, k_0)$ and so on until finally we get $d(j_0, j_1, \ldots, j_{m-1})$, which is equal to $D(j_{m-1}, j_{m-2}, \ldots, j_0)$:

$$D(j_{m-1}, j_{m-2}, \ldots, j_0) = d(j_0, j_1, \ldots, j_{m-1}).
\tag{11.41}$$

Thus the FFT is calculated by employing the m-step operation of recursive formula (11.40) and finally reversing the order of binary digits of the argument.

Let us illustrate this process for the case of $m = 3$. The first step is given by (11.40) as

(i) $\quad d(j_0, k_1, k_0) = \sum_{k_2=0}^{1} E^{k_2 4 j_0}\, d(k_2, k_1, k_0)$

$$= E^0\, d(0, k_1, k_0) + E^{4 j_0}\, d(1, k_1, k_0).$$

The second step yields

(ii) $\quad d(j_0, j_1, k_0) = \sum_{k_1=0}^{1} E^{k_1 2(j_0 + j_1 2)}\, d(j_0, k_1, k_0)$

$$= E^0\, d(j_0, 0, k_0) + E^{2(j_0 + j_1 2)}\, d(j_0, 1, k_0).$$

The third step gives

(iii) $\quad d(j_0, j_1, j_2) = \sum_{k_0=0}^{1} E^{k_0(j_0 + j_1 2 + j_2 4)}\, d(j_0, j_1, k_0)$

$$= E^0\, d(j_0, j_1, 0) + E^{(j_0 + j_1 2 + j_2 4)}\, d(j_0, j_1, 1).$$

Then we obtain the final result

(iv) $\quad D(j_2, j_1, j_0) = d(j_0, j_1, j_2).$

Figure 11.7 shows the above steps from left to right. At the extreme left, $d_k = d(k_2, k_1, k_0)$ is arranged in a bit-reversed order. This order of argument is kept the same throughout the entire set of steps. The arrows with the power of E indicate the result of each step. For example, during the first step, $E^0\, d(0, 0, 0)$ plus $E^4\, d(1, 0, 0)$ will yield the new $d(1, 0, 0)$. Since $d(k_2, k_1, k_0)$ are no longer needed in further computation after completion of the first step, the computer memories holding them can be erased to store $d(j_0, k_1, k_0)$. After the completion of the third step, the order of binary digits in the argument is reversed to produce the final result.

Since $E = \exp(2\pi i/N)$ and $N = 2^3 = 8$, E^0 is 1 and E^4 is -1. Therefore, as shown in Figure 11.7, the first step includes no multiplication but only four additions and subtractions. The second step includes four multiplications and the same total number of additions and subtractions as the first step. The third step includes six multiplications and the same total number of additions and subtractions as the earlier steps. From this result, we may expect that the total number of additions and subtractions is $N \log_2 N$, as mentioned in Section

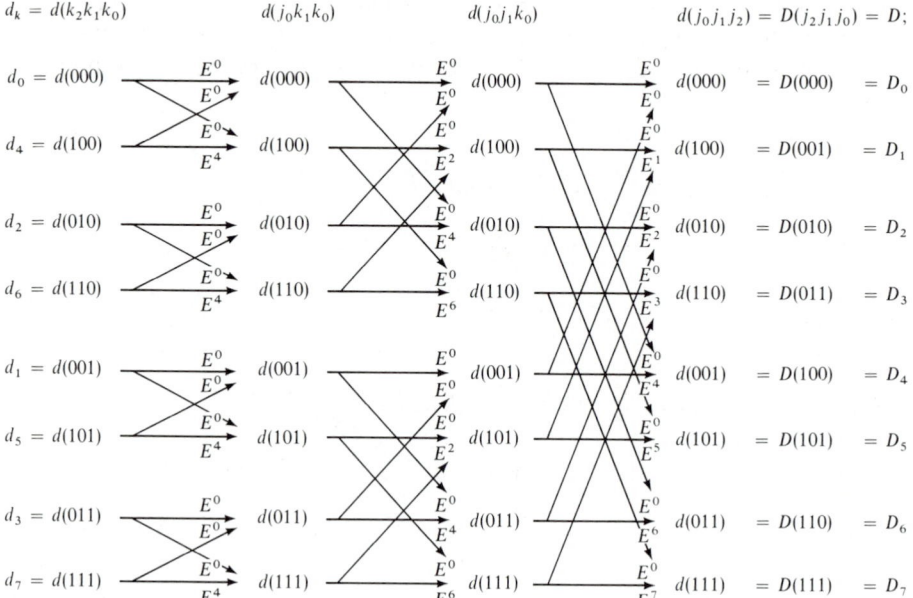

FIGURE **11.7**

Step (i) through (iv) in the Fast Fourier Transform (FFT) for the case of data length $2^3 = 8$.

```
      SUBROUTINE FFT(A,M)
      COMPLEX A(1024),U,W,T
      N = 2**M
      NV2 = N/2
      NM1 = N-1
      J = 1
      DO 7 I = 1, NM1
      IF(I.GE.J) GO TO 5
      T = A(J)
      A(J) = A(I)
      A(I) = T
    5 K = NV2
    6 IF(K.GE.J) GO TO 7
      J = J-K
      K = K/2
      GO TO 6
    7 J = J+K
      PI = 3.1415926538979
      DO 20 L = 1,M
      LE = 2**L
      LE1 = LE/2
      U = (1.0,0.)
      W = CMPLX(COS(PI/LE1),SIN(PI/LE1))
      DO 20 J=1, LE1
      DO 10 I = J,N,LE
      IP = I+LE1
      T = A(IP)*U
      A(IP) = A(I)-T
   10 A(I) = A(I)+T
   20 U = U*W
      RETURN
      END
```

FIGURE **11.8**

Program for computing DFT by the FFT method. [From Cooley et al., 1969.]

11.3.4. The direct calculation by (11.35) requires N^2 multiplications and additions. This means that the FFT is more than a factor of $N/\log_2 N$ faster than the ordinary method. For $N = 1024 = 2^{10}$, the ratio $N/\log_2 N$ is 100, and for $N = 16384 = 2^{14}$ it is more than 1000. Reproduced in Figure 11.8 is a complete operable FFT program in FORTRAN written by Cooley et al. (1969).

11.3.6 Stationary time series

In this section, we shall consider a stochastic model of seismic noise. We shall first generate white noise, i.e., a random sequence of a statistical variable which follows the Gaussian distribution. Then, applying a linear digital filter as discussed in Section 11.3.4 to the white noise, we shall generate a model of seismic noise.

Let x be a statistical variable that obeys the Gaussian distribution with the mean m and the variance σ^2. The probability density function $f(x)$ is given by

$$f(x) = \frac{1}{\sigma\sqrt{2\pi}} \exp\left[-\frac{(x-m)^2}{2\sigma^2}\right] = N(m, \sigma^2). \tag{11.42}$$

The corresponding moment-generating function $\phi(\theta)$ is

$$\phi(\theta) = \int_{-\infty}^{\infty} e^{\theta x} f(x)\, dx = \exp[\tfrac{1}{2}(\theta^2\sigma^2) + \theta m]. \tag{11.43}$$

$\phi(\theta)$ gives moments by definition,

$$\phi(0) = \int_{-\infty}^{\infty} f(x)\, dx = 1,$$

$$\left.\frac{d\phi}{d\theta}\right|_{\theta=0} = \int_{-\infty}^{\infty} xf(x)\, dx = \langle x \rangle,$$

$$\left.\frac{d^2\phi}{d\theta^2}\right|_{\theta=0} = \int_{-\infty}^{\infty} x^2 f(x)\, dx = \langle x^2 \rangle.$$

For the Gaussian distribution,

$$\langle x \rangle = (\theta\sigma^2 + m)\phi\big|_{\theta=0} = m,$$

$$\langle x^2 \rangle = \{\sigma^2 + (\theta\sigma^2 + m)^2\}\phi\big|_{\theta=0} = \sigma^2 + m^2.$$

Consider two variables x_1 and x_2, each following independently the Gaussian distributions $N(m_1, \sigma_1^2)$ and $N(m_2, \sigma_2^2)$, respectively. Then the probability that

the sum $x_1 + x_2$ falls between x and $x + dx$ is given by

$$f(x)\,dx = \iint_{x < x_1 + x_2 < x + dx} f_1(x_1)f_2(x_2)\,dx_1\,dx_2$$

$$= \int_{-\infty}^{\infty} f_1(x - x_1)f_2(x_1)\,dx_1\,dx. \tag{11.44}$$

Since the above operation is a convolution, the corresponding moment-generating functions (just like Fourier transforms) satisfy the following relation:

$$\phi(\theta) = \phi_1(\theta)\phi_2(\theta)$$

$$= \exp[\tfrac{1}{2}\theta^2(\sigma_1^2 + \sigma_2^2) + \theta(m_1 + m_2)]. \tag{11.45}$$

This corresponds to the Gaussian distribution $N(m_1 + m_2, \sigma_1^2 + \sigma_2^2)$. Thus the sum of two Gaussian variables follows the Gaussian distribution. The mean of the sum is equal to the sum of the means, and the variance of the sum is equal to the sum of the variances.

Suppose $x_1, x_2, \ldots, x_n$ are n independent samples from an ensemble following the Gaussian distribution $N(m, \sigma^2)$. Then $x = x_1 + x_2 + \ldots + x_n$ will follow the Gaussian distribution $N(nm, n\sigma^2)$. The sample mean x/n will obey the distribution $N(m, \sigma^2/n)$, because $\langle x \rangle$ is n times $\langle x/n \rangle$ and $\langle x^2 \rangle$ is n^2 times $\langle (x/n)^2 \rangle$. Thus the variance of the sample mean decreases inversely in proportion to the sample size n.

Similarly, an arbitrary linear combination of x's, $a_1x_1 + a_2x_2 + \cdots + a_nx_n$, where the a's are constants, obeys the Gaussian distribution

$$N((a_1 + a_2 + \cdots + a_n)m, (a_1^2 + a_2^2 + \cdots + a_n^2)\sigma^2).$$

Let us define a discrete white noise x_t $(t = \ldots, 0, 1, 2, \ldots)$ as an infinite series of independent samples from the Gaussian ensemble $N(0, \sigma^2)$. If we take a finite length $N\,\Delta t$ of the white noise and compute the discrete Fourier transform by (11.25), we obtain

$$X_k = \sum_{n=0}^{N-1} x_n \exp(ik\,\Delta\omega \cdot n\,\Delta t)\,\Delta t \qquad (k = 0, 1, 2, \ldots, N-1) \tag{11.46}$$

where $\Delta\omega = 2\pi/(N\,\Delta t)$. Separating it into real and imaginary parts, we have

$$X_k = \left(\sum_{n=0}^{N-1} x_n \cos\frac{2\pi kn}{N} + i \sum_{n=0}^{N-1} x_n \sin\frac{2\pi kn}{N} \right)\Delta t. \tag{11.47}$$

The squared absolute value of X_k is then

$$|X_k|^2 = \left(\Delta t \sum_{n=0}^{N-1} x_n \cos\frac{2\pi kn}{N} \right)^2 + \left(\Delta t \sum_{n=0}^{N-1} x_n \sin\frac{2\pi kn}{N} \right)^2. \tag{11.48}$$

Both real and imaginary parts are linear combinations of x_n, and each obeys the Gaussian distribution. Their means are both zero, and the variances for the cosine sum and the sine sum are, respectively,

$$\sigma^2 \Delta t^2 \sum_{n=0}^{N-1} \cos^2 \frac{2\pi kn}{N} \quad \text{and} \quad \sigma^2 \Delta t^2 \sum_{n=0}^{N-1} \sin^2 \frac{2\pi kn}{N}.$$

Therefore, taking the average of (11.48), we have

$$\langle |X_k|^2 \rangle = \sigma^2 \Delta t^2 \sum_{n=0}^{N-1} \left(\cos^2 \frac{2\pi kn}{N} + \sin^2 \frac{2\pi kn}{N} \right) = N\sigma^2 \Delta t^2. \quad (11.49)$$

$\langle |X_k|^2 \rangle$ is the measure of energy carried by the kth Fourier component. As shown in (11.49), this estimate increases proportionally with the record length $N \Delta t$. Therefore, we define the discrete power spectrum P_k by dividing the energy by the time length $N \Delta t$ and making N sufficiently large:

$$P_k = \lim_{N \to \infty} \frac{\langle |X_k|^2 \rangle}{N \Delta t} = \sigma^2 \Delta t. \quad (11.50)$$

Thus the power spectrum of white noise is a constant, independent of frequency Moreover, (11.50) agrees with the discrete Fourier transform of the auto-correlation of x_t, which is σ^2 for zero lag, and vanishes otherwise.

With these preparations, we now construct a model of seismic noise by passing the discrete white noise through a digital filter. Calling the impulse response of the filter f_i and its discrete Fourier transform F_k, the noise n_t can then be expressed either by a convolution,

$$n_t = \sum_i x_{t-i} f_i \, \Delta t, \quad (11.51)$$

or by the corresponding power spectrum $P_k^{(n)}$, the energy per unit time in the kth frequency,

$$P_k^{(n)} = \lim_{N \to \infty} \frac{\langle |X_K \cdot F_K|^2 \rangle}{N \Delta t} = \sigma^2 |F_k|^2 \, \Delta t. \quad (11.52)$$

Thus the power spectral density of the noise n_i is the product of the spectral density of white noise and the squared absolute value of the discrete Fourier transform of the filter impulse response.

The autocorrelation function $r_k = \langle n_t n_{t+k} \rangle = \langle n_t n_{t-k} \rangle$ can be written, using (11.51), as $r_k = \sum_i \sum_j \langle x_{t-i} x_{t+k-j} \rangle f_i f_j (\Delta t)^2$. Since $\langle x_t x_{t+k} \rangle = \sigma^2$ for $k = 0$ and vanishes otherwise, $\langle x_{t-i} x_{t+k-j} \rangle$ vanishes except for $j = i + k$. Thus we have $r_k = \sigma^2 (\Delta t)^2 \sum_i f_i f_{i+k}$. On the other hand, the discrete Fourier transform

of $|F_k|^2$ is equal to $\sum_i f_i f_{i+k} \Delta t$, because, since

$$|F_k|^2 = F_k F_k^*$$

$$= \left(\sum_{n=0}^{N-1} f_n \exp(ik \, \Delta\omega n \, \Delta t) \, \Delta t \right) \left(\sum_{n'=0}^{N-1} f_{n'} \exp(-ik \, \Delta\omega n' \, \Delta t) \, \Delta t \right),$$

$$\sum_{k=0}^{N-1} |F_k|^2 \exp(-ik \, \Delta\omega m \, \Delta t) \frac{\Delta\omega}{2\pi}$$

$$= \sum_{n=0}^{N-1} \sum_{n'=0}^{N-1} f_n f_{n'} \, (\Delta t)^2 \frac{\Delta\omega}{2\pi} \sum_{k=0}^{N-1} \exp[ik \, \Delta\omega(n - m - n') \, \Delta t]$$

$$= \sum_{n'=0}^{N-1} f_{n'} f_{(n'+m)} \, \Delta t,$$

where $(n' + m) = n' + m$ if $0 \leqslant n' + m \leqslant N - 1$ and $(n' + m) = n' + m - N$ if $0 \leq n' + m - N \leq N - 1$. We may call this "circular autocorrelation" in accordance with the circular convolution defined in (11.30). Therefore, comparing (11.52) with the expression for autocorrelation r_k, we find

$$r_m = \sum_{k=0}^{N-1} P_k^{(n)} \exp(-ik \, \Delta\omega m \, \Delta t) \frac{\Delta\omega}{2\pi}.$$

In particular, the noise power is expressed as

$$\langle n_t^2 \rangle = r_0 = \sum_{k=1}^{N-1} P_k^{(n)} \frac{\Delta\omega}{2\pi},$$

justifying the name *power spectral density* for $P_k^{(n)}$.

Since both f_k and F_k are considered as periodic with period N,

$$f_k = f_{N+k} \quad \text{and} \quad F_k = F_{N+k}.$$

On the other hand, both r_k and $|F_k|^2$ are even functions with respect to k. Therefore, it follows that

$$r_k = r_{N-k} \quad \text{and} \quad |F_k|^2 = |F_{N-k}|^2.$$

Because of this property, the frequency $(N - m) \Delta\omega$ should be considered as $-m \Delta\omega$ for $m < N/2$.

Filtering can be applied to the white noise x_i by the recursive formula (11.32):

$$n_i + \sum_{m=1}^{M_1} a_m n_{i-m} = \sum_{l=0}^{M_2} b_l x_{i-l}. \tag{11.53}$$

The power spectral density of output n_i defined by the above equation is given by

$$P_k^{(n)} = \sigma^2 \cdot \frac{\Delta t \left| \sum_{l=0}^{M_2} b_l \exp(ikl \, \Delta\omega \, \Delta t) \right|^2}{\left| 1 + \sum_{m=1}^{M_1} a_m \exp(ikm \, \Delta\omega \, \Delta t) \right|^2} \cdot \tag{11.54}$$

Consider a special case $M_2 = 0$. Setting $M_1 = M$ and $b_0 = 1$, we have

$$n_i + \sum_{m=1}^{M} a_m n_{i-m} = x_i. \tag{11.55}$$

The time series n_i defined by the above equation is called *autoregressive* (Wold, 1938). The filter that generates n_i from the white-noise input has a frequency response $(1 + \sum_{m=1}^{M} a_m \exp(ikm \, \Delta\omega \, \Delta t))^{-1}$. Since this response has no zeros or poles in the upper ω-plane, the filter has minimum delay (Box 5.8). As shown in Box 5.8, the phase response of a minimum-delay filter can be found from its amplitude response. In other words, the total filter response can be found solely from the power response.

Robinson's model of the reflection seismogram, discussed in Section 11.3.3, is precisely the autoregressive time series defined by (11.55). Robinson's deconvolution process consists in finding the white noise x_i from the reflection seismogram n_i by (11.55). Because the filter has minimum delay, one can find the filter coefficients a_i from the autocorrelation function, which contains only amplitude information, as follows.

Since x_i is purely random, the present value of x_i is not related to the past value of n_j. In other words, the correlation $\langle x_i n_j \rangle = 0$ for $j < i$. Multiplying n_j on both sides of (11.55) and averaging, we have

$$\langle n_j n_i \rangle + \sum_{m=1}^{M} a_m \langle n_j n_{i-m} \rangle = 0, \qquad j < i. \tag{11.56}$$

Using the autocorrelation function r_k, we rewrite (11.56) as

$$\sum_{m=1}^{M} a_m r_{k-m} = -r_k, \qquad (k = 1, 2, \ldots, M)$$

or

$$\begin{pmatrix} r_0 & r_1 & \cdots & r_{M-1} \\ r_1 & r_0 & r_1 & \\ \vdots & r_1 & \ddots & r_1 \\ r_{M-1} & \cdots & r_1 & r_0 \end{pmatrix} \begin{pmatrix} a_1 \\ a_2 \\ \vdots \\ a_M \end{pmatrix} = - \begin{pmatrix} r_1 \\ r_2 \\ \vdots \\ r_M \end{pmatrix}.$$

Matrices with the banded structure of the $M \times M$ matrix above are known as Toeplitz matrices; a quick, recursive method for solving the above equation will be given in Section 11.3.8. Thus, knowing the autocorrelation function, the filter coefficients in (11.55) can be determined by the above equation, which can be rewritten as

$$
\begin{pmatrix}
r_0 & \cdots\cdots & r_M \\
\vdots & r_0 & \\
 & & \ddots \\
r_M & & r_0
\end{pmatrix}
\begin{pmatrix}
1 \\
a_1 \\
\vdots \\
a_M
\end{pmatrix}
=
\begin{pmatrix}
\alpha \\
0 \\
\vdots \\
0
\end{pmatrix},
\tag{11.57}
$$

where $\alpha = r_0 + r_1 a_1 + r_2 a_2 + \cdots + r_M a_M$. Multiplying both sides of (11.55) by n_i and averaging, we find that $\alpha = \langle n_i x_i \rangle$. On the other hand, since x_i is purely random, $\langle x_i n_j \rangle = 0$ for $j < i$. Then, multiplying both sides of (11.55) by x_i and averaging, we find

$$
\langle n_i x_i \rangle = \langle x_i^2 \rangle = \sigma^2.
$$

In other words, α is the variance of the white noise.

Equation (11.57) for a_i can also be derived by minimizing the mean square of $(n_i + \sum_{m=1}^{M} a_m n_{i-m})$ with respect to a_i. This is because the least-squares residuals are intended to be randomized, removing any systematic predictable part. The minimum value of the mean square is equal to α.

11.3.7 Measurement of the power spectrum

Let us now consider the problem of estimating the power spectrum when a length $N\Delta t$ of noise data n_i ($i = 0, 1, \ldots, N - 1$) is given. The discrete Fourier transform of n_i can be written as

$$
N_k = F_k X_k,
\tag{11.58}
$$

where F_k is the frequency response of a digital filter and X_k is the discrete Fourier transform of white noise x_k. Note that N_k is a statistical variable because X_k is a statistical variable. Taking the squared absolute value of N_k and using (11.48), we can write

$$
|N_k|^2 = |F_k|^2 \left[\left(\Delta t \sum_{n=0}^{N-1} x_n \cos \frac{2\pi k n}{N} \right)^2 + \left(\Delta t \sum_{n=0}^{N-1} x_n \sin \frac{2\pi k n}{N} \right)^2 \right].
\tag{11.59}
$$

Each term in ( ) is a linear combination of x_n, and therefore obeys the Gaussian distribution. Their means are zero, and when N is large, both the sine term and the cosine term share the same variance $N\sigma^2 \Delta t^2/2$, as shown in (11.49).

Since $|N_k|^2$ is the sum of two squares of Gaussian variables, the normalized value with respect to the variance,

$$\chi_2^2 = \frac{2|N_k|^2}{|F_k|^2 N\sigma^2 \, \Delta t^2},$$ (11.60)

obeys the chi-squared distribution with two degrees of freedom.

In general, the mean of χ_n^2, which obeys the chi-squared distribution with n degrees of freedom, is n, and the RMS deviation of χ_n^2 from its mean is $\sqrt{2n}$. The ratio of RMS error to mean is, therefore, $\sqrt{2}/\sqrt{n}$, which is as large as 1 for $n = 2$.

We can reduce this error in estimating $|N_k|^2$ by averaging over the neighboring frequencies. Taking the sum over $2m$ neighbor points, over which $|F_k|^2$ may be assumed to vary slowly,

$$\chi_{2(2m+1)}^2 = \frac{2 \displaystyle\sum_{l=-m}^{m} |N_{k+l}|^2}{|F_k|^2 N\sigma^2 \, \Delta t^2}$$ (11.61)

will obey the chi-squared distribution with $2(2m + 1)$ degrees of freedom. Then the fractional RMS error of $\chi_{2(2m+1)}^2$—i.e., the ratio of RMS deviation of $\chi_{2(2m+1)}^2$ to its mean—is $1/\sqrt{2m + 1}$. The same fractional RMS error is applicable to the power spectrum estimate (11.52) after smoothing over $2m + 1$ points,

$$\frac{1}{N \, \Delta t} \cdot \frac{1}{2m + 1} \sum_{l=-m}^{m} |N_{k+l}|^2.$$

For a given fractional error $\varepsilon = 1/\sqrt{2m + 1}$, the spectral estimate must be smoothed over the frequency interval $\pm \Delta f$ given by

$$\Delta f = \frac{\Delta\omega}{2\pi} (m + \tfrac{1}{2}) = \frac{m + \tfrac{1}{2}}{N \, \Delta t} = \frac{1}{2N \, \Delta t \varepsilon^2}.$$ (11.62)

For a stable estimate of the power spectrum, ε must be small, and the estimate must be smoothed over a wide frequency range. The reliability of the estimate is improved at the expense of frequency resolution. For a given reliability, the frequency resolution improves linearly with the record length. On the other hand, for a given frequency resolution, the accuracy of the estimate improves linearly as the square root of the record length.

For a given record length $N \, \Delta t$, the resolvable frequency width is ultimately limited to $\Delta\omega = 2\pi/(N \, \Delta t)$. However, if the time series is of the autoregressive

type, as defined by (11.55), it is possible to gain the resolution beyond this limit. The property of the filter that generates the autoregressive time series is determined by a small number of filter coefficients a_i ($i = 1, \ldots, M$). Furthermore, the filter coefficients can be determined by (11.57), using a finite portion of the autocorrelation function $r_k = \langle n_j n_{j+k} \rangle$ for k from 0 up to only $M - 1$. Once the a's are determined, the power spectrum can be calculated by (11.54), where $M_2 = 0$ for the autoregressive time series.

The above process may be more clearly visualized by considering the following experiment. First, we put noise n_i through a filter and observe the output. We determine the filter coefficients in such a way that the output x_i is whitened, or purely randomized, or minimized in the least-squares sense. We then construct its inverse filter and feed in sinusoidal inputs. The squared amplitude of the output will be the desired power spectrum. This is the basis of the so-called *maximum entropy method* (Burg, 1967) for estimating the power spectrum, and it can attain an excellent resolution from a short length of data when the data are well approximated by an autoregressive time series. Andersen (1974) describes the practical steps for fast application of the maximum entropy method.

The autoregressive time series is rather common in nature. For example, the case $M = 1$, in which the autocorrelation function decays exponentially with the lag, may be found in the Brownian motion of a particle in a viscous fluid. From (11.55) we find for this case

$$n_i + a_1 n_{i-1} = x_i.$$

Multiplying by n_{i-1} and averaging over i, we see that

$$r_1 + a_1 r_0 = 0.$$

Multiplying by n_{i-j} ($i > j$) and averaging over i gives

$$r_j + a_1 r_{j-1} = 0$$

or

$$r_j/r_{j-1} = r_1/r_0 = -a_1.$$

Thus we obtain an autocorrelation function that decays exponentially. The corresponding power spectrum is given by (11.54) as $\sigma^2 \, \Delta t / |1 + a_1 \exp(ik \, \Delta\omega \, \Delta t)|^2$. The case $M = 2$ corresponds to an autocorrelation function with the form of a damped oscillation. LaCoss (1971) compared several methods of estimating power spectra and demonstrated the high resolution attained by the maximum-entropy method using time series with known spectra.

11.3.8 *Signal filtering by the least-squares method*

Consider that the seismogram d_t is the sum of signal s_t and noise n_t, both of which are stationary time series, as discussed in Section 11.3.6:

$$d_t = n_t + s_t. \tag{11.63}$$

The objective of this section is to design a linear filter that will best eliminate noise from the seismogram. Let the impulse response of the filter be f_i $(i = 0, 1, 2 \dots)$. Passing the data d_t through the filter, the output will be $\sum_{i=0}^{\infty} d_{t-i} f_i$. We want this output to be the best estimate of s_t. Let us consider repeated experiments in which we take out samples of time series n_t and s_t from their ensembles. Writing the samples in the kth experiment as

$$d_t^{(k)} = n_t^{(k)} + s_t^{(k)},$$

we shall choose the filter response f_i that will, on the average, minimize the square of the difference between filter output and signal, i.e.,

$$\sum_{k=1}^{N} \left(\sum_i d_{t-i}^{(k)} f_i - s_t^{(k)} \right)^2.$$

Taking derivatives of this square sum with respect to the parameters f_i and setting them at zero, we have

$$\frac{\partial}{\partial f_j}: \quad \sum_k \left(\sum_i d_{t-i}^{(k)} f_i - s_t^{(k)} \right) d_{t-j}^{(k)} = 0, \quad (j = 0, 1, 2, \dots)$$

or

$$\sum_i f_i \sum_k d_{t-i}^{(k)} d_{t-j}^{(k)} = \sum_k s_t^{(k)} d_{t-j}^{(k)}. \quad (j = 0, 1, 2, \dots) \tag{11.64}$$

The sums with respect to k can be identified as the autocorrelation function r_τ of data and the cross-correlation function g_τ between data and signal:

$$r_\tau = \lim_{N \to \infty} \frac{1}{N} \sum_{k=1}^{N} d_t^{(k)} d_{t-\tau}^{(k)},$$

$$\tag{11.65}$$

$$g_\tau = \lim_{N \to \infty} \frac{1}{N} \sum_{k=1}^{N} s_t^{(k)} d_{t-\tau}^{(k)}.$$

We can then rewrite (11.64) as

$$\sum_{i=0}^{\infty} f_i r_{j-i} = g_j, \quad (j = 0, 1, 2, \dots). \tag{11.66}$$

This equation gives the desired filter coefficient f_i. Such a least-squares filter was first studied by Wiener (1949).

An efficient method of solving the above equation is well known. Truncating the filter length to n, equation (11.66) can be written as

$$
\begin{pmatrix}
r_0 & r_1 & \cdots & r_{n-1} \\
r_1 & r_0 & r_1 & \\
\vdots & r_1 & \ddots & r_1 \\
r_{n-1} & \cdots & r_1 & r_0
\end{pmatrix}
\begin{pmatrix}
f_0 \\ f_1 \\ \vdots \\ f_{n-1}
\end{pmatrix}
=
\begin{pmatrix}
g_0 \\ g_1 \\ \vdots \\ g_{n-1}
\end{pmatrix},
\tag{11.67}
$$

where all f_i and g_i are considered as the components of column vectors. The matrix formed by the autocorrelation functions is a Toeplitz matrix; it appeared previously in the determination of filter coefficients for an autoregressive time series (11.56). A quick recursive method of solving the linear equation involving this matrix was found by Levinson (1949). The method was later extended by Wiggins and Robinson (1965) to the case of multiple time series.

To solve (11.67), we shall first solve (11.57). The latter equation was formulated in the problem of representing a stationary time series by the filtered white noise and finding the filter response from the autocorrelation function under the assumption of minimum delay. Equation (11.57) corresponds to the special case of the present problem in which the signal is absent.

The recursive method of solving (11.57) starts with the solution of

$$
\begin{pmatrix}
r_0 & r_1 \\
r_1 & r_0
\end{pmatrix}
\begin{pmatrix}
1 \\
a_1^{(1)}
\end{pmatrix}
=
\begin{pmatrix}
\alpha^{(1)} \\
0
\end{pmatrix}
$$

and uses the result in obtaining the solution of

$$
\begin{pmatrix}
r_0 & r_1 & r_2 \\
r_1 & r_0 & r_1 \\
r_2 & r_1 & r_0
\end{pmatrix}
\begin{pmatrix}
1 \\
a_1^{(2)} \\
a_2^{(2)}
\end{pmatrix}
=
\begin{pmatrix}
\alpha^{(2)} \\
0 \\
0
\end{pmatrix},
$$

and so on. It is sufficient to show how to go from the nth step to the $(n+1)$th step. Suppose we know the solution $a_i^{(n)}$ of the following equation:

$$
\begin{pmatrix}
r_0 & r_1 & \cdots & r_n \\
r_1 & r_0 & & \\
\vdots & r_1 & \ddots & r_1 \\
r_n & \cdots & r_1 & r_0
\end{pmatrix}
\begin{pmatrix}
1 \\
a_1^{(n)} \\
\vdots \\
a_n^{(n)}
\end{pmatrix}
=
\begin{pmatrix}
\alpha^{(n)} \\
0 \\
\vdots \\
0
\end{pmatrix}.
\tag{11.68}
$$

We then increase the dimension by rewriting (11.68) as

$$
\begin{pmatrix}
r_0 & r_1 & \cdots & r_{n+1} \\
r_1 & r_0 & & \\
\vdots & & \ddots & r_1 \\
r_{n+1} & \cdots & r_1 & r_0
\end{pmatrix}
\begin{pmatrix}
1 \\
a_1^{(n)} \\
\vdots \\
a_n^{(n)} \\
0
\end{pmatrix}
=
\begin{pmatrix}
\alpha^{(n)} \\
0 \\
\vdots \\
0 \\
\beta^{(n)}
\end{pmatrix},
\tag{11.69}
$$

where $\beta^{(n)} = r_{n+1} + r_n a_1^{(n)} + \cdots + r_1 a_n^{(n)}$. Using the symmetry of the matrix, we find that (11.69) is identical to

$$
\begin{pmatrix}
r_0 & & \cdots & r_{n+1} \\
\vdots & & & \vdots \\
r_{n+1} & & \cdots & r_0
\end{pmatrix}
\begin{pmatrix}
0 \\
a_n^{(n)} \\
\vdots \\
a_1^{(n)} \\
1
\end{pmatrix}
=
\begin{pmatrix}
\beta^{(n)} \\
0 \\
\vdots \\
0 \\
\alpha^{(n)}
\end{pmatrix}.
\tag{11.70}
$$

Multiplying (11.70) by a constant $k^{(n)}$ and adding to (11.69) gives

$$
\begin{pmatrix}
r_0 & \cdots & r_{n+1} \\
\vdots & \ddots & \vdots \\
r_{n+1} & \cdots & r_0
\end{pmatrix}
\begin{pmatrix}
1 \\
a_1^{(n)} + k^{(n)} a_n^{(n)} \\
a_2^{(n)} + k^{(n)} a_{n-1}^{(n)} \\
\vdots \\
k^{(n)}
\end{pmatrix}
=
\begin{pmatrix}
\alpha^{(n)} + k^{(n)} \beta^{(n)} \\
0 \\
\vdots \\
0 \\
\beta^{(n)} + k^{(n)} \alpha^{(n)}
\end{pmatrix}.
\tag{11.71}
$$

If we choose $k^{(n)} = -\beta^{(n)}/\alpha^{(n)}$, the bottom component on the right-hand side vanishes, and the above equation will have the desired form (11.68). Thus the $(n + 1)$th solution $a_i^{(n+1)}$ is expressed by $a_i^{(n)}$ and $\alpha^{(n)}$ as

$$
a_i^{(n+1)} = a_i^{(n)} + k^{(n)} a_{n+1-i}^{(n)}, \qquad 1 \le i \le n
$$

and

$$
a_{n+1}^{(n+1)} = k^{(n)},
$$

where

$$
k^{(n)} = -\frac{1}{\alpha^{(n)}} (r_{n+1} + r_n a_1^{(n)} + \cdots + r_1 a_n^{(n)}).
\tag{11.72}
$$

The recursive solution for (11.67) can be obtained immediately using the above result. Starting with the solution $f_i^{(n)}$ of the equation

$$\begin{pmatrix} r_0 & \cdots & r_{n-1} \\ \vdots & \ddots & \vdots \\ r_{n-1} & \cdots & r_0 \end{pmatrix} \begin{pmatrix} f_0^{(n)} \\ \vdots \\ f_{n-1}^{(n)} \end{pmatrix} = \begin{pmatrix} g_0 \\ \vdots \\ g_{n-1} \end{pmatrix}, \tag{11.73}$$

we increase the dimension by writing

$$\begin{pmatrix} r_0 & \cdots & r_n \\ \vdots & \ddots & \vdots \\ r_n & \cdots & r_0 \end{pmatrix} \begin{pmatrix} f_0^{(n)} \\ \vdots \\ f_{n-1}^{(n)} \\ 0 \end{pmatrix} = \begin{pmatrix} g_0 \\ \vdots \\ g_{n-1} \\ g^{(n)} \end{pmatrix}, \tag{11.74}$$

where

$$g^{(n)} = f_0^{(n)} r_n + f_1^{(n)} r_{n-1} + \ldots + f_{n-1}^{(n)} r_1.$$

On the other hand, by definition (11.68), we have

$$\begin{pmatrix} r_0 & \cdots & r_n \\ \vdots & \ddots & \vdots \\ r_n & \cdots & r_0 \end{pmatrix} \begin{pmatrix} a_n^{(n)} \\ \vdots \\ a_1^{(n)} \\ 1 \end{pmatrix} = \begin{pmatrix} 0 \\ \vdots \\ 0 \\ \alpha^{(n)} \end{pmatrix}. \tag{11.75}$$

Multiplying (11.75) by $h^{(n)}$ and adding to (11.74), we can express the result in the desired form of (11.73) if we choose $h^{(n)}$ which satisfies

$$g^{(n)} + h^{(n)} \alpha^{(n)} = g_n \quad \text{or} \quad h^{(n)} = \frac{g_n - g^{(n)}}{\alpha^{(n)}}. \tag{11.76}$$

Then

$$f_i^{(n+1)} = f_i^{(n)} + h^{(n)} a_{n-i}^{(n)} \qquad 0 \le i \le n-1,$$
$$f_n^{(n+1)} = h^{(n)}. \tag{11.77}$$

will satisfy the following equation (which is indeed the equation we wanted to solve):

$$\begin{pmatrix} r_0 & \cdots & r_n \\ \vdots & \ddots & \vdots \\ r_n & \cdots & r_0 \end{pmatrix} \begin{pmatrix} f_0^{(n+1)} \\ \vdots \\ f_n^{(n+1)} \end{pmatrix} = \begin{pmatrix} g_0 \\ \vdots \\ g_n \end{pmatrix}. \tag{11.78}$$

In obtaining the above least-squares filter, we assumed that the filter is physically realizable and operates on the past data only. If we relax this constraint and allow the filter coefficients f_i to be two-sided, the suffix j in (11.66) extends to the negative range, and g_τ becomes the simple convolution of r_τ and f_τ. Then the Fourier transform of f_τ will be the ratio of the Fourier transform of g_τ to that of r_τ. If the signal and noise are uncorrelated, r_τ will be the sum of signal and noise autocorrelation functions, and g_τ will be the signal autocorrelation function. The frequency response of the least squares filter then becomes simply $P_S/(P_S + P_N)$, where P_S and P_N are the signal and noise spectral densities. Thus the filter will pass the data without change when noise power is negligible, but in general it attenuates the data proportionally to the ratio of signal power to total power.

11.4 Analysis of Data from a Seismic Array

A seismic array is a set of seismographs distributed over an area of the Earth's surface at spacings narrow enough so that the signal waveform may be correlated between adjacent seismographs. Such an array is useful for studying detailed characteristics of wave propagation across the array, as well as for obtaining signal enhancement on the basis of differences in the characteristics of wave propagation between signal and noise. Putting the array on the xy-plane, a given wave field $f(x, y, t)$ may be decomposed into plane waves by the Fourier transform

$$f(x, y, t) = \int\!\!\!\int\!\!\!\int_{-\infty}^{\infty} f(k_x, k_y, \omega) \exp(+ik_x x + ik_y y - i\omega t) \frac{d\omega \, dk_x \, dk_y}{8\pi^3},$$

where $f(k_x, k_y, \omega)$ is the frequency-wavenumber spectrum. This expresses the amplitude and phase of plane waves that propagate with an apparent velocity c in the xy-plane, and in the direction specified by azimuthal angle ϕ:

$$k_x = \frac{\omega \cos \phi}{c},$$

$$k_y = \frac{\omega \sin \phi}{c}.$$

(11.79)

If we can estimate $f(k_x, k_y, \omega)$ from $f(x_i, y_i, t)$ observed at many seismograph sites (x_i, y_i), we can interpret a given wave field by a superposition of simple plane waves. In estimating the frequency-wavenumber spectrum using an array, we are using the Earth immediately below the array as a part of the measuring device. If that part of the Earth is homogeneous, the measured spectrum will clearly represent the nature of incident waves. Unfortunately, this is usually not the case, and the "plane waves" cannot maintain a plane wavefront as they

propagate through the Earth under the array. An example of an irregular wavefront observed at Montana LASA is shown in Figure 13.10 (giving fluctuations in phase).

To remedy this unfortunate situation, we introduce the so-called *station correction*. The arrival time of a wave at the ith station is expressed as

$$t_i = t_0 + \frac{\cos \phi}{c} (x_i - x_0) + \frac{\sin \phi}{c} (y_i - y_0) + \tau_i, \qquad (11.80)$$

where t_0 is the arrival time at a reference point (x_0, y_0) and τ_i is the station residual ($-\tau_i$ is the station correction). For a number of stations greater than three, the three parameters t_0, c, and ϕ can be determined by the least-squares method, minimizing $\sum_i \tau_i^2$. Then τ_i may be determined as the observed time minus the calculated, using the parameters determined by the least-squares method.

To prepare for a general discussion of estimating wavenumber spectra, we shall describe some signal-enhancement techniques using the array data.

One of the simplest but most effective methods of signal enhancement is to shift each record in time by the amount given by (11.80) so that the signal-arrival time becomes common at all the stations. A simple algebraic sum of all the records is then computed at each instant. This process is called *beam-forming*.

This method requires an advance knowledge of c, ϕ, and station correction, $-\tau_i$, which usually depends strongly on both c and ϕ. Therefore, one cannot enhance the signal unless the propagation velocity and azimuth of approach are known in advance. This problem may be avoided if one forms all possible beams and selects the one that gives the clearest signal. Six hundred pre-steered beams were once formed in real time at the Montana LASA.

The reciprocal of apparent velocity c is called $dT/d\Delta$, or ray parameter p, for teleseismic body waves. As shown in the next chapter, $dT/d\Delta$ plays an important part in travel-time inversion to obtain Earth structure.

Let us now go into the detail of signal-enhancement techniques using data from a seismic array.

11.4.1   The multi-channel least-squares filter

Let us first consider the straightforward extension of the least-squares filter discussed in Section 11.3.8 to the case of multi-channel time series. We shall represent the seismogram $d_{t,i}$ at the ith station as the sum of noise $n_{t,i}$ and signal $s_{t,i}$:

$$d_{t,i} = n_{t,i} + s_{t,i} \qquad (i = 1, 2, \ldots, M), \qquad (11.81)$$

where t is the discretized time.

Our problem is to find a multi-channel filter $f_{T,ij}$ that will operate on all the past data and give the best estimate of signal at the ith station at time t,

$$\hat{s}_{t,i} = \sum_{T=0}^{\infty} \sum_{j=1}^{M} f_{T,ij} d_{t-T,j}. \tag{11.82}$$

Assuming that both signal and noise are stationary time series, the process of obtaining the best filter goes parallel to the one given in Section 11.3.8. Considering many samples of data $d_{t,i}^{(k)}$ and signal $s_{t,i}^{(k)}$, we determine the filter coefficients $f_{T,ij}$ that minimize, on the average, the squares of difference between the signal estimate $\hat{s}_{t,i}^{(k)}$ and the signal sample. Since the time series are all stationary, the filter coefficient determined for a certain t applies to any time. Designating the average over samples by $\langle \ \rangle$, the filter coefficient that minimizes

$$\left\langle \left(\sum_{T} \sum_{j=1}^{M} f_{T,ij} d_{t-T,j} - s_{t,i} \right)^2 \right\rangle$$

can be obtained by solving

$$\sum_{T=0}^{\infty} \sum_{j=1}^{M} f_{T,ij} \langle d_{t-T,j} d_{t-T',k} \rangle = \langle s_{t,i} d_{t-T',k} \rangle. \tag{11.83}$$

Defining

$$r_{T,ij} = \langle d_{t,i} d_{t-T,j} \rangle,$$

$$g_{T,ij} = \langle s_{t,i} d_{t-T,j} \rangle,$$

we can rewrite (11.83) as

$$\sum_{T=0}^{\infty} \sum_{j=1}^{M} f_{T,ij} r_{T'-T,jk} = g_{T',ik}. \tag{11.84}$$

We consider $\mathbf{f}_{T,i}$ and $\mathbf{g}_{T,i}$ as column vectors with M elements, and $\mathbf{r}_T$ as an $M \times M$ matrix such that

$$\mathbf{f}_{T,i} = \begin{pmatrix} f_{T,i1} \\ f_{T,i2} \\ \vdots \\ f_{T,iM} \end{pmatrix}, \qquad \mathbf{g}_{T,i} = \begin{pmatrix} g_{T,i1} \\ g_{T,i2} \\ \vdots \\ g_{T,iM} \end{pmatrix},$$

and

$$\mathbf{r}_T = \begin{pmatrix} r_{T,11} & r_{T,12} & \cdots & r_{T,1M} \\ r_{T,21} & r_{T,22} & \cdots & \vdots \\ \vdots & \vdots & & \vdots \\ r_{T,M1} & \cdots & \cdots & r_{T,MM} \end{pmatrix}.$$

Then we can further simplify the equation to

$$\sum_{T=0}^{\infty} \mathbf{r}_{T'-T}\mathbf{f}_{T,i} = \mathbf{g}_{T',i}. \tag{11.85}$$

This is the same form as (11.66), except that $\mathbf{f}_{T,i}$ and $\mathbf{g}_{T',i}$ are vectors and $\mathbf{r}_{T'-T}$ is the matrix defined above. The same recursive method of solving the equation that was discussed in Section 11.3.8 can be used to solve (11.85) by replacing the scalar sum and multiplication by the appropriate matrix sum and multiplication (Wiggins and Robinson, 1965).

An actual example of the data autocorrelation $r_{T,ij}$ obtained by Claerbout (1964) from the Uinta Basin, Utah, array station (seismograph spacing about 1 km) is shown in Figure 11.9, where the value of $r_{T,ij}$ is drawn as a stepwise continuous function of T for each station pair (i, j). Note that, by definition, $r_{T,ji} = r_{-T,ij}$. Using this set of autocorrelation matrices, Claerbout obtained a least-squares filter that best predicts the value of noise at a future point when there is no signal. The difference between the actual data and the predicted value, which is called the *prediction-error filter output*, enhanced a teleseismic signal, as shown at the bottom of Figure 11.10. Both the arrival time and the sense of first motion are clearly detected in the output, although the signal is barely seen in the actual data shown at the top. Usual band-pass filters can enhance the signal amplitudes, but fail to define the arrival time and sense of first motion, as shown in Figure 11.10. This filter, however, distorts the signal waveform. The least-squares filter without signal distortion is called the *maximum-likelihood filter*, and will be discussed in Section 11.4.2.

The simple interpretation of the least-squares filter in the frequency domain discussed in Section 11.3.8 may be extended to the frequency-wavenumber domain. The summation with respect to the station j in (11.84) may be considered as a spatial convolution if the seismographs are uniformly spaced and properly numbered. Allowing for both-sided operations, and assuming that signal and noise are uncorrelated, the response $f(k_x, k_y, \omega)$ of the least-squares filter in the frequency-wavenumber domain may be written as the ratio of signal power density to the sum of the signal and noise power density,

$$f(k_x, k_y, \omega) = \frac{P_S(k_x, k_y, \omega)}{P_S(k_x, k_y, \omega) + P_N(k_x, k_y, \omega)}, \tag{11.86}$$

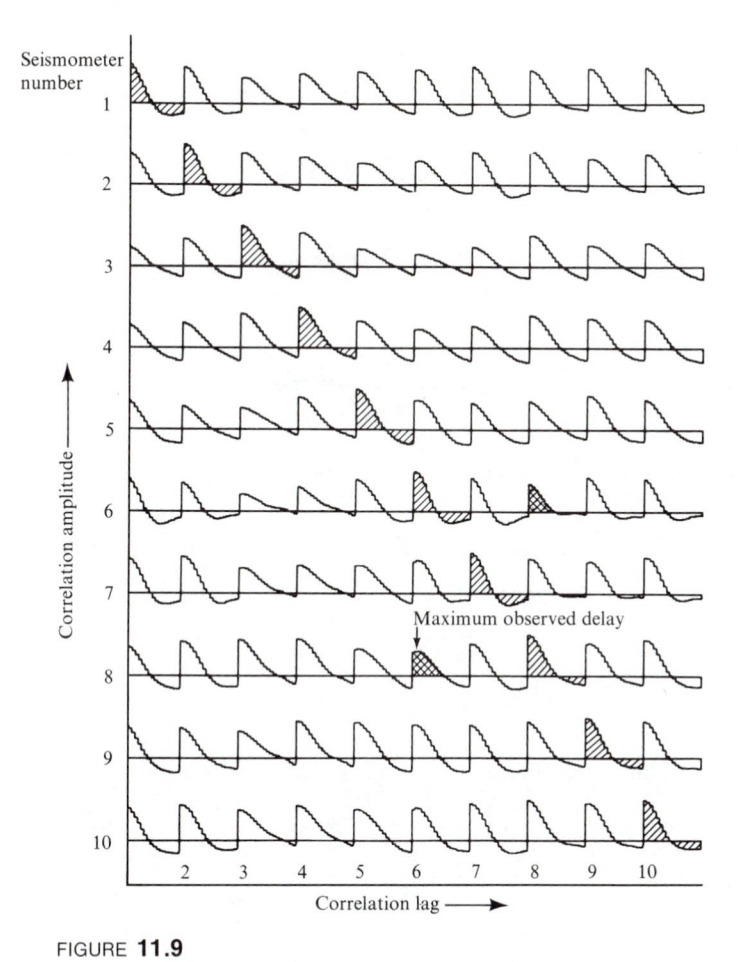

FIGURE **11.9**

The autocorrelation function $r_{T,ij}$ obtained for seismic noise at the
Uinta Basin array station. $r_{T,ij}$ are drawn as a stepwise continuous
function of T for each pair (i,j). Diagonal elements are shaded. By
definition, $r_{-T,ij} = r_{T,ji}$. [From Claerbout, 1964.]

which is a natural extension of one-dimensional results. As Burg (1964) points
out, the frequency-wavenumber approach may give a better physical insight
(if the medium below an array is homogeneous), but the multi-channel approach
is more practical for filter design. A detailed illustration of different approaches
was given by Schneider et al. (1964) for a relatively simple two-channel problem
of eliminating ghost arrivals from reflection seismograms.

Successful separation of fundamental and several higher modes of Rayleigh
waves by Laster and Linville (1966), using the multi-channel least-squares
filter, was made on the seismograms obtained for laboratory models.

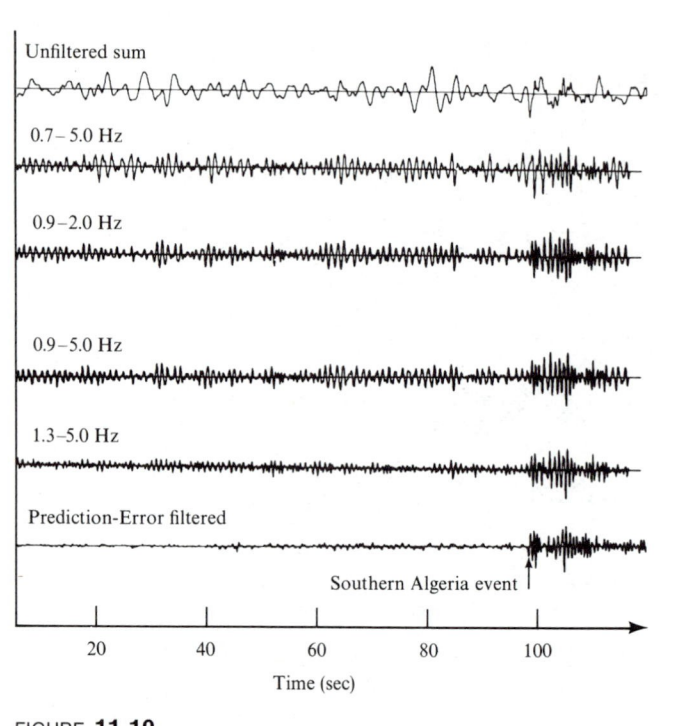

FIGURE **11.10**

The top trace is the unfiltered seismogram. The bottom trace
is the output of a multi-channel least-squares filter. The first
motion of a teleseismic event hidden in noise in the unfiltered
record is clearly detected by the least-squares filter. Usual
band-pass filters can enhance the signal amplitude but fail to
define the first motion, which supplies the most important
information. [From Claerbout, 1964.]

11.4.2   Common waveform model of the signal

In the previous section, we assumed that the signal is a stationary time series
and that we know the cross-correlation function $g_{T,ij}$ between the data and the
signal. These assumptions may not always be realistic. Another simple, but
probably more realistic, assumption for teleseismic P-waves, is that the signal
waveform may be the same at all the stations but have different arrival times,
as determined by (11.80). Using the arrival time t_i given by (11.80), we shift the
data in time and write

$$d_{t,i} = s_t + n_{t+t_i,i},\tag{11.87}$$

where s_t is the common signal, which is unknown, but not a statistical variable
anymore.

For simplicity, let us first consider the data at only one time point. Writing

$$d_i = s + n_i \qquad (i = 1, 2, \ldots, M), \tag{11.88}$$

we assume that d_i obeys the Gaussian distribution with the mean value s and the covariance matrix

$$\rho_{ij} = \langle n_i n_j \rangle. \tag{11.89}$$

The probability density function $f(d_1, d_2, \ldots, d_M)$ may be written as

$$f(d_1, d_2, \ldots, d_M) = \frac{|\phi|^{1/2}}{(2\pi)^{M/2}} \exp\left[-\frac{1}{2} \sum_{i,j=1}^{M} \phi_{ij}(d_i - s)(d_j - s) \right], \tag{11.90}$$

where ϕ_{ij} is the (ij) element of the inverse matrix of the covariance matrix ρ_{ij} and $|\phi|$ is its determinant (Papoulis, 1965). Subscripts i and j here denote different stations. The inverse relation between ϕ_{ij} and ρ_{ij} can be proved in the following way.

Since $\int \ldots \int\int f(d_1, d_2, \ldots, d_M) \, dd_1 dd_2 \ldots dd_M = 1$, we have

$$|\phi|^{-1/2} = \frac{1}{(2\pi)^{M/2}} \int \ldots \int \exp\left\{ -\frac{1}{2} \sum_{i,j=1}^{M} \phi_{ij}(d_i - s)(d_j - s) \right\} dd_1 \ldots dd_M.$$

Differentiating both sides by ϕ_{ij}, we get

$$|\phi|^{-3/2} \times \text{(cofactor of } \phi_{ij})$$

$$= \frac{1}{(2\pi)^{M/2}} \int \ldots \int (d_i - s)(d_j - s) \exp\left[-\frac{1}{2} \sum_{i,j=1}^{M} \phi_{ij}(d_i - s)(d_j - s) \right] dd_1 \ldots dd_M.$$

Since the right-hand side is $|\phi|^{-1/2}\langle(d_i - s)(d_j - s)\rangle$ and the left-hand side is $|\phi|^{-1/2} \cdot [(ij) \text{ element of } \phi^{-1}]$, we get

$$[(ij) \text{ element of } \phi^{-1}] = \langle(d_i - s)(d_j - s)\rangle = \rho_{ij},$$

or $\phi = \rho^{-1}$.

The probability (11.90) that a given set of data occurred is maximized if we choose s to minimize

$$\sum_{i,j}^{M} \phi_{ij}(d_i - s)(d_j - s).$$

Taking the derivative with respect to s and setting the result equal to zero, we

find that the minimum occurs when

$$s = \frac{\sum\limits_{i,j}^{M} \phi_{ij}d_i}{\sum\limits_{i,j}^{M} \phi_{ij}}. \tag{11.91}$$

This is a weighted average of data, with weight proportional to the sum of the elements of the inverse matrix of the covariance matrix. This is the *maximum-likelihood estimate* of the signal when the data consists of only one time point.

If the noises are uncorrelated among the stations, ρ_{ij} will be a diagonal matrix, with diagonal elements equal to the noise power. The weight ϕ_{ij} will be also diagonal with elements inversely proportional to the noise power. Thus the filter puts greater weight on the low-noise station.

The maximum-likelihood estimate (11.91) may be obtained by the least-squares-filter approach described previously under a certain constraint. We shall apply a linear filter f_i on the data d_i to find the best estimate of signal s. We shall minimize the mean square of $(\sum_i f_i d_i - s)$ under the constraint that when the data consist entirely of signal, the filter will pass it without distortion. This constraint can be described by $\Sigma f_i s = s$ or $\Sigma f_i = 1$. Then we obtain

$$\left\langle \left(\sum_i f_i d_i - s \right)^2 \right\rangle = \left\langle \left(\sum_i f_i(s + n_i) - s \right)^2 \right\rangle$$

$$= \left\langle \left(\sum_i f_i n_i \right)^2 \right\rangle$$

$$= \sum_i \sum_j f_i f_j \rho_{ij}.$$

We want to minimize this under the constraint $\Sigma f_i = 1$. Using the Lagrange multiplier λ, we shall minimize

$$\sum_i \sum_j f_i f_j \rho_{ij} - \lambda \left(\sum_i f_i - 1 \right)$$

with respect to f_i and λ. Then we find

$$\sum_j \rho_{ij} f_j = \lambda \qquad (i = 1, 2, \ldots, M)$$

and

$$\sum_i f_i = 1.$$

Remembering that $\sum_j \rho_{ij}\phi_{jk} = \delta_{ik}$, we see that $f_j = \lambda \sum_k \phi_{jk}$ will satisfy the first equations. The second equation is satisfied by $\lambda = 1/\sum_j \sum_k \phi_{jk}$. Thus we have

$$f_j = \frac{\sum_k \phi_{jk}}{\sum_j \sum_k \phi_{jk}},$$

which gives the same estimate of signal as the maximum-likelihood estimate given in (11.91).

Thus we may say that the maximum-likelihood estimate of the signal can be obtained by a least-squares filter under the constraint that the signal is not distorted.

Let us now generalize this result to data with a finite length N. The probability density function for $M \times N$ variables $d_{t,i}$ can be written as

$$f = \frac{|\mathbf{\Phi}|^{1/2}}{(2\pi)^{MN/2}} \exp\left[-\frac{1}{2} \sum_{i,j=1}^{M} \sum_{k,l=1}^{N} \Phi_{ij}^{kl}(d_{k,i} - s_k)(d_{l,j} - s_l) \right], \quad (11.92)$$

where Φ_{ij}^{kl} is an element of the $MN \times MN$ matrix that is the inverse matrix of the covariance matrix, whose element is

$$\rho_{ij}^{kl} = \langle (d_{k,i} - s_k)(d_{l,j} - s_l) \rangle. \quad (11.93)$$

Subscripts i and j refer to stations, and superscripts k and l refer to times. The above relation can be obtained in a manner analogous to the case of $N = 1$.

The maximum-likelihood estimate of signal s_t is given by minimizing the factor in [ ] in (11.92) with respect to s_l. Taking derivatives with respect to s_l and equating them to zero, we obtain

$$\sum_{k=1}^{N} s_k \sum_{i,j=1}^{M} \Phi_{ij}^{kl} = \sum_{k=1}^{N} \sum_{i,j=1}^{M} d_{k,i}\Phi_{ij}^{kl} \quad (l = 1, \ldots, N). \quad (11.94)$$

The solution s_k gives the maximum-likelihood estimate of the signal, which again corresponds to the least-squares filter output without signal distortion.

If the noise is purely random in both space and time, Φ_{ij}^{kl} will vanish except for $i = j$ and $k = l$. In addition, if the noise is stationary in space and time, Φ_{ii}^{kk} will be a constant independent of (i, k). Then (11.94) reduces to

$$s_l = \frac{1}{M} \sum_{i=1}^{M} d_{l,i}. \quad (11.95)$$

Since $d_{t,i}$ is already time-shifted according to (11.87), the above formula gives

the simple beam-forming output discussed earlier. In this case, the noise amplitude will be reduced inversely in proportion to $\sqrt{M}$, where M is the number of seismographs.

The maximum-likelihood method based on (11.94) has been applied to both the short-period and long-period data from the Montana LASA, as summarized in Capon et al. (1968, 1969). The short-period noises are uncorrelated between the subarrays (Fig. 11.2), and the maximum-likelihood method offers little advantage over the beam-forming method. When the method is applied to short-period data within a subarray, it primarily works over microseismic noise with frequency around 0.2 Hz and reduces it. But the great predominance of the microseismic noise power at 0.2 Hz tends to obscure the estimates at other frequencies. The performance may be improved by prefiltering, which reduces the microseisms, but again such prefiltering introduces an undesirable distortion of the signal, which may be practically impossible to correct. In general, the maximum-likelihood processing over a small array (~ 0.5 km spacing) gave a result comparable to that of beam-forming over a larger array (e.g., 3-km spacing); in either case, the noise reduction is about a factor of $\sqrt{M}$.

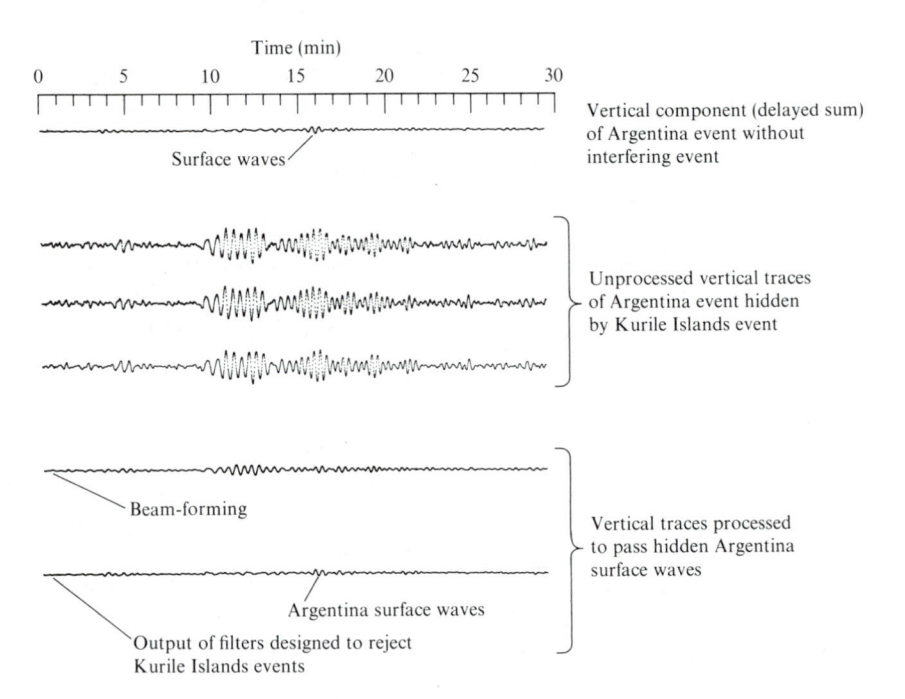

FIGURE **11.11**

Testing the maximum-likelihood method with artificial data. The signal (at the top) is mixed with noise, as shown in the second through fourth traces. The bottom trace shows the successful recovery of signal by the maximum-likelihood method. Beam-forming is unable to identify the signal, as shown in the second trace from the bottom. [From Capon et al., 1969.]

The maximum-likelihood method works efficiently when the noise is highly coherent across the array. One of the most successful results was obtained when long-period Rayleigh waves from a region other than the source under study were considered as noise. Rayleigh waves from small events in Argentina were added to those from the Kurile Islands, and processed by the maximum-likelihood method. As shown in Figure 11.11, the noise (Kurile Rayleigh wave) was strongly suppressed (by a factor of 10 in amplitude), so that the Argentina event became identifiable. However, the process of beam-forming (designated as delayed sum output in Fig. 11.11) was unable to identify the Argentina event.

11.4.3 Frequency-wavenumber power spectrum

The frequency-wavenumber power spectrum will give us a direct physical insight into the wave field because it will tell us the amount of power distributed among different wave velocities and directions of approach. As we did in Sections 11.3.6 and 11.3.7, where we treated the seismic noise as a stationary time series and discussed the measurement of power spectra, we shall assume that the seismic noise is stationary in both time t and two spatial coordinates x, y.

There are two basic ways of estimating the power spectrum. One is to estimate the autocorrelation function and then do the Fourier transformation. The other is to calculate directly the Fourier transform of the noise and then do the absolute-value squaring and averaging.

Let us write these two approaches more explicitly. Writing the noise field as $n(x, y, t)$, the autocorrelation function as $R(\xi, \eta, \tau)$, and the power spectral density as $P(k_x, k_y, \omega)$, we have

$$R(\xi, \eta, \tau) = \langle n(x, y, t)n(x + \xi, y + \eta, t + \tau) \rangle \qquad (11.96)$$

and

$$P(k_x, k_y, \omega) = \int\int\int_{-\infty}^{\infty} R(\xi, \eta, \tau) \exp[i(\omega\tau - k_x\xi - k_y\eta)] \, d\tau \, d\xi \, d\eta. \qquad (11.97)$$

In the second approach, we shall introduce the discrete power spectrum analogous to the one described in Section 11.3.6. We may define the power spectrum P_{lmk} by

$$P_{lmk} = \lim_{LMK \to \infty} \frac{\langle |F_{lmk}|^2 \rangle}{L \, \Delta x M \, \Delta y K \, \Delta t}, \qquad (11.98)$$

where F_{lmk} is the discrete Fourier transform of the digitized noise field $n(l \, \Delta x, m \, \Delta y, k \, \Delta t)$ at the points in the three-dimensional space located at an interval Δx in the x-direction, Δy in the y-direction, and Δt in the t-direction.

The lengths of the data are $L \, \Delta x$, $M \, \Delta y$, and $K \, \Delta t$ in the x-, y-, and t-direction, respectively.

$$F_{lmk} = \sum_{l'=0}^{L-1} \sum_{m'=0}^{M-1} \sum_{k'=0}^{K-1} n(l' \, \Delta x, m' \, \Delta y, k' \, \Delta t)$$

$$\times \exp\left(-i \frac{2\pi ll'}{L} - i \frac{2\pi mm'}{M} + i \frac{2\pi kk'}{K}\right) \Delta x \, \Delta y \, \Delta t \qquad (11.99)$$

The above methods are directly applicable to the data obtained more or less continuously in space. When the seismic noise is completely stationary, a mobile seismograph can be used repeatedly to cover any desired spatial point, making the measurement of the autocorrelation function continuous in space. Such is the case of ground motion caused by traffic in a big busy city.

In most problems, however, the seismic noise is not completely stationary, and is caused by atmospheric and oceanographic disturbances, which are transient. The seismic arrays designed for the study of noise are immobile and make the continuous spatial coverage of the autocorrelation function difficult. Thus the basic methods described earlier cannot apply to most data.

In practice, several approximate methods have been developed. The simplest method is to combine beam-forming with a power-spectrum estimate for the beam output. From (11.79) and (11.80), the time shift required for beam-forming for the point (k_x, k_y, ω) is

$$t_i = t_0 + \frac{k_x}{\omega}(x_i - x_0) + \frac{k_y}{\omega}(y_i - y_0) + \tau_i, \qquad (11.100)$$

where τ_i is the station residual introduced in (11.80). Expressing the noise time-series at the ith station as $n_i(t)$, the beam output can be written as

$$b(k_x/\omega, k_y/\omega, t) = \frac{1}{N} \sum_{i=1}^{N} n_i(t + t_i). \qquad (11.101)$$

The power spectrum of $b(k_x/\omega, k_y/\omega, t)$ as a time series can be obtained by calculating the autocorrelation and then performing the Fourier transform. The result is

$$\hat{P}(k_x, k_y, \omega) = \int \exp(i\omega\tau) \, d\tau \, \frac{1}{N^2} \left\langle \sum_{i=1}^{N} n_i(t + t_i) \sum_{j=1}^{N} n_j(t + \tau + t_j) \right\rangle. \qquad (11.102)$$

By definition (11.96), we can write this result as

$$\hat{P}(k_x, k_y, \omega) = \int \exp(i\omega\tau) \, d\tau \, \frac{1}{N^2} \sum_{i,j=1}^{N} R(x_j - x_i, y_j - y_i, t_j - t_i + \tau). \qquad (11.103)$$

Introducing a weight function

$$W(\kappa_x, \kappa_y) = \frac{1}{N^2} \sum_{i,j=1}^{N} \exp[-i\kappa_x(x_i - x_j) - i\kappa_y(y_i - y_j) + i\omega(\tau_i - \tau_j)],$$

(11.104)

we shall show that our simple estimate $\hat{P}$ is a weighted average of the true power spectrum according to the formula

$$\hat{P}(k_x, k_y, \omega) = \int_{-\infty}^{\infty}\!\!\int W(\kappa_x - k_x, \kappa_y - k_y)P(\kappa_x, \kappa_y, \omega)\, d\kappa_x\, d\kappa_y. \quad (11.105)$$

Inserting (11.104) into (11.105), we find

$$\hat{P}(k_x, k_y, \omega) = \sum_{i,j} \int_{-\infty}^{\infty}\!\!\int \frac{1}{N^2} \exp[+i\kappa_x(x_j - x_i) + i\kappa_y(y_j - y_i)]$$
$$\times \exp[-ik_x(x_j - x_i) - ik_y(y_j - y_i) - i\omega(\tau_j - \tau_i)]$$
$$\times P(\kappa_x, \kappa_y, \omega)\, d\kappa_x\, d\kappa_y. \quad (11.106)$$

On the other hand, from the inverse transform of (11.97),

$$R(x, y, \tau) = \int_{-\infty}^{\infty}\!\!\!\int\!\!\int P(\kappa_x, \kappa_y, \omega) \exp(-i\omega\tau + i\kappa_x x + i\kappa_y y)\, d\omega\, d\kappa_x\, d\kappa_y/8\pi^3$$

or

$$\int_{-\infty}^{\infty} R(x, y, \tau') \exp(i\omega\tau')\, d\tau' = \int_{-\infty}^{\infty}\!\!\int P(\kappa_x, \kappa_y, \omega) \exp(+i\kappa_x x + i\kappa_y y)\, d\kappa_x\, d\kappa_y/4\pi^2$$

Putting this result into (11.106), we have

$$\hat{P}(k_x, k_y, \omega) = \frac{1}{N^2} \sum_{i,j} \int_{-\infty}^{\infty} \exp(i\omega\tau')\, d\tau' R(x_j - x_i, y_j - y_i, \tau')$$
$$\times \exp[-ik_x(x_j - x_i) - ik_y(y_j - y_i) - i\omega(\tau_j - \tau_i)].$$

Rewriting $\tau' - \dfrac{k_x}{\omega}(x_j - x_i) - \dfrac{k_y}{\omega}(y_j - y_i) - (\tau_j - \tau_i)$ as τ,

$$\hat{P}(k_x, k_y, \omega) = \frac{1}{N^2} \sum_{i,j} \int_{-\infty}^{\infty} \exp(i\omega\tau)\, d\tau\, R(x_j - x_i, y_j - y_i, \tau + t_j - t_i),$$

which agrees with (11.103). Therefore, the power spectrum of the beam output is a weighted average of the true power spectrum. The weight function $W(\kappa_x, \kappa_y)$ can be calculated by (11.104) once the station distribution (x_i, y_i) is known.

Two subsets of Montana LASA seismographs and their corresponding weight functions are shown in Figure 11.12, reproduced from LaCoss et al. (1969). If the weight function is a delta function centered at $k_x = k_y = 0$, our estimate gives the exact value of the true spectrum. The figure, however, shows a spread (6 db down from the peak at the center) of about ± 0.035 km^{-1} for the array of diameter 22 km, and about ± 0.025 km^{-1} for the array of diameter 30 km. Examples of actual wavenumber spectra for various frequencies are

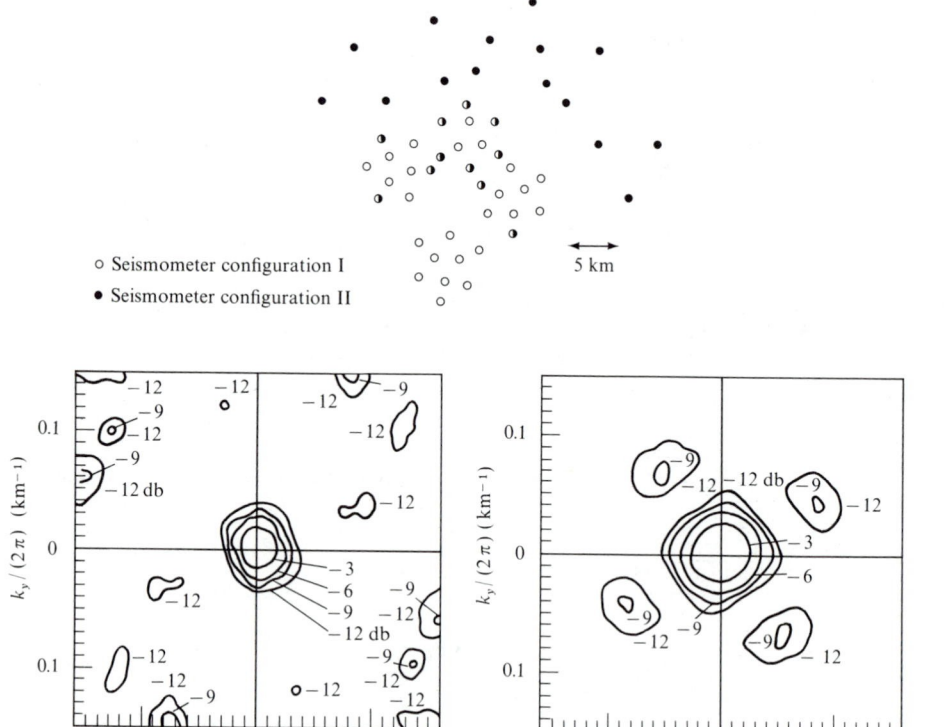

FIGURE **11.12**

Examples of the weight function $W(k_x, k_y)$ for two different array configurations; configuration I (array diameter 22 km) is on the right; configuration II (diameter 30 km), on the left. [From LaCoss et al., 1969.]

shown in Figure 11.13. LaCoss and his co-workers studied the mode structure
of seismic noise recorded at Montana LASA and found that noise at frequencies
higher than 0.3 Hz is primarily compressional waves that probably originate
beneath large storms at sea; the noisiest band between 0.2 to 0.3 Hz consists
of both body waves and higher-mode Rayleigh waves. At frequencies lower than
0.15 Hz, vertical-component microseisms consist primarily of fundamental-
mode Rayleigh waves. They detected an appreciable amount of Love-wave
energy at these low frequencies.

In Section 11.3.7, we have discussed the method of estimating the power
spectrum with high resolution for time series of the autoregressive type. The
same method could be extended for a high-resolution-wavenumber spectral
analysis. However, a direct extension is difficult because of nonuniform spacing
of seismograph locations.

Another method, called the *maximum-likelihood estimator*, developed by
Capon (1969), is also claimed to have a higher resolution than the conventional
method. In order to illustrate this method, we shall go back to equation (11.92)
in Section 11.4.2 and consider the simple case of one station ($M = 1$). The
probability density function for N variables d_t can be written as

$$f = \frac{|\Phi|^{1/2}}{(2\pi)^{N/2}} \exp\left[-\frac{1}{2} \sum_{k,l=1}^{N} \Phi^{kl}(d_k - s_k)(d_l - s_l) \right], \qquad (11.107)$$

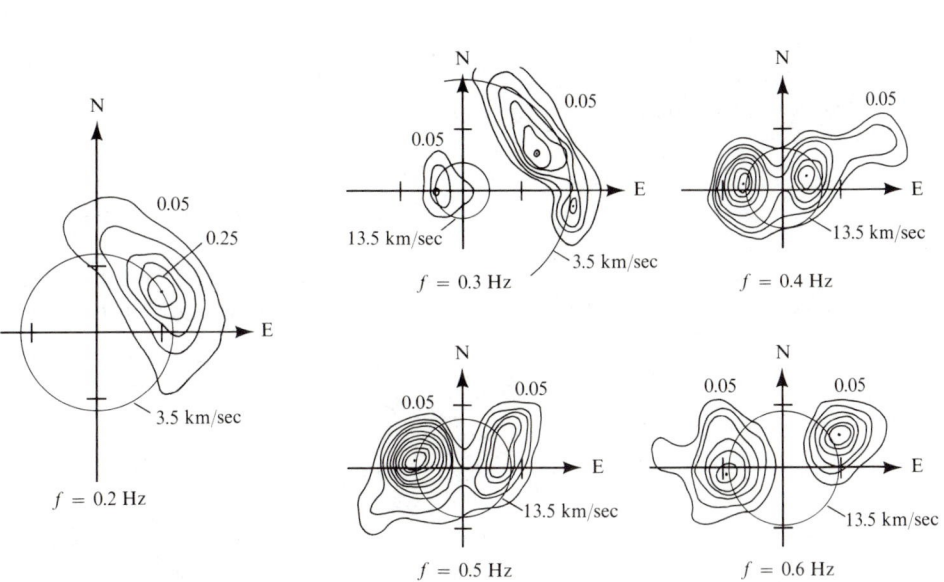

FIGURE **11.13**

Wavenumber spectra of microseisms for five different frequencies observed at the Montana
LASA. [From LaCross et al., 1969.]

where Φ^{kl} is an element of $N \times N$ matrix $\mathbf{\Phi}$, which is the inverse matrix of the covariance matrix ρ whose element is

$$\rho^{kl} = \langle (d_k - s_k)(d_l - s_l) \rangle. \tag{11.108}$$

We assume that the signal s_k has the known shape f_k $(k = 1, 2, \ldots, N)$ but that its amplitude contains an unknown factor c:

$$s_k = cf_k. \tag{11.109}$$

Let us first ask the question: What is the maximum-likelihood estimate of c when the data d_k $(k = 1, 2, \ldots, N)$ are given? Using the matrix notation for brevity, the exponent in (11.107) can be written as $-\frac{1}{2}(\mathbf{d} - c\mathbf{f})^T \mathbf{\Phi}(\mathbf{d} - c\mathbf{f})$, where $\mathbf{d}$ and $\mathbf{f}$ are column vectors with the components d_k and f_k, respectively, and T indicates taking the transpose of the vector.

$$-\tfrac{1}{2}(\mathbf{d} - c\mathbf{f})^T\mathbf{\Phi}(\mathbf{d} - c\mathbf{f}) = -\tfrac{1}{2}[\mathbf{d}^T\mathbf{\Phi}\mathbf{d} - c\,\mathbf{d}^T\mathbf{\Phi}\mathbf{f} - c\mathbf{f}^T\mathbf{\Phi}\mathbf{d} + c^2\mathbf{f}^T\mathbf{\Phi}\mathbf{f}].$$

Taking the derivative with respect to c, setting the result equal to zero, and observing $\mathbf{d}^T\mathbf{\Phi}\mathbf{f} = \mathbf{f}^T\mathbf{\Phi}\mathbf{d}$ because $\mathbf{\Phi}$ is symmetric, we find the maximum likelihood estimate to be

$$\hat{c} = \frac{\mathbf{d}^T\mathbf{\Phi}\mathbf{f}}{\mathbf{f}^T\mathbf{\Phi}\mathbf{f}}. \tag{11.110}$$

The corresponding estimate of $\mathbf{s} = c\mathbf{f}$ is

$$\hat{\mathbf{s}} = \hat{c}\mathbf{f} = \frac{\mathbf{d}^T\mathbf{\Phi}\mathbf{f}}{\mathbf{f}^T\mathbf{\Phi}\mathbf{f}}\,\mathbf{f},$$

which is equal to $c\mathbf{f}$ when $\mathbf{d}$ is equal to $c\mathbf{f}$. In other words, it does not distort the signal. We can find the variance of the estimate c by following Robinson (1963). Rewriting

$$\hat{c} - c = (\mathbf{d} - c\mathbf{f})\mathbf{\Phi}\mathbf{f}(\mathbf{f}^T\mathbf{\Phi}\mathbf{f})^{-1}$$

and noting that $\langle (\mathbf{d} - c\mathbf{f})(\mathbf{d} - c\mathbf{f})^T \rangle = \rho$ and $\rho = \mathbf{\Phi}^{-1}$, we obtain

$$\begin{aligned}
\langle (\hat{c} - c)^2 \rangle &= \langle (\hat{c} - c)^T(\hat{c} - c) \rangle \\
&= (\mathbf{f}^T\mathbf{\Phi}\mathbf{f})^{-1}\mathbf{f}^T\mathbf{\Phi}\langle (\mathbf{d} - c\mathbf{f})(\mathbf{d} - c\mathbf{f})^T \rangle\mathbf{\Phi}\mathbf{f}(\mathbf{f}^T\mathbf{\Phi}\mathbf{f})^{-1} \\
&= (\mathbf{f}^T\mathbf{\Phi}\mathbf{f})^{-1}\mathbf{f}^T\mathbf{\Phi}\rho\mathbf{\Phi}\mathbf{f}(\mathbf{f}^T\mathbf{\Phi}\mathbf{f})^{-1} = (\mathbf{f}^T\mathbf{\Phi}\mathbf{f})^{-1}\mathbf{f}^T\mathbf{\Phi}\mathbf{f}(\mathbf{f}^T\mathbf{\Phi}\mathbf{f})^{-1} \\
&= (\mathbf{f}^T\mathbf{\Phi}\mathbf{f})^{-1} \\
&= (\mathbf{f}^T\rho^{-1}\mathbf{f})^{-1}. \tag{11.111}
\end{aligned}$$

Summarizing the above result, the maximum likelihood estimate of the signal amplitude c is given by $(\mathbf{d}^T \mathbf{\Phi} \mathbf{f})(\mathbf{f}^T \mathbf{\Phi} \mathbf{f})^{-1}$, and the variance of the estimate is equal to $(\mathbf{f}^T \rho^{-1} \mathbf{f})^{-1}$, where ρ is the covariance matrix of noise and $\mathbf{\Phi} = \rho^{-1}$. The maximum-likelihood estimate of the power spectrum used by Capon is the variance of the signal estimate when the signal shape $\mathbf{f}$ is the unit sinusoidal oscillation: $f_k = \exp[i\omega(k - 1)\Delta t]$, where ω is the frequency at which the power spectrum is to be estimated. The power spectral estimate is

$$\frac{1}{\mathbf{f}^T \mathbf{\Phi} \mathbf{f}^*} = \frac{1}{\displaystyle\sum_{k=1}^{N} \sum_{l=1}^{N} \Phi^{kl} \exp[i\omega(k - l)\Delta t]}, \tag{11.112}$$

where a conjugate operation $*$ is included because $\mathbf{f}$ is complex. Equation (11.112) is a reasonable estimate of the power spectrum, because it is the variance of the best estimate of a virtual sinusoid with a given frequency. Since the variance is caused by the noise power in the vicinity of that frequency, this must give a high resolution estimate of the noise power spectrum at that frequency. Note that this estimate does not require equally spaced samples, because f_k can be arbitrarily chosen.

Capon's estimate of the frequency-wavenumber spectrum is given by a natural extension of (11.112) to the two-dimensional case:

$$P(k_x, k_y, \omega) = \left\{ \sum_{i=1}^{N} \sum_{j=1}^{N} \phi_{ij}(\omega) \exp[ik_x(x_i - x_j) + ik_y(y_i - y_j)] \right\}^{-1}, \tag{11.113}$$

where $\phi_{ij}(\omega)$ is an element of the matrix $\boldsymbol{\phi}(\omega)$; (x_i, y_i) indicates the ith seismograph location. $\boldsymbol{\phi}(\omega)$ is the inverse matrix of the Fourier transform of the covariance matrix $\rho_{\tau,ij}$, given by

$$\rho_{\tau,ij} = \langle n_{t,i} n_{t+\tau,j} \rangle,$$

where $n_{t,i}$ is the noise at the ith station.

Capon has extensively used this method in separating various modes of waves propagating across the LASA. He showed, among other things, that late-arriving Rayleigh waves were caused by scattering at some ocean-continent boundaries.

11.5 Some Examples of Effective Analysis Methods

In this chapter, we have covered a general approach for signal enhancement within the linear Gaussian scheme in which the statistical properties of data are completely described by the mean and covariance matrix. Although we showed some successful examples in previous sections, it must be cautioned that seismic data do not always obey a Gaussian distribution.

For example, it is common practice to plot the logarithm of amplitude as a function of epicentral distance in the determination of earthquake magnitude, seismic moment, or attenuation. This implies that the logarithm of amplitude obeys a Gaussian distribution. In Chapter 13, we shall work with the variation of logarithm of amplitude and phase of P-waves observed at Montana LASA sensors, instead of the real and imaginary parts of the Fourier transform, because Chernov's theory of wave scattering predicts the variance of the logarithm of Fourier transform instead of the variance of the Fourier transform. If the noise in a seismogram is "signal-generated," such as that due to scattering, the linear Gaussian scheme used in this chapter may not be appropriate.

This is not a trivial problem, because the linear-inversion method is being applied to the Fourier transform of free oscillations and surface waves to obtain elements of the seismic-moment tensor, using formulas such as (7.149). A large variation in phase introduces underestimation of the seismic moment because both real and imaginary parts of the Fourier transform tend to scatter above and below zero. On the other hand, the skewness in amplitude distribution may cause overestimation when the phase is coherent.

Nevertheless, there are expanding areas in which the analysis method described in the present chapter seems to be working well. For example, the beam-forming method found effective in signal enhancement for teleseismic body waves is a simple process of time shift and sum. The same process is used in common-depth-point (CDP) stacking, one of the most successful data-reduction techniques in reflection seismology. Even simpler is the process of sign-change and sum used in stacking free-oscillation data from WWSSN stations by Mendiguren (1973), who succeeded in obtaining an unambiguous, high-resolution identification of spectral peaks using information about the earthquake source mechanism and initial Earth model. The stacking method was later extended by Nolet (1975) to find the phase velocities of a number of higher-mode Rayleigh waves. We shall describe these successful methods briefly here, along with a few other methods that have proved to be useful. One thing common to these successful techniques is the efficient manner by which *a priori* knowledge can be incorporated into the analysis process. For example, the beam-forming worked well because of the accurate station corrections available for various directions of incident waves. The more one can utilize pre-existing knowledge about the Earth and the seismic source along with the raw data, the more refined is the information that can be extracted from the data.

11.5.1 *Common-depth-point stacking*

CDP stacking was introduced by W. H. Mayne (patent application in 1950) in order to enhance the reflection signal from a given reflection point by com-

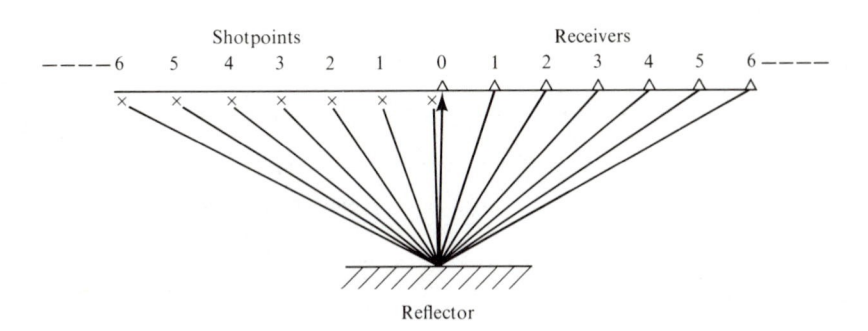

FIGURE **11.14**
Shot-receiver pairs for a CDP stack.

bining many records obtained by a multiplicity of shotpoint and geophone locations. The method has become exceedingly popular since the advent of digital processing. As shown in Figure 11.14, the shotpoints and receiver sites are chosen at equal distances from the reflector point. The records are shifted in time in such a way that the arrival of reflected signals at all the receivers are coincident with one vertically above the reflection point. The CDP stack is the algebraic sum of these time-shifted records, which will enhance the reflection as long as the same waveform persists for the reflection in the range of the horizontal distance covered and the travel times are accurately known.

Suppose that the medium above the reflector is uniform with compressional velocity α. The travel time between the reflection point and the receiver is $\sqrt{x^2 + h^2}/\alpha$, where x is the horizontal travel distance and h is the depth of the reflector. The time shift required for the CDP stack is $2\,\Delta t$, where $\Delta t = \sqrt{(x^2 + h^2)}/\alpha - h/\alpha$. We can rewrite this in the form

$$(t_0 + \Delta t)^2 - t_0^2 = x^2/\alpha^2, \tag{11.114}$$

where $t_0 = h/\alpha$.

When the medium is layered, as shown in Figure 11.15, the travel time t and the distance x can be written as

$$t = px + \sum_i d_i \sqrt{(1/\alpha_i)^2 - p^2},$$

$$x = \sum_i d_i \tan \theta_i = \sum_i d_i \frac{p}{\sqrt{(1/\alpha_i)^2 - p^2}}.$$

Expanding the travel time into a Taylor series with respect to x,

$$t = t_0 + \left(\frac{dt}{dx}\right)_0 x + \frac{1}{2}\left(\frac{d^2 t}{dx^2}\right)_0 x^2 + \cdots,$$

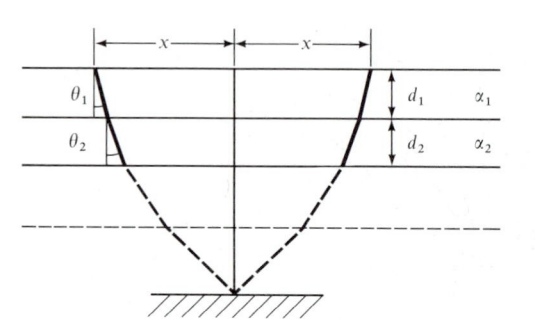

FIGURE **11.15**

A stack of homogeneous layers above the reflector. d_i, α_i, and θ_i are the thickness, velocity, and angle between ray and vertical for the ith layer, respectively.

we find

$$\frac{dt}{dx} = p = 0 \qquad \text{at } x = 0,$$

$$\frac{d^2t}{dx^2} = \frac{dp}{dx} = \frac{1}{\displaystyle\sum_i \frac{d_i\alpha_i}{(1 - \alpha_i^2 p^2)^{3/2}}} = \frac{1}{\Sigma d_i\alpha_i} \qquad \text{at } x = 0.$$

Thus we have

$$t = t_0 + \frac{x^2}{2\Sigma d_i\alpha_i} + \cdots.$$

By squaring and neglecting the terms higher than x^2, we find

$$t^2 = t_0^2 + \frac{t_0 x^2}{\Sigma d_i\alpha_i}. \tag{11.115}$$

Comparing this with (11.114), we find

$$\alpha = \left(\frac{\Sigma d_i\alpha_i}{t_0}\right)^{1/2}$$

$$= \left(\frac{\Sigma \alpha_i^2 \, \Delta t_i}{\Sigma \, \Delta t_i}\right)^{1/2}, \tag{11.116}$$

where $\Delta t_i = d_i/\alpha_i$ is the vertical travel time in the ith layer. This shows that the appropriate constant velocity to be used in the calculation of time shift $2 \, \Delta t$ by (11.114) should be the weighted RMS velocity (11.116) rather than the simple average velocity. A glossary of terms used in exploration geophysics was published by Sheriff (1968, 1969). Thus the horizontal distance $2x$ is called

the *offset* and the time shift 2 Δt is called the *normal move-out*. When the receiver site shows a variation in elevation and in the thickness of the low-velocity surificial layer, a station correction must be applied to the move-out. The station correction is called a *static*. In exploration geophysics, the CDP stack is a stack of records corrected for statics and normal move-out.

Although *a priori* knowledge of velocity distribution is essential for a successful CDP stack, one can improve the stack in several ways, starting with an approximate velocity distribution. For example, Schneider and Backus (1968) compute CDP stacks using data for different offsets on the basis of an initial-velocity model. The differences between the stacks, due to error in the initial model, may be determined by a cross-correlation computation and used for correcting the error in the initial model. Another method is to use several trial velocity values and determine the optimal velocities by maximizing the amplitude of the resultant CDP stack, as done by Carotta and Michon (1967).

11.5.2   *Identification of free-oscillation spectral peaks when the earthquake source mechanism is known*

Consider a free oscillation excited in a spherically symmetric nonrotating Earth by an earthquake. From Chapter 8, we know that the oscillation consists of normal modes with distinct frequencies, expressed by spherical harmonics of different angular and azimuthal order numbers and depth-dependent eigenfunctions specified by another order number. If we know the epicenter, focal depth, and fault-plane solution of the earthquake, we can calculate the motion due to a normal mode of a given frequency by the method described in Chapter 8. Assuming a point source, the shape of the time function for each mode will be common at all the stations on the Earth, but the amplitude and sign will change from one station to another.

Suppose that we collect many seismograms of an earthquake with known source parameters from the WWSSN and correct the sign according to the theoretical prediction for a particular mode based on the source parameters (if the theoretical sign is plus, the record is unchanged, but if it is minus, the sign of the record is reversed.) We then sum all these corrected seismograms algebraically with no time shift. The result will enhance the signal associated with the particular mode and will tend to reduce the other modes. The Fourier transform of the stacked record will more clearly show the spectral peak of the wanted mode. The same result is obtained by first making Fourier transforms of the records, correcting the phase by π for a sign difference, and then stacking them.

The method was first applied by Mendiguren (1973) to the WWSSN data for a large deep-focus earthquake in Columbia. The method not only improved the resolution of spectral analysis, but also made possible a unique, unambiguous identification of many overtones, because, as shown in Figure 11.16,

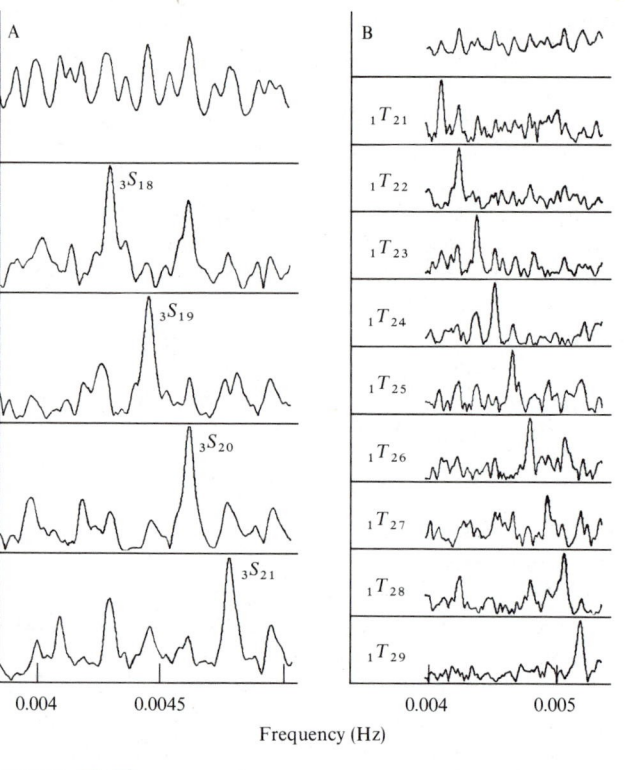

Frequency (Hz)

FIGURE **11.16**

Spectral peaks of free oscillation for the third higher
spheroidal modes $_3S$ in the colatitudinal component (**a**),
and the first higher torsional modes $_1T$ in the azimuthal
component (**b**). The spectra at the top are the sum of the
absolute amplitudes of the spectra at all stations. The
remaining spectra are the result of stacking with sign
correction. [From Mendiguren; copyright 1973 by the
American Association for the Advancement of Science.]

the method enhances the mode with given order numbers and suppresses the
neighboring modes. Subsequently, many hundreds of new modes have been
identified by Gilbert and Dziewonski (1975) using a similar technique.

11.5.3   Measurement of higher-mode dispersion

In discussing the general problems of surface-wave analysis (Section 11.3.1),
we emphasized the importance of separating different modes and mentioned
several possible techniques for separation. An application was made by Nolet

(1975, 1976), who successfully isolated higher modes of Rayleigh waves, up to the sixth higher mode, using the network of long-period stations in Western Europe.

Nolet's method starts with an approximate wavenumber spectrum based on the idea of beam-forming introduced in Section 11.4.3. Let the seismogram at the jth station be $d_j(t)$ and its Fourier transform $D_j(\omega)$. Reducing the problem to one space dimension by using the epicentral distance Δ_j and neglecting any station correction, the Fourier transform of beam output for wave number k (along the great-circle path) is given by

$$\hat{D}(k, \omega) = \frac{1}{N} \sum_{j=1}^{N} D_j(\omega) \exp(-ik\Delta_j). \tag{11.117}$$

By a proper choice of the portion of each seismogram, we may assume that $D_j(\omega)$ consists of a small number of modes each having the form of (11.11):

$$D_j(\omega) = \sum_n F_n(\omega) \exp[ik_n(\omega)\Delta_j + \phi_n(\omega)], \tag{11.118}$$

where $k_n(\omega) = \omega/c_n(\omega)$. For simplicity, the amplitude $F_n(\omega)$ and the source phase $\phi_n(\omega)$ are assumed to be independent of the station. Inserting (11.118) into (11.117), we obtain

$$\hat{D}(k, \omega) = \sum_n F_n(\omega) \exp[i\phi_n(\omega)]H(k_n(\omega) - k), \tag{11.119}$$

where $H(k) = \sum_j \exp(ik\Delta_j)$ is called the "array response." $|H(k)|^2$ is nothing but the weight function W introduced in (11.104). We showed in Section 11.4.3 that the power spectrum $\hat{P}$ of the beam output is the weighted average of the true power spectrum with W as the weight function. In the case of surface waves of the form (11.118), the wavenumber spectra consists of δ-functions in k for a given ω, and the beam-output spectrum $\hat{D}(k, \omega)$ becomes a sum of weighted and shifted array responses.

If there are many stations distributed uniformly and densely along the great circle, the array response $H(k)$ will be close to the desirable form of a δ-function. When the number of stations is small, $H(k)$ will have large side lobes close to the main lobe, producing spurious peaks that interfere with the true peaks. To overcome this problem, we first use the group-velocity window. Putting the window function $W(U, t)$, which opens only for a time interval around the time of arrival of the wave with group velocity U, we compute the windowed spectrum

$$D_j(\omega, U) = \int_{-\infty}^{\infty} d_j(t)W(U, t) \exp(i\omega t) \, dt \tag{11.120}$$

and then the beam output for wavenumber k,

$$D(k, \omega, U) = \frac{1}{N} \sum_{j=1}^{N} D_j(\omega, U) \exp(-ik\Delta_j). \qquad (11.121)$$

An example of beam-output power spectrum $|D(k, \omega, U)|^2$ is shown on the left side of Figure 11.17 as a function of angular order ($kr_\oplus - \frac{1}{2}$; $r_\oplus$ is the radius of the Earth) and U for a given ω. The number of stations used is 13. The datum in this case is a synthetic signal composed of the fundamental and seven higher Rayleigh modes. The contour interval is 3 db. The true locations of these modes are denoted by closed triangles. We find many spurious peaks in this diagram.

The spurious peaks can be cleaned up by the following procedure. We first specify ω and U and analyze $\hat{D}(k, \omega, U)$ as a function of k alone. Next we find the value of k at which $|\hat{D}|$ is the maximum and call it k_{max}. We assume that this maximum peak is not due to side lobes, and that a true mode exists here. We then subtract the contribution of this mode from $\hat{D}$. To assure stability in the iteration process to follow, we introduce a loop-gain factor γ ($0 < \gamma \le 1$) and subtract $\gamma \hat{D}(k_{max}, \omega, U) H(k_{max} - k)$ from $\hat{D}$. The above process is repeated with the residual, and we record $k_{max}^{(i)}$ and $D(k_{max}^{(i)}, \omega, U)$ at each iteration. The iteration is stopped when the amplitude of the residual integrated over the

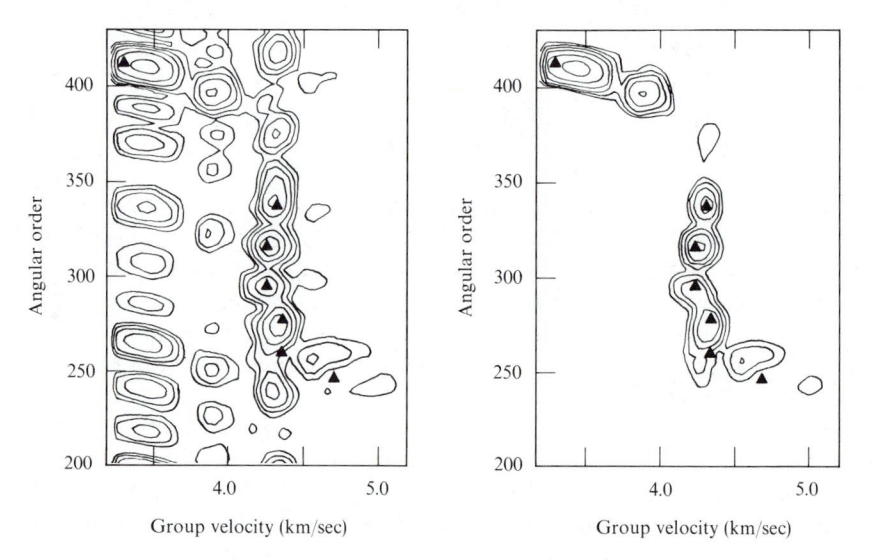

FIGURE **11.17**

An example of beam-output spectrum $|\hat{D}(k, \omega, U,)|^2$ is shown on the left-hand side as a function of angular order ($kr_\oplus - 1/2$, where $r_\oplus$ is the Earth's radius) and group velocity U for a given ω. The result of cleaning the beam-output spectrum is shown on the right-hand side. Triangles indicate the true locations of modes. [From Nolet (1976); copyrighted by the American Geophysical Union.]

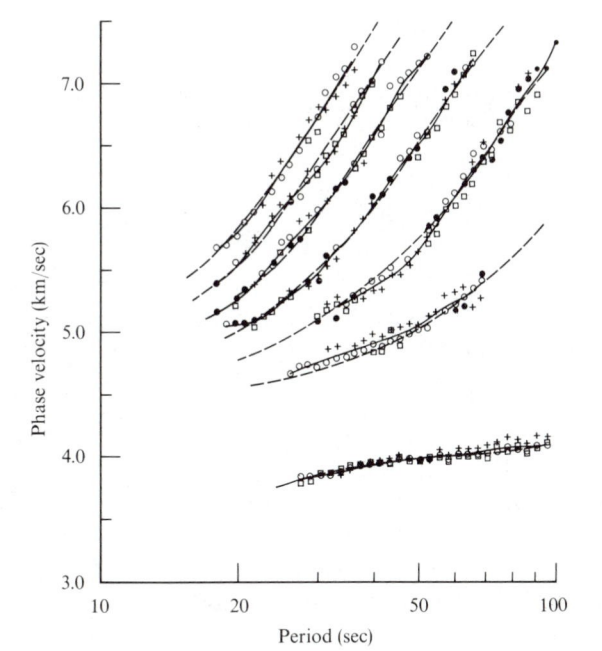

FIGURE **11.18**

The phase velocities of fundamental mode Rayleigh waves and the first six higher modes in Western Europe. Different symbols correspond to different events. The broken curve represents the theoretical values for the Gutenberg model, and the continuous one is obtained by smoothing data. [From Nolet, 1975.]

whole range of k becomes less than a certain fraction of the starting value. In reconstructing the wavenumber spectrum, only the main lobe $H_0(k)$ of the array response is used, and $\sum_i \gamma \hat{D}(k_{max}^{(i)}, \omega, U) H_0(k_{max}^{(i)} - k)$ is added to the final residual.

The result of applying the above cleanup process to the data on the left-hand side of Figure 11.17, using $\gamma = 0.4$, is shown in the right-hand side of the same figure. Most of the spurious peaks were indeed cleaned up.

Using this method, Nolet determined Rayleigh-wave phase velocity over an array of stations in Western Europe for several events in Eastern Asia, as shown in Figure 11.18. The solid curves are obtained by smoothing data, and the broken curves represent the theoretical values for the Gutenberg model.

11.5.4   The cross-correlation filter

Consider that the data consist of noise n_t and signal s_t,

$$d_t = n_t + s_t,$$

where n_t is a stationary time series as defined in (11.51) with the discrete power density $P_k^{(n)}$ and s_t as a transient signal. The filter that has the impulse response of the same shape as s_t but reversed in time is called a *cross-correlation filter*,

or a *matched filter*, and is very often used successfully in both exploration and earthquake seismology.

Let us first estimate how much gain can be obtained in the signal-to-noise ratio by such a filter. As shown in Section 11.3.6, the noise power $\langle n^2 \rangle$ was expressed in terms of discrete power density $P_k^{(n)}$ as

$$\langle n^2 \rangle = \sum_{k=1}^{N-1} P_k^{(n)} \frac{\Delta \omega}{2\pi}, \tag{11.122}$$

where $\Delta \omega = 2\pi/(N \, \Delta t)$. Noting that $P_{N-k}^{(n)}$ refers to the frequency $-k \, \Delta \omega$, the summation is actually taken over the frequency range $-N/(2N \, \Delta t) < f < N/(2N \, \Delta t)$. Assuming that the noise power density is constant P over the range $k_1 < |k| < k_2$, we write

$$\begin{aligned} \langle n^2 \rangle &= 2P(k_2 - k_1) \, \Delta \omega/2\pi \\ &= 2PW, \end{aligned} \tag{11.123}$$

where $W = (k_2 - k_1) \Delta \omega/2\pi$ is the frequency bandwidth.

Passing the noise through the filter whose impulse response is s_{-t}, we get the output n_t' given by

$$n_t' = \sum_i n_{t-i} s_{-i} \, \Delta t. \tag{11.124}$$

Then the output noise power is given by

$$\langle n'^2 \rangle = \sum_{k=1}^{N-1} P_k^{(n)} |S_k|^2 \frac{\Delta \omega}{2\pi}, \tag{11.125}$$

where S_k is the discrete Fourier transform of the filter response s_k. Taking the bandwidth $W = f_2 - f_1$ to cover the frequency range where S_k exists significantly, and using (11.123), we can rewrite (11.125) as

$$\langle n'^2 \rangle = P \sum_{k=1}^{N-1} |S_k|^2 \frac{\Delta \omega}{2\pi}. \tag{11.126}$$

Using the discrete equivalent of the Parseval theorem,

$$\sum_{k=0}^{N-1} |S_k|^2 \frac{\Delta \omega}{2\pi} = \sum_{t=0}^{N-1} s_t^2 \, \Delta t, \tag{11.127}$$

we can write

$$\langle n'^2 \rangle = P \, \Delta t \sum_{i=0}^{N-1} s_i^2. \tag{11.128}$$

The noise power was increased by a factor

$$\frac{\langle n'^2 \rangle}{\langle n^2 \rangle} = \frac{\Delta t}{2W} \sum_{i=0}^{N-1} s_i^2. \tag{11.129}$$

On the other hand, the filter response to the input signal s_t is $\Sigma s_{t-i} s_{-i} \, \Delta t$, which has the maximum value $\sum_{i=0}^{N-1} s_i^2 \, \Delta t$ at $t = 0$, in which case the signal amplitude will be improved by a factor $\Delta t \Sigma s_i^2 / \max(s_i)$. Therefore, the gain in signal-to-noise power ratio is given by

$$\text{gain} = \frac{\Sigma s_i^2}{[\max(s_i)]^2} \cdot 2W \, \Delta t. \tag{11.130}$$

For example, for a so-called *chirp filter*, which is a sinusoidal oscillation of unit amplitude and duration $N \, \Delta t$ with frequency changing linearly with time from f_1 to f_2, we find, approximately,

$$\sum_{i=0}^{N-1} s_i^2 / [\max(s_i)]^2 \sim N/2.$$

Thus the gain in signal-to-noise ratio is the product of the frequency bandwidth ($W \sim f_2 - f_1$) and the duration of signal. Capon et al. (1969) discussed the signal-to-noise gain of the chirp filter in more detail when they used it for the enhancement of dispersed Rayleigh-wave signals. Typically, teleseismic Rayleigh waves from a medium-sized earthquake recorded on a standard long-period seismograph last about 600 sec with frequency varying from 0.025 to 0.05 Hz. The corresponding time-bandwidth product is equal to a factor of 15 or 12 db gain in signal-to-noise power ratio. Actual Rayleigh-wave trains are not of uniform amplitude, making the value of $\sum_{i=0}^{N-1} s_i^2 / [\max(s_i)]^2$ less than $N/2$. The actual gains observed by Capon et al. ranged from 6 to 10 db.

The same principle of signal enhancement is successfully used in the VIBROSEIS method of reflection seismology, in which a hydraulic vibrator is used as an energy source to generate a sinusoidal wave train of varying frequency, and the records are cross-correlated with the signal waveform. The force level of the vibrator is moderate, on the order of 10^4 kg. With the signal duration of a few seconds and the frequency range 10 to 100 Hz, the gain in signal-to-noise power ratio can be more than a factor of 100. Repeated sweeps will further enhance the signal. The VIBROSEIS method has been increasingly replacing the explosion method, because it does not require the expensive and time-consuming drilling of shotholes.

Another important application of the cross-correlation filter is the preprocessing of data for a distortionless time-domain window, as discussed in detail by Dziewonski and Hales (1972). Suppose that the phase spectrum of the signal is approximately known in advance. We first obtain the Fourier transform of the entire length of data without windowing. The approximately known

phase $\phi_0(\omega)$ is subtracted from the observed phase, and then its inverse Fourier transform is taken. The above process is equivalent to applying a cross-correlation filter whose frequency response has the unit spectral density and the given phase spectrum $\phi_0(\omega)$ with a minus sign.

If the phase correction is accurate, the filtered signal will be concentrated around zero lag and have the shape of an even function with respect to lag. Then we apply a symmetric window with its center at zero lag. Because the cross-correlation filter concentrates the signal energy near zero lag, the length of window can be shortened, thereby decreasing the contribution from noise. Further, since no phase distortion is introduced by a symmetric window over a symmetric signal, the damaging effect of signal distortion by windowing is minimized.

We then recalculate the Fourier transform of windowed data and obtain the correction to the approximate phase spectrum $\phi_0(\omega)$. The process can be iterated until no further correction is required.

The phase correction or equalization by a cross-correlation filter was suggested by Tukey (1959) as a means of eliminating the effect of propagation from surface waves in order to find their radiation pattern at the source. The method was successfully applied to Love and Rayleigh waves by Aki (1960), demonstrating the usefulness of surface waves for studying earthquake mechanisms.

11.5.5 Spectral ratios

As discussed in Section 11.3, seismograms are often considered as an output of a series of linear filters, each expressing the effect of separate factors. Taking the ratio of the spectra of two seismograms, therefore, one can eliminate the effect of certain common factors.

For example, Berckhemer (1962) successfully isolated the source-scale effect on the earthquake spectrum by taking the spectral ratio between two seismograms recorded at the same station for two different earthquakes with common epicenter but different magnitudes. Likewise, Ben-Menahem's (1961) directivity function, the spectral ratio between two stations at opposite azimuths from the source, was intended to isolate the effect of fault-rupture propagation.

There are several spectral ratios intended to eliminate the effect of the seismic source. The attenuation and dispersion of world-encircling surface waves were obtained from the spectral ratio between wave packets observed at a single station (Satô, 1958). The spectral ratio of surface waves at two stations along the wave path has been used extensively to measure the attenuation and dispersion along the path. The spectral ratio between vertical-component and horizontal-component displacement due to teleseismic body waves has been used in the study of crustal structure (Phinney, 1964). There are many other attempts to use the spectral ratio in isolating certain factors and eliminating others.

The spectral ratio, however, is a very dangerous quantity when noise is present, because of its enormous variability. Suppose that we make a Fourier transform of a finite sample of a time series and label the square of its absolute value as energy spectral density. As discussed in Section 11.3.7, the energy spectral density obeys the χ^2 distribution with two degrees of freedom. Therefore, its RMS deviation from the mean is equal to the mean itself. The ratio of two such energy spectra sharing the same mean will obey the F distribution with 2×2 degrees of freedom. We find that the probability that this ratio will lie in the range between $1/19$ and 19 is 90%.

Now let us consider the problem of obtaining the best estimate of the spectral ratio between signals at two stations from n independent observations. The signal may be different for different observations, but the spectral ratio between the pair of signals is common. At a given frequency ω, we write the common ratio as $re^{i\theta}$, where r is the amplitude ratio and θ is the phase difference. Both r and θ are functions of ω.

Assuming that the noise at two stations obeys Gaussian distributions independently from each other, then the cosine and sine transform of data will also obey the Gaussian distribution, as discussed in Section 11.3.6. Writing the cosine transform and the sine transform of the kth data at station i as $C_k^{(i)}$ and $S_k^{(i)}$, respectively, we express their means as $\langle C_k^{(i)} \rangle$ and $\langle S_k^{(i)} \rangle$. Assuming that noise has zero mean, we have

$$\frac{\langle C_k^{(2)} \rangle + i \langle S_k^{(2)} \rangle}{\langle C_k^{(1)} \rangle + i \langle S_k^{(1)} \rangle} = re^{i\theta} \tag{11.131}$$

or

$$\langle C_k^{(2)} \rangle = r \cos \theta \langle C_k^{(1)} \rangle - r \sin \theta \langle S_k^{(1)} \rangle,$$

$$\langle S_k^{(2)} \rangle = r \cos \theta \langle S_k^{(1)} \rangle + r \sin \theta \langle C_k^{(1)} \rangle.$$

Then the joint probability density function for given data can be written as

$$\frac{1}{(2\pi\sigma_1^2)^n} \exp \sum_{k=1}^{n} \frac{(C_k^{(1)} - \langle C_k^{(1)} \rangle)^2 + (S_k^{(1)} - \langle S_k^{(1)} \rangle)^2}{2\sigma_1^2}$$

$$\times \frac{1}{(2\pi\sigma_2^2)^n} \exp \sum_{k=1}^{n} \left[\frac{(C_k^{(2)} - r \cos \theta \langle C_k^{(1)} \rangle + r \sin \theta \langle S_k^{(1)} \rangle)^2}{2\sigma_2^2} \right.$$

$$\left. + \frac{(S_k^{(2)} - r \cos \theta \langle S_k^{(1)} \rangle - r \sin \theta \langle C_k^{(1)} \rangle)^2}{2\sigma_2^2} \right],$$

where σ_1^2 and σ_2^2 are the variance of cosine and sine transforms of the noise at stations 1 and 2. The maximum-likelihood estimate of r and θ can be obtained

by maximizing the above probability with respect to parameters. Following Pisarenko (1970), we find after some algebra that the maximum likelihood estimates of θ and r are

$$\hat{\theta} = \tan^{-1} G/F,$$

$$\hat{r} = \frac{V - ZU}{2(F^2 + G^2)^{1/2}} + \sqrt{\left(\frac{(V - ZU)^2}{4(F^2 + G^2)} + Z \right)}, \qquad (11.132)$$

where

$$U = \frac{1}{n} \sum_{k=1}^{n} [(C_k^{(1)})^2 + (S_k^{(1)})^2],$$

$$V = \frac{1}{n} \sum_{k=1}^{n} [(C_k^{(2)})^2 + (S_k^{(2)})^2],$$

$$F = \frac{1}{n} \sum_{k=1}^{n} [C_k^{(1)}C_k^{(2)} + S_k^{(1)}S_k^{(2)}],$$

$$G = \frac{1}{n} \sum_{k=1}^{n} [C_k^{(1)}S_k^{(2)} - C_k^{(2)}S_k^{(1)}],$$

and $Z = \sigma_2^2/\sigma_1^2$. Lowes (1971) pointed out that the maximum-likelihood estimate $\hat{\theta}$ can be rewritten as

$$\tan \hat{\theta} = \frac{\Sigma a_k A_k \sin \theta_k}{\Sigma a_k A_k \cos \theta_k}, \qquad (11.133)$$

where

$$C_k^{(1)} + iS_k^{(1)} = a_k \exp(i\lambda_k),$$

$$C_k^{(2)} + iS_k^{(2)} = A_k \exp(i\Lambda_k)$$

and $\theta_k = \Lambda_k - \lambda_k$. In other words, $\tan \hat{\theta}$ is the ratio of the weighted average of $\sin \theta_i$ to that of $\cos \theta_i$, with weight proportional to the product of the two amplitudes. Figure 11.19 shows a simple geometric interpretation of $\hat{\theta}$.

Pisarenko's formula (11.132) for the maximum-likelihood estimate of amplitude ratio and phase difference has been used by Patton (1978) in a study of dispersion and attenuation of Rayleigh waves across Eurasia. He found that the method works well with periods longer than about 40 sec, but gives unreliable results for shorter periods. This is probably because fluctuations in long-period waves are caused by ambient noises in the instrument and the Earth, to which the statistical model used for deriving Pisarenko's formula

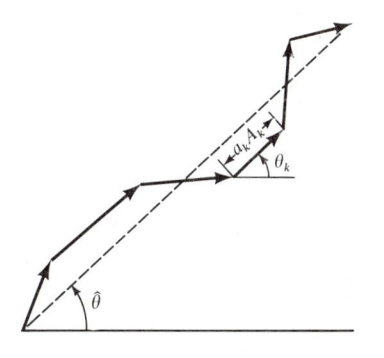

FIGURE **11.19**

A graphical interpretation of the maximum-likelihood estimate of phase difference.

is applicable. On the other hand, fluctuations of waves with periods shorter than about 40 sec (for the propagation distances around 4000 km) may be caused by scattering, multipath interference, and interference with other waves. Such signal-generated noises may not be adequately modeled by a super-position of a Gaussian noise. For these short waves, more reliable results are obtained by matching the observed and theoretical spectra in the domain of logarithmic amplitude and linear phase angle.

SUGGESTIONS FOR FURTHER READING

Beauchamp, K. G. *Signal Processing Using Analog and Digital Techniques*. London: George Allen and Unwin Ltd., 1973.

Beauchamp, K. G. *Exploitation of Seismograph Networks*. Leiden: Nordhoff International Publishing, 1975.

Blackman, R. B., and J. W. Tukey. *The Measurement of Power Spectra*. New York: Dover Publishers, 1958.

Brigham, E. O., *The Fast Fourier Transform*. Englewood Cliffs, New Jersey: Prentice-Hall, 1974.

Claerbout, J. F. *Fundamentals of Geophysical Data Processing With Applications to Petroleum Prospecting*. New York: McGraw-Hill, 1976.

Dziewonski, A. M., and A. L. Hales. Numerical analysis of dispersed seismic waves. *In* B. A. Bolt (editor), *Seismology: Surface Waves and Earth Oscillations* (Methods in Computational Physics, Vol. 11). New York: Academic Press, 1972.

Gold, B., and C. M. Rader. *Digital Processing of Signals*. New York: McGraw-Hill, 1969.

Peacock, K. L., and S. Treitel. Predictive deconvolution, theory and practice. *Geophysics*, **34**, 155–169, 1969.

Robinson, E. R. *Statistical Communication and Detection*. London: Charles Griffin and Co., 1967.

PROBLEMS

11.1 Plan an experiment to measure phase velocity of surface waves for a wave path between two stations. Suppose we use an earthquake of magnitude $M_s = 3\frac{1}{2}$ at $\Delta = 90°$ during the quiet period of ground noise; what accuracy can you achieve in the phase-velocity measurement? Use the formula given in (11.18). Consult Figures 10.9, 10.10, and 10.11 for the spectral data on surface waves and ground noise.

11.2 Consider a layered half-space with various impedance contrasts $\beta_1\rho_1/\beta_2\rho_2$ between the layer and the half-space. Formulate a recursive filter that will generate the normally incident plane wave from observed seismograms at the surface using equation (11.24). Obtain the corresponding frequency-response function by (11.33). Use the result of Section 11.3.6 to find the autocorrelation and power spectra for the surface motion when the incident wave is white noise.

11.3 Consider a recursive filter with the z-transform response given by

$$F(z) = \frac{(z - 1)(z + 1)}{(z - c)(z - c^*)}$$

where $c = (1 + h)e^{i\omega_0\Delta t}$. Write down the filter in the forms of (11.32) and (11.33). Show that the center frequency and band-width of the filter are determined by ω_0 and $h\omega_0$, respectively.

11.4 Obtain the array response, i.e., the weight function $W(\xi, \eta)$ defined in (11.104) for a square array with seismometers equally spaced in x and y. Discuss the resolution and the effect of side lobes as a function of the number of seismometers and spacing.

11.5 In constructing the Jeffreys-Bullen travel-time table, the residual is considered to obey a statistical distribution with the probability density given by

$$f(t) = \frac{1 - \varepsilon}{\sigma\sqrt{2\pi}} \exp\left[-\frac{(t - \bar{t})^2}{2\sigma^2} \right] + \varepsilon g(t),$$

where the first term is the Gaussian distribution given in (11.42) and $g(t)$ is a smoothly varying function representing a very gradual decay in the travel-time distribution; ε is assumed to be small. Show that the maximum-likelihood estimate (Section 11.4.2) of the mean is given by a weighted average $\bar{t} = \Sigma w_i t_i/\Sigma w_i$, and that the weight w_i for observed time t_i is given by

$$w_i = \frac{1}{1 + \mu \exp[(t_i - \bar{t})/2\sigma^2]},$$

where μ is the ratio of the tail amplitude to the peak value of $f(t)$.

CHAPTER 12

Inverse Problems in Seismology

One of the main jobs of a seismologist is to determine the Earth's interior structure from data obtained at the Earth's surface. Ultimately we want to find a method that will give structure and source parameters by processing the whole seismogram. So far in this book, we have developed solutions to the "forward problem." That is, for models of Earth structure and seismic source we have found out how to compute properties of seismic motion that are observable, such as travel times; dispersion curves; spectra of near-field, strong ground motion; free-oscillation periods; far-field waveforms; and complete seismograms. The "inverse problem" is to compare these computed properties with the data, in order to learn about Earth structure and seismic source. Whenever the solution to the forward problem is known, then trial and error provides one method of inversion: the parameters of a model are readjusted on some ad hoc basis until some acceptable agreement between observables and data is discovered. This method immediately brings out the main features that we must consider in *any* inverse method: is the model *adequate* to explain the data; is it *unique*; and, if not unique, what knowledge about the model do we in fact gain from a particular data set? To answer this last question, which is crucial, trial and error is usually poor, because so many trials must be made, as in the Monte Carlo method. Therefore, far more sophisticated inversion methods have been developed for certain limited data sets. The first two are

(i) the travel times of seismic rays transmitted through a structure, and (ii) the reflection seismograms scattered back from the structure. The remaining inversion methods we shall explore are applicable to (iii) any set of data for which small changes in the underlying model produce small changes in the observable and for which an adequate starting model is known. Data sets meeting these criteria are, again, a limited subset of the data in whole seismograms. So far, such invertible data sets (iii) have included dispersion of surface waves on a local or regional scale; free-oscillation eigenfrequencies on a global scale; and, in a preliminary fashion, waveforms from selected portions of certain seismograms. In this chapter, we shall inquire as to what can be uncovered about the structure of the Earth's interior from each of these three types of data set, and derive formulas for the data inversion.

We shall first consider the classic problem of travel-time inversion for body waves propagated through an Earth model in which the seismic velocity depends only on depth or distance from the Earth's center. The velocity distribution with depth can be uniquely determined by the formula due to Herglotz (1907) and Wiechert (1910) in the absence of a low-velocity layer. We shall also derive formulas applicable to the case including low-velocity layers, following Gerver and Markushevitch (1966).

We shall then ask what can be uncovered about the Earth's interior from a reflection seismogram. The answer to this fundamental question is known only for the simplest case of one-dimensional waves in one-dimensional media, i.e., the plane waves incident on a medium whose properties vary only in the direction of wave propagation. Although the case is too simple to have direct practical application, it nevertheless gives some insight into the physics involved.

We shall find that a reflection seismogram cannot give elastic constants and density as a function of depth, but only impedance as a function of travel time. We shall derive the inversion formulas for continuous and discrete cases, following Claerbout (1968) and Ware and Aki (1969).

Few analytical solutions are known for the inversion of dispersion data for surface waves in vertically heterogeneous media. For example, using the WKBJ method, applicable only to short waves, Takahashi (1955) reduced the relation between the Love-wave phase-velocity curve and the shear-velocity depth function to the form of Abel's integral equation.

Thus we conclude that the applicability of existing analytic solutions of inverse problems is very much restricted, because of their oversimplified assumptions about both the medium and the theory of wave propagation. In the absence of formulas for a direct inversion of data, the usual approach is the trial and error method. In this case, we start with an initial guess as to the wave source and media, predict what should be observed, and compare the prediction with what actually is observed. The process is repeated for various choices of model parameters, describing the wave source and media until a satisfactory fit between the predicted and the observed is obtained. With the use of the computer, a systematic search for satisfactory models has become

practical by uniformly or randomly scanning plausible regions of the parameter space.

The above process of trial and error can be done quickly and systematically if the small perturbation in model parameters is linearly related to the resultant change in the observable. In that case, by requiring that the residual between the predicted and the observed data be minimal, the correction to the parameters of an initial model can be given as the solution of a system of linear equations whose coefficients are functions of the initial model. A satisfactory model may be obtained by repeating the above process until the solution converges.

The property of the linear system is given by the matrix that relates the data and model parameters. We shall follow Lanczos (1961) in the general discussion of interplay between the data and model spaces in order to understand the interrelationship between observations and parameters.

A large number of parameters are often required to describe the detail of a model. On the other hand, the uncertainty of each individual parameter increases with the number of parameters to be determined from given data. Thus a compromise must be found between the requirements for resolution and accuracy. A quantitative discussion of trade-off between resolution and accuracy is possible when the statistical properties of errors in data are known. This problem has been analyzed by Backus and Gilbert in a series of papers that aroused broad interest in the inverse problem.

In addition to the known statistical properties of data, it may happen that the statistical properties of model parameters are known, or we may be interested only in the smoothed properties of the model in a specified way. Following Franklin (1970), we shall discuss the optimal way of inverting data when the statistical properties of both data and model are known.

12.1 Travel-time Inversion

12.1.1 The Herglotz-Wiechert formula

The travel time data are arrival times $T(X)$ of the onset of a wave packet as a function of distance X between the seismic source and receiver measured along the Earth's surface. Let us first consider the case of a flat Earth in which the velocity $c(z)$ varies only with depth z. From (4.45a), we know that the ray parameter p defined by

$$p = \frac{\sin i}{c(z)}, \tag{12.1}$$

where i is the angle between the ray path and the vertical (see Figure 12.1), is a constant along a given ray path. The maximum depth $Z(p)$ penetrated by the

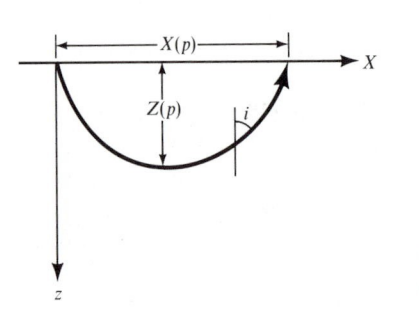

FIGURE **12.1**

Ray path in a flat Earth model.

ray with parameter p is given by $c(Z(p)) = 1/p$, because $i = \pi/2$ at the deepest point. The horizontal distance $X(p)$ that the ray with parameter p travels before reaching the surface (Figure 12.1) is given by

$$X(p) = 2 \int_0^{Z(p)} \tan i \, dz$$

$$= 2 \int_0^{Z(p)} \frac{p \, dz}{\sqrt{(c(z))^{-2} - p^2}}, \qquad (12.2)$$

where we used (12.1). Likewise, the travel time $T(p)$ for the ray with parameter p is given by

$$T(p) = 2 \int_0^{Z(p)} \frac{(c(z))^{-2} \, dz}{\sqrt{(c(z))^{-2} - p^2}}. \qquad (12.3)$$

Since $dT/dX = \sin i_0/c(0)$, where $c(0)$ and i_0 are the values of $c(z)$ and i, respectively, at the surface, the ray parameter p can be determined directly from the travel-time data $T(X)$ as

$$p = \frac{dT}{dX}. \qquad (12.4)$$

Thus both functions $X(p)$ and $T(p)$ are observable quantities at the Earth's surface. The travel-time inverse problem is to find $c(z)$ from observed $X(p)$ or $T(p)$.

In the case of a spherically symmetric Earth, the ray parameter is no longer related to the horizontal phase velocity but to the angular velocity. As shown in (4.45b), the ray parameter p defined by

$$p = \frac{r \sin i}{c(r)} \qquad (12.5)$$

is constant along a ray path for a spherically symmetric Earth, where i is the angle between the ray path and the radial direction from the Earth's center (Fig. 12.2). For both the source and the receiver on the surface, the angular distance $\Delta(p)$ between the source and the point at which the ray with parameter p emerges can be written as

$$\Delta(p) = 2 \int_{r_p}^{r_\oplus} \tan i \, \frac{dr}{r}$$

$$= 2p \int_{r_p}^{r_\oplus} \frac{1}{\sqrt{(r/c)^2 - p^2}} \frac{dr}{r}, \tag{12.6}$$

where $r_\oplus$ is the Earth's radius and r_p is the radial distance to the deepest point of the ray (Fig. 12.2). The corresponding travel time $T(p)$ is given by

$$T(p) = 2 \int_{r_p}^{r_\oplus} \frac{(r/c)^2}{\sqrt{(r/c)^2 - p^2}} \frac{dr}{r}. \tag{12.7}$$

Corresponding to the relation (12.4) for the flat Earth, we have

$$p = dT/d\Delta \tag{12.8}$$

for the spherical Earth.

The above equations (12.5) through (12.8) for the spherical Earth can be obtained from the corresponding ones (12.1) to (12.4) for the flat Earth simply by replacing $X(p)$ by $r_\oplus \, \Delta(p)$, depth z by $r_\oplus \ln(r_\oplus/r)$, $c(z)$ by $c(r) \, r_\oplus/r$, and the ray parameter p by $p/r_\oplus$. It therefore suffices to solve the inverse problem for the flat Earth. The solution for spherical Earth problems can be obtained immediately by the above changes of variables (see also Box 9.9).

The formulas expressing the travel distance in terms of the velocity-depth function, (12.2) and (12.6), can be reduced to the form of Abel's integral equation, for which the inverse problem has been solved.

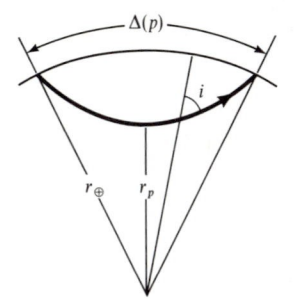

FIGURE **12.2**
Ray path in a spherical Earth model.

We shall first rewrite equation (12.2) using $(c(z))^{-2}$ as the integration variable:

$$\frac{X(p)}{2p} = \int_0^{Z(p)} \frac{dz}{\sqrt{(c(z))^{-2} - p^2}}$$

$$= \int_{c_0^{-2}}^{p^2} \frac{dz/d(c^{-2})}{\sqrt{c^{-2} - p^2}} d(c^{-2}), \tag{12.9}$$

where c_0 is the velocity at $z = 0$. The upper limit of integration $c^{-2} = p^2$ follows from the fact that $Z(p)$ is the depth of a ray's deepest point, at which the velocity c is the reciprocal of ray parameter p. Equation (12.9) is identical to (9) of Box 12.1 if we replace $X(p)/2p$ by $t(x)$, p^2 by x and $(c(z))^{-2}$ by ξ. Therefore,

---

BOX **12.1**

Abel's problem

Abel's (1826) problem was to determine the shape of a hill when the time taken by a particle to go up and return is known as a function of the particle's initial velocity. The particle is constrained to the surface of the hill without friction and moves under the force of gravity. From conservation of energy, we can find the maximum height x it reaches with the initial velocity v_0 by $gx = \frac{1}{2}v_0^2$. Consider the movement along a path from the point of maximum height $P(x, y)$ down to the starting point 0, as shown in Figure A. The potential energy of the particle at height ξ is equal to $-mg(x - \xi)$, and the kinetic energy is $\frac{1}{2}m(ds/dt)^2$, where m is the particle mass and s is the distance measured along the path. Since there is no frictional loss, we have

$$\left(\frac{ds}{dt}\right)^2 = 2g(x - \xi). \tag{1}$$

Taking the square root of both sides and integrating, we have

$$t(x) = \int_0^x \frac{ds/d\xi}{\sqrt{2g(x - \xi)}} d\xi. \tag{2}$$

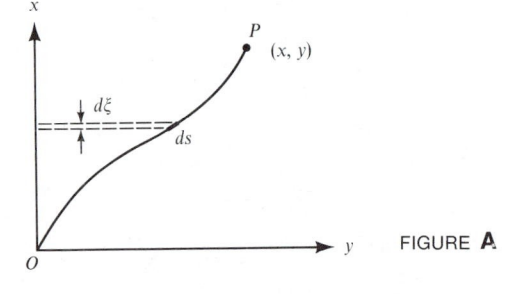

FIGURE **A**

Rewriting this in the form of Abel's integral equation,

$$t(x) = \int_0^x \frac{f(\xi)}{\sqrt{x - \xi}} \, d\xi, \tag{3}$$

our inverse problem is to find $f(\xi)$ when $t(x)$ is known.

The solution is obtained by the following steps. First, multiply both sides of (3) by $dx/\sqrt{\eta - x}$ and integrate with respect to x from 0 to η:

$$\int_0^\eta \frac{t(x) \, dx}{\sqrt{\eta - x}} = \int_0^\eta \frac{dx}{\sqrt{\eta - x}} \int_0^x \frac{f(\xi)}{\sqrt{x - \xi}} \, d\xi$$

$$= \int_0^\eta f(\xi) \, d\xi \int_\xi^\eta \frac{dx}{\sqrt{\eta - x}\sqrt{x - \xi}} \tag{4}$$

by changing the order of integration. The integration limits are also changed to cover the shaded area (Figure C), which is the original area of integration (Figure B). The integral with respect to x in (4) is seen to be a constant by making the variable change $x = \xi \cos^2 \theta + \eta \sin^2 \theta$, giving

$$\int_\xi^\eta \frac{dx}{\sqrt{\eta - x}\sqrt{x - \xi}} = \int_0^{\pi/2} 2 \, d\theta = \pi. \tag{5}$$

Then equation (4) can be written as

$$\int_0^\eta \frac{t(x) \, dx}{\sqrt{\eta - x}} = \pi \int_0^\eta f(\xi) \, d\xi. \tag{6}$$

Differentiating (6) with respect to η, we obtain

$$\frac{d}{d\eta} \int_0^\eta \frac{t(x) \, dx}{\sqrt{\eta - x}} = \pi f(\eta). \tag{7}$$

Replacing η by ξ, we have the standard form

$$f(\xi) = \frac{1}{\pi} \frac{d}{d\xi} \int_0^\xi \frac{t(x) \, dx}{\sqrt{\xi - x}} \tag{8}$$

as the solution of integral equation (3).

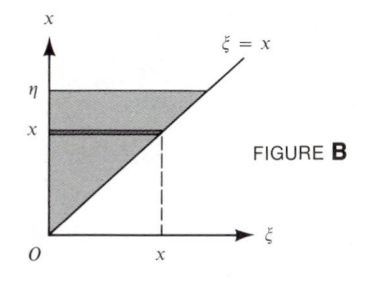

FIGURE **B**

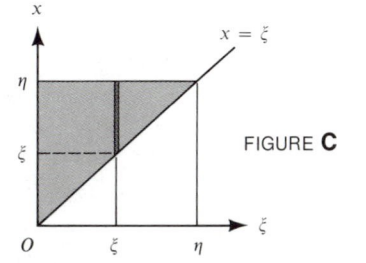

FIGURE **C**

According to Bôcher (1909), the necessary and sufficient conditions that Abel's integral equation (3) has the continuous solution (8) are the following:

i) $t(x)$ must be continuous,
ii) $t(0) = 0$, and
iii) $t(x)$ has a derivative that is finite with at most a finite number of discontinuities.

The most serious restriction is the exclusion of a discontinuity in $t(x)$ (which would imply two different return times for a given velocity at 0). This happens if the hill is undulating with peaks and troughs. In the case of travel-time inversion, this corresponds to the presence of low-velocity channels.

Finally, changing the variables in (3) and (8) such that $\xi \rightarrow a - \xi$ and $x \rightarrow a - x$, we obtain the formulas more suitable to our problem:

$$t(x) = \int_x^a \frac{f(\xi)}{\sqrt{\xi - x}}\, d\xi \tag{9}$$

and

$$f(\xi) = -\frac{1}{\pi}\frac{d}{d\xi}\int_\xi^a \frac{t(x)\, dx}{\sqrt{x - \xi}}. \tag{10}$$

the solution of the inverse problem corresponding to equation (10) of Box 12.1 can be written as

$$z(c) = -\frac{1}{\pi}\int_{c_0^{-2}}^{c^{-2}} \frac{X(p)/2p}{\sqrt{p^2 - c^{-2}}}\, d(p^2)$$

$$= -\frac{1}{\pi}\int_{c_0^{-1}}^{c^{-1}} \frac{X(p)}{\sqrt{p^2 - c^{-2}}}\, dp, \tag{12.10}$$

where we have dropped the derivative with respect to c^{-2} from both sides. Equation (12.10) gives the depth z for a given velocity c from observed travel-time data $X(p)$. It can be written in the following forms by integration by parts:

$$z(c) = -\frac{1}{\pi}\int_{\cosh^{-1}(c/c_0)}^{0} X(p)\, d(\cosh^{-1}(pc))$$

$$= \frac{1}{\pi}\int_0^{X(c^{-1})} \cosh^{-1}(pc)\, dX, \tag{12.11}$$

where we used $X(c_0^{-1}) = 0$. The result for a spherical Earth is obtained from (12.11) by variable changes mentioned earlier, giving

$$\ln[r_\oplus/r(c)] = \frac{1}{\pi}\int_0^{\Delta(r/c)} \cosh^{-1}(pc/r)\, d\Delta. \tag{12.12}$$

In applications, a particular value of the ratio r/c is selected. Via the $\Delta = \Delta(p)$ curve, the upper limit of integration in (12.12) is then known, and the integral itself is carried out (using general $p \geq r/c$) to give a value of $\ln(r_\oplus/r)$. Hence r is known for a value of $r/c(r)$, leading to the determination of one point in the velocity profile $c = c(r)$. Equations (12.10) through (12.12) are called Herglotz-Wiechert formulas.

The conditions for the validity of Herglotz-Wiechert formulas may be obtained from Bôcher's conditions, discussed in Box 12.1. Translating them into the travel-time problem, we find that the formulas are applicable to cases in which the derivative of $X(p)$ is discontinuous, but not to cases in which $X(p)$ itself is discontinuous. Cases of the latter type occur when velocity decreases with depth (for a spherical Earth, an increase in $c(r)/r$ with r). Thus the Herglotz-Wiechert formula cannot be used in cases involving low-velocity channels. It can be applied, however, to the case of a rapid velocity increase that causes a triplication in the travel-time curve.

Figure 12.3 shows an example of the triplication. For ray paths bottoming above the zone of rapid velocity increase, the rays penetrating deeper usually emerge farther away from the source point. However, the strong refraction

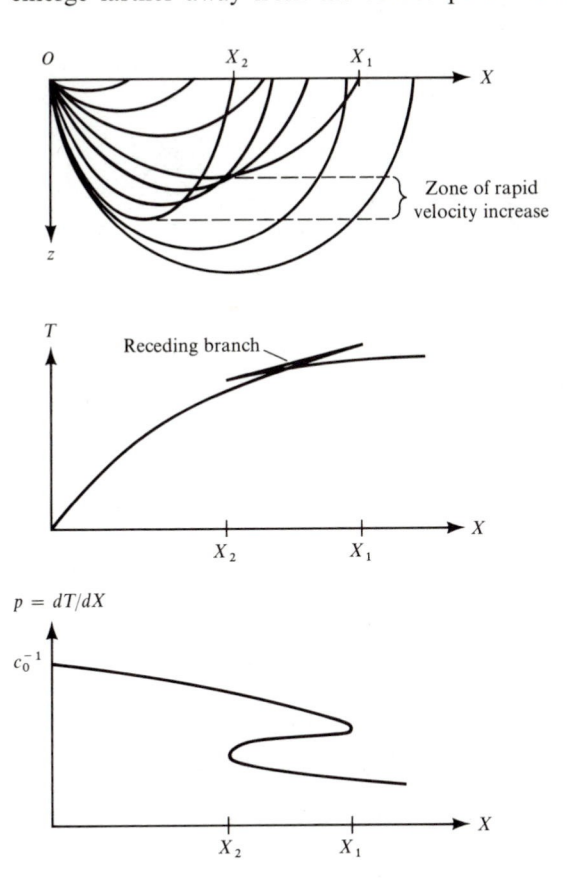

FIGURE **12.3**

Triplication of travel-time curve due to a rapid increase in velocity with depth.

by a transition zone can cause the ray to emerge closer to the source, as depicted in Figure 12.3. If the ray bottoms below the transition zone, the behavior of the travel-time curve returns to normal. The travel time $T(X)$ and ray parameter $p = dT/dX$ are also shown schematically as functions of X. The travel-time curve is downward concave on the normal branches, but upward concave at the receding branch because $d^2T/dX^2 = dp/dX > 0$ there. As shown at the bottom of Figure 12.3, $X(p)$ is clearly a single-valued continuous function of p, and therefore the Herglotz-Wiechert formula (12.10) is applicable in this case.

Let us apply the formula to an extreme example due to Slichter (1932), consisting of a homogeneous layer overlying a homogeneous half-space, in which velocity increases discontinuously as shown in Figure 12.4. The direct waves traveling with velocity c_0 in the upper layer are designated as P. The refracted and reflected waves from the half-space are designated as P_n and P_MP, respectively. The travel-time branches that correspond to them meet at a distance where the total reflection first shows up at the critical angle i_n, which satisfies $\sin i_n = c_0/c_1$. The corresponding $X - p$ relation is shown at the bottom.

Since the travel time $T(X)$ for the reflection branch P_MP is given by

$$T(X) = \frac{2}{c_0}\left[h^2 + \left(\frac{X}{2}\right)^2\right]^{1/2},$$

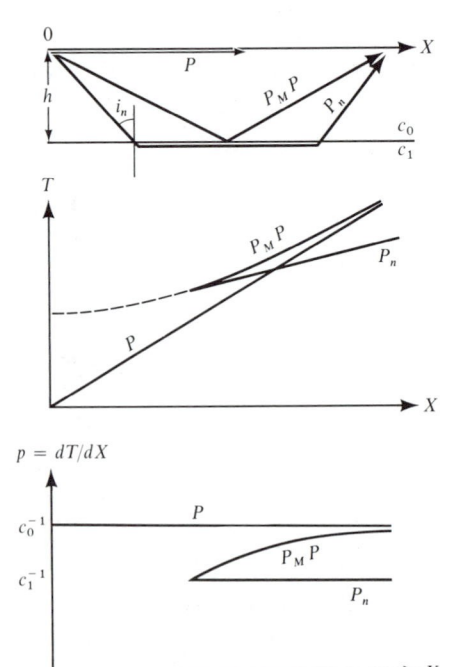

FIGURE **12.4**

Ray paths and travel times for a homogeneous layer over a half-space.

we can easily obtain the explicit form of $X(p)$ as

$$X(p) = \frac{2hp}{\sqrt{1/c^2 - p^2}} \qquad \text{for } c_1^{-1} < p < c_0^{-1}$$

for that branch. Inserting this result into the inversion formula (12.10), we find that

$$z(c) = -\frac{1}{\pi} \int_{c_0^{-1}}^{c^{-1}} \frac{2hp \, dp}{\sqrt{c_0^{-2} - p^2}\sqrt{p^2 - c^{-2}}}$$
$$= h \qquad \text{for } c_0^{-1} > c^{-1} > c_1^{-1}. \qquad (12.13)$$

Thus the correct depth h is recovered for velocity in the range from c_0 to c_1.

For a successful inversion of travel-time data with a triplication, it is necessary to identify the normal and receding branches correctly. This requires the correct identification of not only the first arrivals but also later arrivals. In practice, correct readings of arrival times of later phases are often difficult because of interference from the coda of earlier arrivals.

Suppose only the first arrival is correctly identified and found to consist of two line segments, as shown in Figure 12.5. There are an infinite number of structures that will give the same first arrival. In a study of crustal structure along the coast of California, Healy (1963) considered n-layered structures, for various n, which give the identifical first arrival, while forcing all the arrivals from "masked layers" to pass through the intersection of the two first-arrival lines. The result is reproduced for $n = 1, 2$, and 50 in Figure 12.6. The 50-layer case approaches the formula $c(z) = c_0 \cosh \pi z/X_c$, obtained by Slichter (1932) for a continuous structure for which all rays emerge at the intersection point (X_c is the horizontal distance to the point). Figure 12.6 illustrates the nonuniqueness of interpreting the travel-time data when only the first arrivals are known.

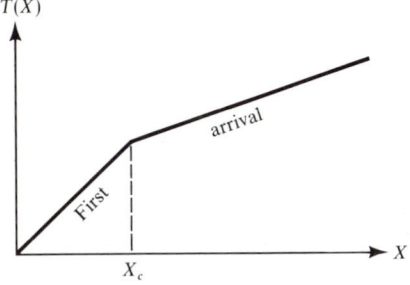

FIGURE **12.5**

Travel time of the first arrival is known as shown here, but those of later arrivals are unknown. The nonuniqueness problem in this case is demonstrated in Figure 12.6.

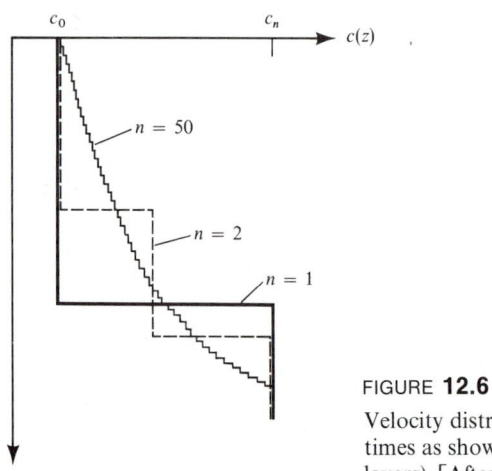

FIGURE **12.6**

Velocity distributions that give the same first-arrival times as shown in Figure 12.5 (n is the number of layers). [After Healy, 1963.]

12.1.2   Travel-time inversion for structures including low-velocity layers

If the velocity $c(z)$ starts decreasing with depth z at a certain depth z_1 in the flat Earth model (or the angular velocity $c(r)/r$ starts to decrease with decreasing r in the case of the spherical Earth model), no ray can have its deepest point below this depth until the velocity comes back to the value it had at z_1, as illustrated in Figures 12.7 and 9.14. In this case, a shadow zone occurs with a gap in the travel-time curve, and a discontinuity appears in the $X - p$ relation, violating a condition for applicability of the Herglotz-Wiechert inversion formula. What can we learn about the Earth's structure from such discontinuous travel-time curves?

Slichter (1932) analyzed this problem and drew some interesting conclusions. Since there are no rays bottoming within the low-velocity layer (LVL), the velocity distribution throughout the LVL affects the travel time, which is always

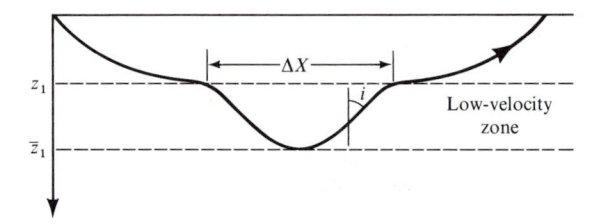

FIGURE **12.7**

Ray paths and travel times for the Earth model containing a low-velocity layer.

integrated over the layer. Suppose we simulate the velocity distribution in the LVL by a stack of playing cards. The cards may be shuffled without changing the integrated travel time. Thus, the ordering of the cards cannot be determined from the travel-time data.

However, Slichter found an upper bound for the thickness of the LVL from observed gaps ΔX and ΔT in the travel-time curve (see Fig. 12.8). Defining upper and lower boundaries of the LVL and the angle i between the ray path and vertical, as shown in Figure 12.7, we find

$$\frac{\Delta X}{2} = \int_{z_1}^{z_1} \tan i \, dz \tag{12.14}$$

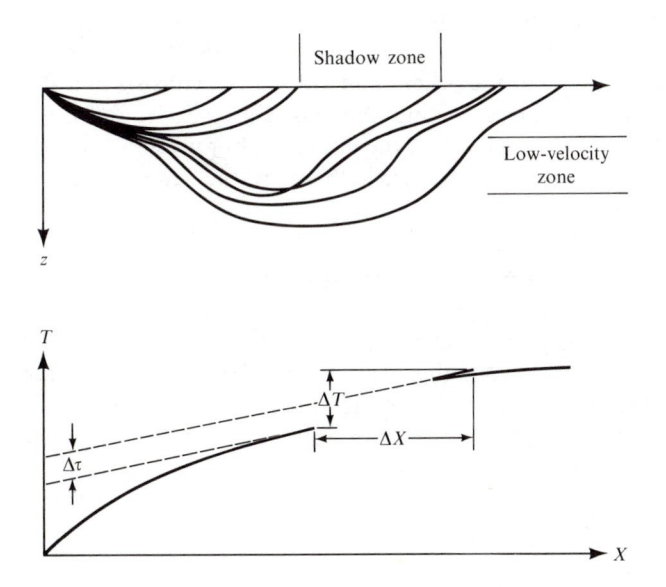

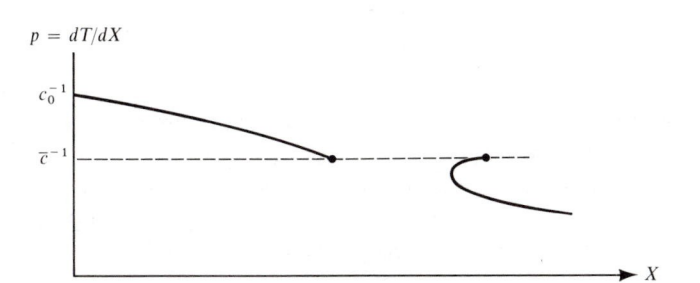

FIGURE **12.8**

Ray paths and travel times for a structure including a low-velocity layer.

and

$$\frac{\Delta T}{2} = \int_{z_1}^{\bar{z}_1} \frac{dz}{c(z) \cos i}$$

$$= \int_{z_1}^{\bar{z}_1} \frac{p_1 \, dz}{\sin i \cos i}$$

$$= p_1 \frac{\Delta X}{2} + p_1 \int_{z_1}^{\bar{z}_1} \cot i \, dz, \tag{12.15}$$

where p_1 is the ray parameter of the ray that grazes at the upper and lower boundaries of the LVL, or $p_1 = c^{-1}(z_1) = c^{-1}(\bar{z}_1)$. For given ΔT and ΔX, the integrals of both $\tan i$ and $\cot i$ over the range $z_1 < z < \bar{z}_1$ are fixed. If the LVL has a uniform velocity, i will be constant and the integrals will become αh and h/α, where $\alpha = \tan i$. On the other hand, if the LVL consists of a stack of homogeneous layers, the corresponding integrals will be $\Sigma \alpha_n h_n$ and $\Sigma h_n/\alpha_n$, where h_n is the thickness and α_n is $\tan i$ for the nth layer. For given ΔT and ΔX, these integrals are given and their product is therefore the same for the two cases,

$$\alpha h \cdot h/\alpha = h^2 = (\Sigma h_n/\alpha_n)(\Sigma \alpha_n h_n).$$

Since the right-hand side is greater than $(\Sigma h_n)^2$ unless the layers all have the same value α_n, we conclude that the thickness of the LVL is maximum for a uniform velocity. In other words, for a given ΔT and ΔX, there exists a bound for the thickness of the LVL. This bound can be obtained by putting $c(z) = c_L$, for $z_1 < z < \bar{z}_1$, in (12.14) and (12.15).

$$\frac{\Delta X}{2} = \frac{p_1(\bar{z}_1 - z_1)_{\max}}{(c_L^{-2} - p_1^2)^{1/2}},$$

$$\frac{\Delta T}{2} = \frac{c_L^{-2}(\bar{z}_1 - z_1)_{\max}}{(c_L^{-2} - p_1^2)^{1/2}}. \tag{12.16}$$

By eliminating c_L from the above equations, we find the maximum estimate of thickness of the LVL from observed gaps in the travel-time curve.

A more general and complete analysis of the problem has been made by Gerver and Markushevitch (1966). The above equations (12.16) for the upper bound of thickness of the LVL can also be derived using their method.

Let us first recall the function $\tau(p)$ defined by

$$\tau(p) = T(p) - pX(p), \tag{12.17}$$

which was introduced (for spherical Earth problems) in (9.22). The physical meaning of $\tau(p)$ is clear; $\tau(p)$ is the intercept with $X = 0$ of a line tangent to the travel-time curve $T = T(X)$, since p is the slope of the line. From (12.2) and (12.3), we have

$$\tau(p) = 2 \int_0^{Z(p)} \sqrt{c^{-2}(z) - p^2}\ dz, \qquad (12.18)$$

which monotonically increases with $Z(p)$ and decreases with p. The observed $\tau(p)$ for P- and S-waves in the Earth are shown in Figure 12.9.

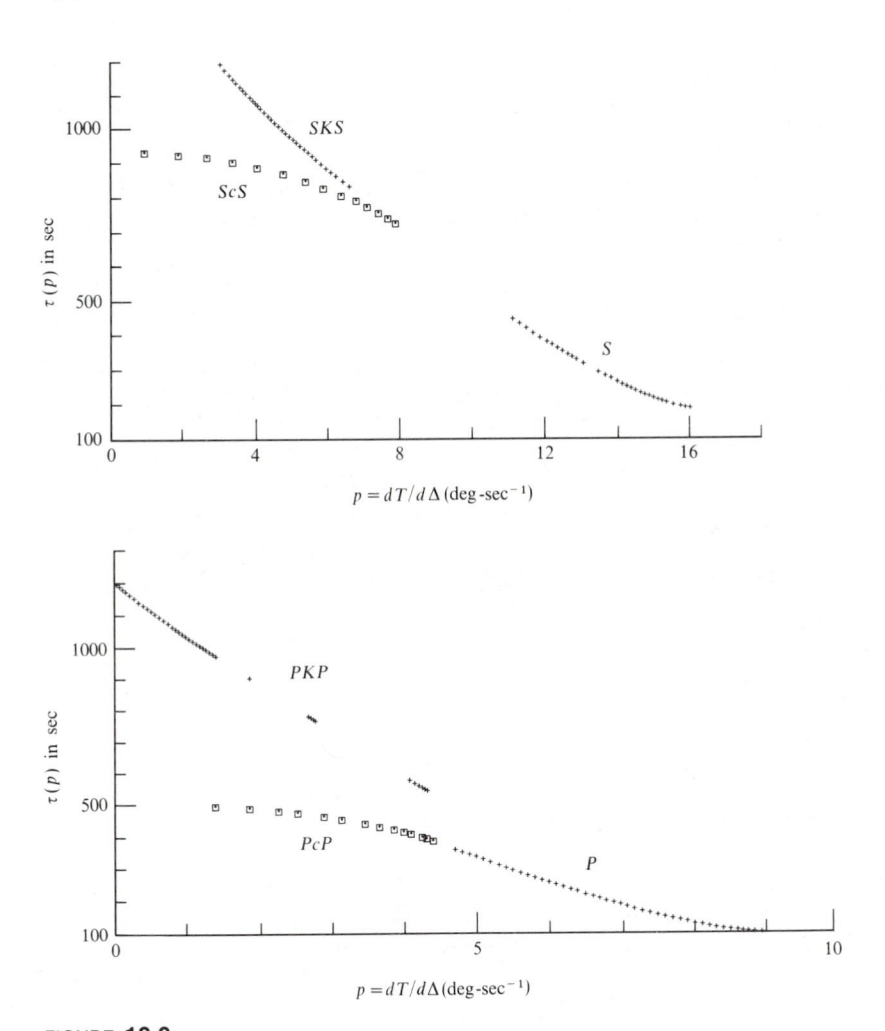

FIGURE **12.9**

Top. $\tau(p)$ for S, SKS, and ScS. Bottom. $\tau(p)$ for P, PKP, and PcP. [From Johnson and Gilbert, 1972.]

The jump in $\tau(p)$ due to a low-velocity layer can be written as

$$\Delta\tau = \Delta T - p_1 \Delta X. \tag{12.19}$$

Note that $\tau(p)$ alone contains all the information available from the travel-time data. Since $dT/dX = p$,

$$\frac{d\tau}{dp} = \frac{dT}{dp} - p\frac{dX}{dp} - X(p) = -X(p). \tag{12.20}$$

BOX **12.2**

Measurement of $\tau(p)$

Bessonova et al. (1974) used a simple and reliable method of obtaining $\tau(p)$ from a set of travel-time data (T_i, X_i). The method is based on an extremum property of $T(p_0) - pX(p_0)$ as p_0 is varied with p fixed.

First, we note that $[T(p_0) - pX(p_0)]$ at $p_0 = p$ is just $\tau(p)$. Second, note that $\partial/\partial p_0[T(p_0) - pX(p_0)] = (p_0 - p)\,dX/dp_0$, so that $[T(p_0) - pX(p_0)]$ has an extremum at $p_0 = p$.

The method of Bessonova et al. proceeds as follows. For a given set of travel-time data (T_i, X_i), plot $T_i - pX_i$ for a fixed p as a function of X_i, as shown in the figure. It will take an extreme value at $X(p)$, and the extreme value of $T_i - pX_i$ is equal to $\tau(p)$.

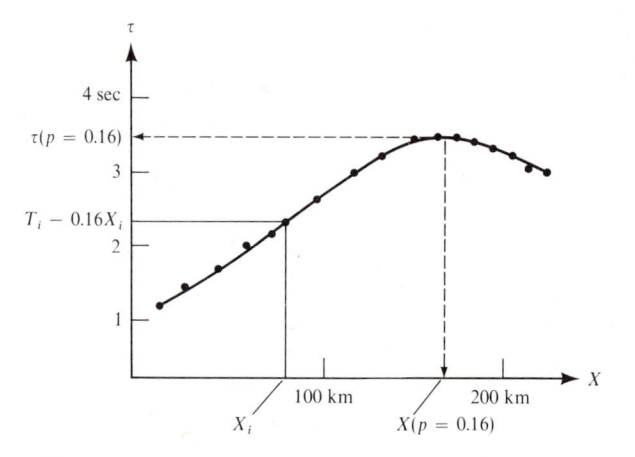

The inversion formulas obtained for $X(p)$ can be written for $\tau(p)$ according to the relation $d\tau(p) = -X(p)\,dp$. For example, equation (12.10) becomes simply

$$Z(p) = \frac{1}{\pi}\int_0^{\tau(p)} \frac{d\tau}{\sqrt{v^2(\tau) - p^2}}, \tag{4}$$

where $v(\tau)$ and $\tau(p)$ are mutually inverse functions and $z(c) = Z(p)|_{p=1/c}$.

We can determine $X(p)$, and then $T(p)$, if $\tau(p)$ alone is known. The gaps ΔT and ΔX in the travel-time curve due to a LVL can also be found if $\tau(p)$ alone is known, because the difference in $d\tau/dp$ between $p_1 - 0$ and $p_1 + 0$ gives ΔX. Then ΔT can be determined by putting known ΔX and $\Delta\tau$ into (12.19).

Let us now obtain the inversion formula applicable to the velocity-depth distribution that includes low-velocity layers. We shall start with the following identity:

$$Z(p) = \int_0^{Z(p)} dz \cdot \frac{1}{\pi} \int_{p^2}^{[c(z)]^{-2}} \frac{d(q^2)}{\sqrt{[c(z)^{-2} - q^2](q^2 - p^2)}}. \qquad (12.21)$$

This is easily verified by substituting $q^2 = p^2 \cos^2\theta + c^{-2}\sin^2\theta$.

The area of the above integration is shown by a shaded region in the $q - z$ diagram in Figure 12.10. The integral is taken for a fixed z with respect to q from p to $1/c(z)$, and then the result is integrated with respect to z from 0 to $Z(p)$. The kth low-velocity zone is specified by the ray parameter p_k at two depths z_k and $\bar{z}_k$ corresponding to the top and bottom of the zone, respectively.

Since $Z(p)$ is the depth at which a ray with ray parameter p bottoms and p is equal to the reciprocal velocity at the bottom, $Z(p)$ as a function of p is equivalent to $1/c(z)$ as a function of z except within the low-velocity zone, where there are no turning points. For the kth low-velocity zone, we shall assume that the curve $Z(p)$ is a straight line connecting the two points (p_k, z_k) and $(p_k, \bar{z}_k)$. By this definition, $Z(p)$ does not increase with p.

Changing the order of integration, we can rewrite (12.21) as

$$Z(p) = \frac{2}{\pi} \int_p^{1/c_0} \frac{q\, dq}{\sqrt{q^2 - p^2}} \int_0^{Z(q)} \frac{dz}{\sqrt{[1/c(z)]^2 - q^2}} + I(p), \qquad (12.22)$$

where $I(p)$ is the integral over the densely shaded area in Figure 12.10. Since from (12.2),

$$X(q) = 2 \int_0^{Z(q)} \frac{q\, dz}{\sqrt{[1/c(z)]^2 - q^2}},$$

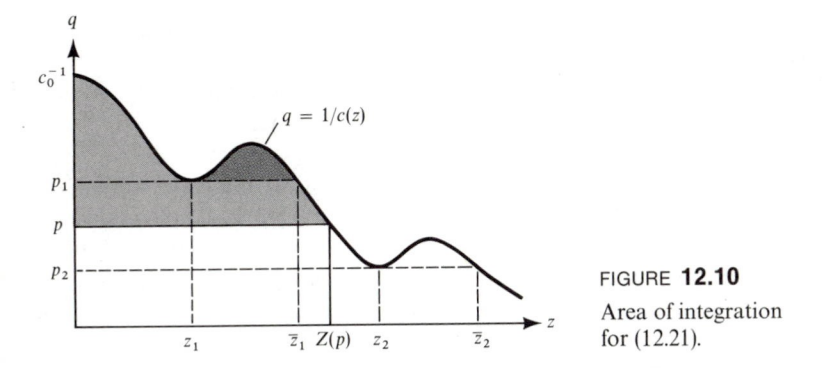

FIGURE **12.10**

Area of integration for (12.21).

we can rewrite (12.22):

$$Z(p) = \frac{1}{\pi} \int_p^{1/c_0} \frac{X(q)\,dq}{\sqrt{q^2 - p^2}} + I(p). \qquad (12.23)$$

The first term of the right-hand side of the above equation is nothing but the Herglotz-Wiechert formula, and is determined uniquely from the observed travel-time data. The contribution $I(p)$ from the low-velocity layers can be obtained as

$$I(p) = \sum_{i=1}^{k} \frac{2}{\pi} \int_{z_i}^{\bar{z}_i} dz \int_{p_i}^{1/c(z)} \frac{q\,dq}{\sqrt{[1/c(z)]^2 - q^2}\,\sqrt{q^2 - p^2}}$$

$$= \sum_{i=1}^{k} \frac{2}{\pi} \int_{z_i}^{\bar{z}_i} \tan^{-1} \sqrt{\frac{[1/c(z)]^2 - p_i^2}{p_i^2 - p^2}}\,dz \qquad \text{for } p_k > p > p_{k+1}. \quad (12.24)$$

The above equations (12.23) and (12.24), obtained by Gerver and Markushevitch (1966), represent the extension of the Herglotz-Wiechert formula to include low-velocity layers.

For $p > p_1$, $I(p) = 0$. Therefore, the Herglotz-Wiechert term gives the solution $Z(p)$, which gives the velocity-depth function uniquely for $z < z_1$.

For $p < p_1$, the Herglotz-Wiechert term is determined from the observed travel-time data. However, through the second term $I(p)$, an arbitrary velocity-depth function $v(z)$ may be assigned to the low-velocity layer subject to some constraints described below. Except for the upper boundary of the first low-velocity layer, the boundary depths $z_k, \bar{z}_k$ are also unknown. The first constraint from observation on $c(z)$, z_k, and $\bar{z}_k$ is given by the discontinuity in $\tau(p)$:

$$\Delta \tau_k = 2 \int_{z_k}^{\bar{z}_k} \sqrt{[1/c(z)]^2 - p_k^2}\,dz. \qquad (12.25)$$

The second is, by definition, that the calculated $Z(p)$ should not increase with p. The third is that $Z(p_k + 0)$ and $Z(p_k - 0)$ must agree, respectively, with the depths of the lower and upper boundaries, z_k and $\bar{z}_k$, of the kth layer. As shown in Figure 12.11, Gerver and Markushevich gave a "giraffe-like" area in which the plots of all possible solutions $c(z)$ lie for the case of two low-velocity zones. The upper bound for $c(z)$ corresponds to the Herglotz-Wiechert term. Figure 12.11 also shows the existence of an upper bound for the thickness of the LVL given earlier by (12.16).

The special methods we have described above for inverting travel-time data are closely associated with the special properties of an Abel integral equation— namely, that a method of *construction* is known for obtaining the inverse. Other methods of inverting travel-time data are of course possible, and Johnson and Gilbert (1972) have used more general methods, which we describe in Section

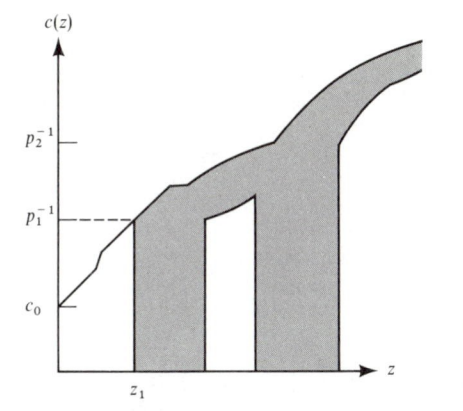

FIGURE **12.11**

A solution obtained by Gerver
and Markushevitch (1966).

12.3. Such methods were also used by Crosson (1976a,b) with local earthquake data. In this case, hypocenters must be determined simultaneously with inversion for structure.

12.2 Inverse Problem for a Reflection Seismogram

After an impulsive seismic wave is sent downward, a seismograph on the surface will record echoes coming back from the Earth's interior. What can we recover from this record, called a reflection seismogram, about the Earth's interior? The solution of this inverse problem is known for the one-dimensional case in which the source wave is a plane wave propagating vertically and the elastic properties of the Earth also vary only vertically. It will be shown that we cannot recover the seismic velocity as a function of depth, as we did in the case of travel-time inversion discussed in the preceding section. Instead, we recover the impedance, the product of seismic velocity and density, as a function of vertical travel time. Thus there is a fundamental difference between travel-time data and the reflection seismogram in the nature of the information they contain.

Let us start with the equation of motion for the one-dimensional case:

$$\rho(x)\frac{\partial^2 u}{\partial t^2} = \frac{\partial}{\partial x}\left[E(x)\frac{\partial u}{\partial x}\right],\tag{12.26}$$

where $\rho(x)$ is density, $E(x)$ is an elastic constant, and $u(x, t)$ may represent longitudinal displacement in the x-direction due to compressional waves when $E(x) = \lambda(x) + 2\mu(x)$, or transverse displacement perpendicular to the x-direction due to shear waves when $E(x) = \mu(x)$, where $\lambda(x)$ and $\mu(x)$ are Lamé's constants. We use x for the depth here to keep z for use in the z-transform.

We shall consider simultaneously an easier problem in which the medium consists of discrete homogeneous layers, as shown in Figure 12.12. For the

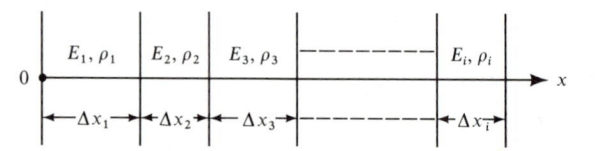

FIGURE **12.12**

A stack of homogeneous layers. Δx_i are chosen such that
the travel time for each layer is the same.

ith layer, the equation of motion is given by

$$\rho_i \frac{\partial^2 u_i}{\partial t^2} = E_i \frac{\partial^2 u_i}{\partial x^2}, \tag{12.27}$$

where E_i and ρ_i are constants. Using an arbitrary function $f_i(t)$, the solution of
(12.27) can be written as

$$u_i = f_i(t \pm x/c_i), \tag{12.28}$$

where $c_i = \sqrt{E_i/\rho_i}$. Following Goupillaud (1961), we shall choose the layer
thickness Δx_i in such a way that the travel time Δt_i across the layer is constant
for all layers:

$$\Delta t_i = \frac{\Delta x_i}{c_i} = \Delta t = \text{const.} \tag{12.29}$$

It is expected intuitively that any realistic variation in $E(x)$ and $\rho(x)$ can be well
approximated by such a stack of layers if Δt is taken sufficiently small.

The above discrete medium is convenient, because if the wave source is an
impulse applied at an interface (at $t = 0$, say), the subsequent seismic motions
at any interface can be expressed as a sequence of pulses existing only at discrete
points separated by the interval $2 \Delta t$.

If we write such a sequence as

$$s(t) = \sum_n s_n \delta(t - 2n \Delta t), \tag{12.30}$$

then its Fourier transform $S(\omega)$ is

$$S(\omega) = \int_{-\infty}^{\infty} s(t) \exp(i\omega t) \, dt$$

$$= \sum_n s_n \exp(2in\omega \Delta t). \tag{12.31}$$

Thus both the time-domain solution and the frequency response are written in terms of s_n. Putting $z = \exp(2i\omega\,\Delta t)$, we rewrite (12.31) as

$$S(z) = \sum_n s_n z^n \qquad (12.32)$$

and call $S(z)$ the *z-transform*.

Now let us consider what can be recovered from a reflection seismogram obtained at the surface ($x = 0$) of such a medium.

12.2.1 Inversion of reflection seismograms

Consider the reflection and transmission at the interface of the ith layer and $(i + 1)$th layer, as shown in Figure 12.13. When the unit impulse is incident from the ith layer, we designate the reflected and transmitted impulses as r_i and t_i, respectively. When the impulse is incident from the $(i + 1)$th layer, we designate them as $\bar{r}_i$ and $\bar{t}_i$. Applying the conditions for continuity of displacement $u_i = u_{i+1}$ and continuity of traction $E_i\,\partial u_i/\partial x = E_{i+1}\,\partial u_{i+1}/\partial x$ at the interface, we obtain the following relations among the amplitudes of reflected and transmitted pulses:

$$1 + r_i = t_i,$$

$$1 + \bar{r}_i = \bar{t}_i, \qquad (12.33)$$

$$r_i = -\bar{r}_i = \frac{c_i\rho_i - c_{i+1}\rho_{i+1}}{c_i\rho_i + c_{i+1}\rho_{i+1}}.$$

These formulas can be obtained also by putting the ray parameter $p = 0$ in (5.39) and taking into account the opposite sign convention for reflected P-waves there.

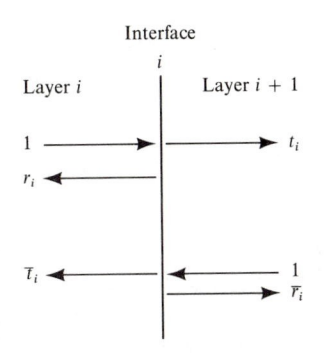

FIGURE **12.13**

Defining reflection and transmission coefficients.

Now we formulate the simplest inverse problem. When we send an impulsive plane wave at $x = 0$ into the layered medium of Figure 12.12, we shall later receive at $x = 0$ reflections from many interfaces in the medium. Given the reflection seismogram, what can we say about the medium?

A simple graphical solution is given in Figure 12.14 following Gerver (1970). In Figure 12.14, we draw the wave path in the $x - t$ diagram. At any interface, the pulses arrive with a time separation of $2 \Delta t$. The pulse amplitude at each wave path is indicated using the reflection and transmission coefficients defined in (12.33). The reflection seismogram is obtained at $x = 0$ as a function of t.

It is easy to find that the reflection coefficient r_1 is directly observed at $t = 2 \Delta t$. Then, from (12.33), we can find $t_1, \bar{t}_1$ and $\bar{r}_1$. Therefore, the reflection amplitude observed at $t = 4 \Delta t$, which is $t_1 r_2 \bar{t}_1 + r_1 \bar{r}_0 r_1$, will determine the only unknown, r_2. The reflection seismogram at $t = 6 \Delta t$ will likewise determine r_3, and so on.

Once $r_1, r_2, \ldots, r_n$ are known, and the surface value of impedance $c_1 \rho_1$ is known, $c_i \rho_i \, (i > 1)$ can be determined iteratively using the relation

$$c_{i+1} \rho_{i+1} = \frac{1 - r_i}{1 + r_i} c_i \rho_i. \tag{12.34}$$

Therefore, from the reflection seismogram of length $2n \, \Delta t$, we can recover the impedance $c_i \rho_i$ from $i = 2$ up to $i = n + 1$. We recover the impedance not as a function of depth x but as a function of i, which is proportional to the travel time of reflected waves.

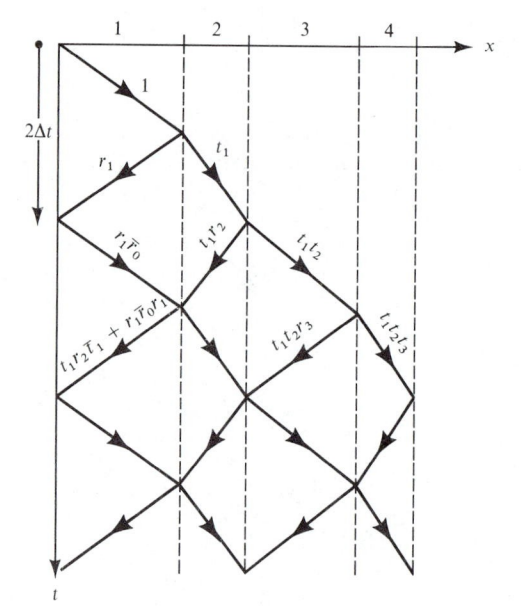

FIGURE **12.14**

Paths of multiply reflected and transmitted waves in the xt-plane.

The same conclusion is obtained for a medium in which $E(x)$ and $\rho(x)$ are continuous functions of x. Equation (12.26) may be reduced to a one-dimensional Schrödinger equation by the following change of variables. First, we change the independent variable x to ζ by

$$dx = c(x)\,d\zeta,$$

$$\zeta = \int^x \frac{dx}{c(x)},\tag{12.35}$$

where $c(x) = [E(x)/\rho(x)]^{1/2}$, so that ζ is the travel time. Equation (12.26) can then be written as

$$\frac{\partial^2 u(\zeta, t)}{\partial \zeta^2} - \frac{\partial^2 u(\zeta, t)}{\partial t^2} = -\frac{d\ln[\rho c]}{d\zeta}\frac{\partial u(\zeta, t)}{\partial \zeta}.$$

Second, we change the dependent variable u to ϕ by

$$\phi = [\rho c]^{1/2}u.\tag{12.36}$$

Then we have

$$\frac{\partial^2 \phi}{\partial \zeta^2} - \frac{\partial^2 \phi}{\partial t^2} = q(\zeta)\phi,$$

where

$$q(\zeta) = \frac{1}{\eta}\frac{d^2\eta}{d\zeta^2} \quad \text{and} \quad \eta = (\rho c)^{1/2}.\tag{12.37}$$

For the time dependence $\exp(-i\omega t)$, we have

$$\frac{d^2\phi}{d\zeta^2} - q(\zeta)\phi + \omega^2\phi = 0,\tag{12.38}$$

which is the one-dimensional Schrödinger equation for potential $q(\zeta)$.

As will be shown in a later section, under certain conditions $q(\zeta)$ can be uniquely determined from the observed reflection seismogram. From $q(\zeta)$, we can obtain the impedance η^2 as a function of the travel time ζ, as shown in (12.37), but no more.

Thus in both continuous and discrete cases, the reflection seismogram gives us impedance as a function of travel time. The next question is how to calculate the impedance as a function of travel time when the reflection seismogram is given. Let us first consider the discrete medium.

12.2.2   Inverse formula for the discrete case

The rigorous derivation of the inversion formula for continuous media is difficult. We shall therefore describe the discrete case in detail, following Claerbout (1968), and then briefly mention analogous results for the continuous medium. It will be shown that the basis of the derivation of the inversion formula is essentially the same in both cases. The causality principle—namely, that a cause precedes an effect—is crucial in both cases.

In order to formulate the inverse problem for the discrete medium shown in Figure 12.12, we need to obtain a general formula to calculate reflected and transmitted waves at the ith interface from those at the $(i - 1)$th interface. As shown in Figure 12.15, we designate the waves propagating in the x-direction by $D_i \exp[-i\omega(t - x/c_i)]$, and call them downgoing waves. The upgoing waves are designated by $U_i \exp[-i\omega(t + x/c_i)]$. D_i and U_i are functions of ω. Likewise, we define D_i' and U_i' as shown in Figure 12.15. Using the reflection and transmission coefficients defined in Figure 12.13, we can write

$$D_{i+1} = t_i D_i' + \bar{r}_i U_{i+1},$$
$$U_i' = r_i D_i' + \bar{t}_i U_{i+1}. \tag{12.39}$$

Rearranging the above equations, we find

$$\begin{pmatrix} U_i' \\ D_i' \end{pmatrix} = \begin{pmatrix} \bar{t}_i - \dfrac{r_i \bar{r}_i}{t_i} & \dfrac{r_i}{t_i} \\[2ex] -\dfrac{\bar{r}_i}{t_i} & \dfrac{1}{t_i} \end{pmatrix} \begin{pmatrix} U_{i+1} \\ D_{i+1} \end{pmatrix}. \tag{12.40}$$

From (12.33),

$$r_i = -\bar{r}_i \quad \text{and} \quad \bar{t}_i - \frac{r_i \bar{r}_i}{t_i} = \frac{1}{t_i}. \tag{12.41}$$

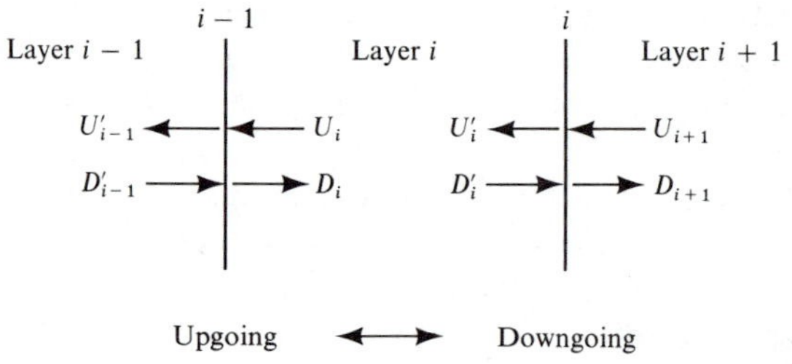

FIGURE **12.15**

Defining upgoing and downgoing waves.

Therefore, we have a very simple relation,

$$\begin{pmatrix} U_i' \\ D_i' \end{pmatrix} = \frac{1}{t_i} \begin{pmatrix} 1 & r_i \\ r_i & 1 \end{pmatrix} \begin{pmatrix} U_{i+1} \\ D_{i+1} \end{pmatrix}. \tag{12.42}$$

On the other hand,

$$\begin{aligned} U_i &= \exp(i\omega\,\Delta t)U_i', \\ D_i &= \exp(-i\omega\,\Delta t)D_i'. \end{aligned} \tag{12.43}$$

Combining (12.42) and (12.43), we find

$$\begin{pmatrix} U \\ D \end{pmatrix}_i = \frac{1}{t_i} \begin{pmatrix} \exp(i\omega\,\Delta t) & r_i\exp(i\omega\,\Delta t) \\ r_i\exp(-i\omega\,\Delta t) & \exp(-i\omega\,\Delta t) \end{pmatrix} \begin{pmatrix} U \\ D \end{pmatrix}_{i+1}.$$

Writing, as before, $z = \exp(2i\omega\,\Delta t)$ and introducing $w = \exp(i\omega\,\Delta t)$, we have

$$\begin{pmatrix} U \\ D \end{pmatrix}_i = \frac{1}{wt_i} \begin{pmatrix} z & zr_i \\ r_i & 1 \end{pmatrix} \begin{pmatrix} U \\ D \end{pmatrix}_{i+1}. \tag{12.44}$$

By iteration of the above process, we can find the upgoing and downgoing waves at the surface in terms of those at the nth interface:

$$\begin{pmatrix} U \\ D \end{pmatrix}_1 = \frac{1}{w^n} \cdot \frac{1}{t_1 t_2 \cdots t_n} \cdot \begin{pmatrix} z & zr_1 \\ r_1 & 1 \end{pmatrix} \begin{pmatrix} z & zr_2 \\ r_2 & 1 \end{pmatrix} \cdots \begin{pmatrix} z & zr_n \\ r_n & 1 \end{pmatrix} \begin{pmatrix} U \\ D \end{pmatrix}_{n+1}.$$

We can now show that the above matrix product has the following form:

$$\begin{pmatrix} U \\ D \end{pmatrix}_1 = \frac{1}{w^n} \begin{pmatrix} z^n F(1/z) & z^n G(1/z) \\ G(z) & F(z) \end{pmatrix} \begin{pmatrix} U \\ D \end{pmatrix}_{n+1}, \tag{12.45}$$

where

$$F(z) = \frac{1}{t_1 t_2 \cdots t_n}(1 + Fz + F_2 z^2 + \cdots + F_{n-1}z^{n-1}),$$

$$G(z) = \frac{1}{t_1 t_2 \cdots t_n}(r_n + G_1 z + G_2 r^2 + \cdots + G_{n-1}z^{n-1})$$

are polynomials up to the power $n - 1$. The inverse z-transforms of $F(z)$ and $G(z)$ are therefore limited in the time range $0 \le t \le 2(n - 1)\,\Delta t$, and vanish outside the range.

The above result can be proved by the method of mathematical induction. Assume that (12.45) is valid for the first n layers, and show that the same form applies to the next interface. For this purpose we shall attach a subscript to F and G. The matrix relating

$$\begin{pmatrix} U \\ D \end{pmatrix}_1 \quad \text{with} \quad \begin{pmatrix} U \\ D \end{pmatrix}_{n+2}$$

is

$$\frac{1}{w^n} \begin{pmatrix} z^n F_n(1/z) & z^n G_n(1/z) \\ G_n(z) & F_n(z) \end{pmatrix} \frac{1}{wt_{n+1}} \begin{pmatrix} z & zr_{n+1} \\ r_{n+1} & 1 \end{pmatrix}$$

$$= \frac{1}{t_{n+1}w^{n+1}} \begin{pmatrix} z^n F_n(1/z)z + z^n G_n(1/z)r_{n+1} & z^n F_n(1/z)zr_{n+1} + z^n G_n(1/z) \\ G_n(z)z + F_n(z)r_{n+1} & G_n(z)zr_{n+1} + F_n(z) \end{pmatrix}$$

$$= \frac{1}{w^{n+1}} \begin{pmatrix} z^{n+1} F_{n+1}(1/z) & z^{n+1} G_{n+1}(1/z) \\ G_{n+1}(z) & F_{n+1}(z) \end{pmatrix}. \tag{12.46}$$

The above result shows that $F_{n+1}(z)$ and $G_{n+1}(z)$ are polynomials up to the term z^n with the constant terms

$$\frac{1}{t_1 t_2 \cdots t_{n+1}} \quad \text{and} \quad \frac{r_{n+1}}{t_1 t_2 \cdots t_{n+1}},$$

respectively. Since equation (12.45) holds for $n = 1$, it must hold for an arbitrary n. The recursive relation in (12.46) is convenient for the computation of $F(z)$ and $G(z)$.

Our inverse problem is to obtain the reflection coefficients at all the interfaces (or impedance of all the layers) from the reflection seismogram observed at the surface. We shall do this by assuming that the number of layers is finite and that the $(n + 1)th$ layer is a half-space containing no seismic sources. By solving this problem we shall find that the reflection coefficients up to the nth interface can be recovered from the reflection seismogram for the recording time up to $2n \Delta t$. Since no reflection can arrive prior to $t = 2(n + 1) \Delta t$ from the interfaces beyond the nth, the reflection seismogram up to the time $2n \Delta t$ is independent of the medium beyond the nth interface. Therefore, the assumption of a finite number of layers does not restrict the applicability of an inversion formula obtained under the assumption.

We shall call the upgoing waves observed at the surface the "reflection seismogram" and denote its z-transform as $R(z)$. The corresponding downgoing wave consists of the source impulse applied at $t = 0$ and the reflection of upgoing waves $R(z)$ by the free surface. The z-transform of the source impulse is 1. From (12.33), the appropriate reflection coefficient at the free surface is

+1. Therefore, the boundary condition at the free surface is given by

$$\begin{pmatrix} U \\ D \end{pmatrix}_1 = \begin{pmatrix} R(z) \\ 1 + R(z) \end{pmatrix}. \tag{12.47}$$

We shall call the downgoing waves in the $(n + 1)$th layer the "transmission seismogram" and denote its z-transform as $T(z)$. Since we assume no seismic sources beyond the nth interface, the boundary conditions there are given by

$$\begin{pmatrix} U \\ D \end{pmatrix}_{n+1} = \begin{pmatrix} 0 \\ T(z) \end{pmatrix}. \tag{12.48}$$

Putting these conditions into (12.45) and rearranging, we obtain

$$\begin{pmatrix} R(z) \\ 1 \end{pmatrix} = \frac{1}{w^n} \begin{pmatrix} z^n F(1/z) & z^n G(1/z) \\ G(z) - z^n F(1/z) & F(z) - z^n G(1/z) \end{pmatrix} \begin{pmatrix} 0 \\ T(z) \end{pmatrix}. \tag{12.49}$$

Introducing

$$M(z) = F(z) - z^n G(1/z), \tag{12.50}$$

we find immediately that

$$T(z) = w^n / M(z),$$
$$R(z) = z^n G(1/z)/M(z). \tag{12.51}$$

From the properties of $F(z)$ and $G(z)$ given in (12.45), it can be seen that for the n-layer medium, $M(z)$ is a polynomial of the form

$$M(z) = M_0 + M_1 z + \cdots + M_n z^n, \tag{12.52}$$

where

$$M_0 = \frac{1}{t_1 t_2 \cdots t_n} \quad \text{and} \quad M_n = \frac{-r_n}{t_1 t_2 \cdots t_n}.$$

The physical meaning of $M(z)$ is clear from (12.51). Since $w^n = \exp(in\omega \, \Delta t)$ represents the delay due to the one-way travel time across n layers, $T(z)/w^n$ is the z-transform of the transmission seismogram with the origin of time shifted to the arrival time of first motion at the nth interface. Equation (12.51) shows that $M(z)$ is the z-transform of a deconvolution filter that will give the source

impulse as the output when the time-shifted transmission seismogram is fed as the input. If the source is not an impulse, then the filter will recover the source waveform.

From (12.52), it is clear that $M(z)$ is analytic for $|z| \leq 1$. Since $z = \exp(2i\omega \, \Delta t)$, $M(z)$ is analytic in the upper-half ω-plane, and $1/M(z)$ is also analytic there. Since $1/M(z)$ has no zeros for $|z| \leq 1$, the time-shifted transmission seismogram not only vanishes for $t < 0$, but is also a minimum-delay wavelet (see Box 5.8).

To obtain inversion formulas for the reflection seismogram, we derive the following equation from (12.50) and (12.51):

$$\begin{aligned}[1 + R(z) + R(1/z)]M(1/z) &= F(z)M(1/z)/M(z) + z^{-n}G(z) \\ &= [F(z)M(1/z) + z^{-n}G(z)M(z)]/M(z) \\ &= [F(z)F(1/z) - G(z)G(1/z)]/M(z).\end{aligned}$$

The numerator of the last expression is equal to the determinant of the propagator matrix in (12.45). Since the determinant of a product of matrices is equal to the product of their determinants, and since the determinant of a layer matrix is

$$\det \frac{1}{wt_i} \begin{pmatrix} z & zr_i \\ r_i & 1 \end{pmatrix} = \frac{1 - r_i}{t_i},$$

we find

$$[1 + R(z) + R(1/z)]M(1/z) = \left(\prod_{i=1}^{n} \frac{1 - r_i}{t_i} \right) \Big/ M(z). \tag{12.53}$$

As mentioned before, $1/M(z)$ corresponds to the time-shifted transmission seismogram that vanishes for $t < 0$. In the z-transform, this means that the right-hand side of (12.53) has no terms with negative powers of z. Writing the reflection seismogram as

$$R(t) = R_1 \, \delta(t - 2 \, \Delta t) + R_2 \, \delta(t - 4 \, \Delta t) + \cdots + R_n \, \delta(t - 2n \, \Delta t) + \cdots$$

or

$$R(z) = R_1 z + R_2 z^2 + \cdots + R_n z^n + \cdots$$

and putting it into (12.53), together with (12.52), we find that

$$(1 + R_1 z + \cdots + R_1 z^{-1} + \cdots)(M_0 + M_1 z^{-1} + \cdots + M_n z^{-n})$$

should not have negative powers of z. For example, the coefficient of z^{-1} must vanish, i.e.,

$$M_0 R_1 + M_1 + M_2 R_1 + \cdots + M_n R_{n-1} = 0.$$

Considering terms up to the z^{-n}, these equations can be summarized in the following matrix form:

$$
\begin{pmatrix}
1 & R_1 & R_2 & \cdots & R_{n-1} \\
R_1 & 1 & R_1 & & \vdots \\
R_2 & R_1 & 1 & & \vdots \\
& & & \ddots & R_1 \\
R_{n-1} & & \cdots\cdots & R_1 & 1
\end{pmatrix}
\begin{pmatrix}
M_1/M_0 \\
M_2/M_0 \\
\vdots \\
\\
M_n/M_0
\end{pmatrix}
=
\begin{pmatrix}
-R_1 \\
-R_2 \\
\vdots \\
\\
-R_n
\end{pmatrix}.
\tag{12.54}
$$

Since $M_n/M_0 = -r_n$, as shown in (12.52), knowing the reflection seismogram up to the time $2n \, \Delta t$ gives the reflection coefficient at the nth interface. Starting with $n = 1$, and solving (12.54) for each step, we can recover r_i. Then, using (12.34), we can recover the impedance $c_i \rho_i$ as a function of i. [Equations (12.51) and (12.53) show the important result that the reflection seismogram $R(z)$ is one side of the autocorrelation function of the time-shifted transmission seismogram $w^{-n} T(z)$ times a constant $\prod_{i=1}^{n} (1 - r_i)/t_i$. This result was extended to the $P - SV$ problem by Frasier (1970) by replacing scalars r, t, and z^n by 2×2 matrices R, T, and Z_n. Since the transmission seismograms are unchanged if the receiver and source are switched (see the reciprocal theorem in Chapter 2), the above results suggest the possibility of converting surface records of tele-seismic events to reflection seismograms for mapping the crust.]

12.2.3 *Inversion formula for the continuous case*

As shown in Section 12.2.1, the equation of motion for a continuous medium may be reduced to the one-dimensional Schrödinger equation,

$$\frac{d^2\phi}{d\zeta^2} + [\omega^2 - q(\zeta)]\phi = 0, \tag{12.38 again}$$

where the travel time ζ, normalized field ϕ, and potential q are defined by (12.35), (12.36), and (12.37), respectively. The solution of (12.38) is oscillatory if $q(\zeta) < \omega^2$, and damped exponentially if $q(\zeta) > \omega^2$. The point ζ corresponding to $q(\zeta) = \omega^2$ is called a turning point, because the oscillatory wave cannot transmit into the region in which $q(\zeta) > \omega^2$ (Box 9.3). We assume $q(\zeta)$ is bounded,

so that waves with infinite frequency will have no turning point. The travel time ζ physically refers to the waves with infinite frequency.

Let us consider the scattering problem depicted in Figure 12.16. We assume that the inhomogeneous medium is bounded by homogeneous media at both sides. Since $q = 0$ in a homogeneous medium, ϕ will be of the form $e^{\pm i\omega\zeta}$ there. At the top of Figure 12.16, we send a signal $e^{i\omega\zeta}$ from the left to the inhomogeneous medium and observe the reflected wave $S_{12}(\omega)e^{-i\omega\zeta}$ and the transmitted wave $S_{11}(\omega)e^{i\omega\zeta}$. At the bottom of Figure 12.16, we send a signal $e^{-i\omega\zeta}$ from the right and observe the reflected wave $S_{21}(\omega)e^{i\omega\zeta}$ and the transmitted wave $S_{22}e^{-i\omega\zeta}$. From reciprocity and conservation of energy,

$$S_{11} = S_{22},$$

$$|S_{11}|^2 + |S_{12}|^2 = 1, \tag{12.55}$$

$$|S_{22}|^2 + |S_{21}|^2 = 1,$$

and, as noted below, the scattering matrix formed from the four coefficients is unitary.

Our problem is to find the potential $q(\zeta)$ from reflected or transmitted waves observed outside the inhomogeneous medium. This problem would be strongly undetermined if we did not have the constraint that impedance η^2 (12.37) must be positive. In general, the differential operator $[d^2/d\zeta^2 - q(\zeta)]$ has a "discrete spectrum," or "bound states"; i.e., for certain values of ω^2, there are solutions of (12.38) that vanish at the exterior boundary of the inhomogeneous region. As pointed out by Sabatier (1976), however, such solutions cannot exist in our problem because impedance is real and positive. In this case, the solution of our problem is uniquely determined by the Gel'fand-Levitan integral equation. We shall follow Faddeev (1967) to derive the integral with the aid of an analogous derivation for the corresponding discrete problem, given by Ware and Aki

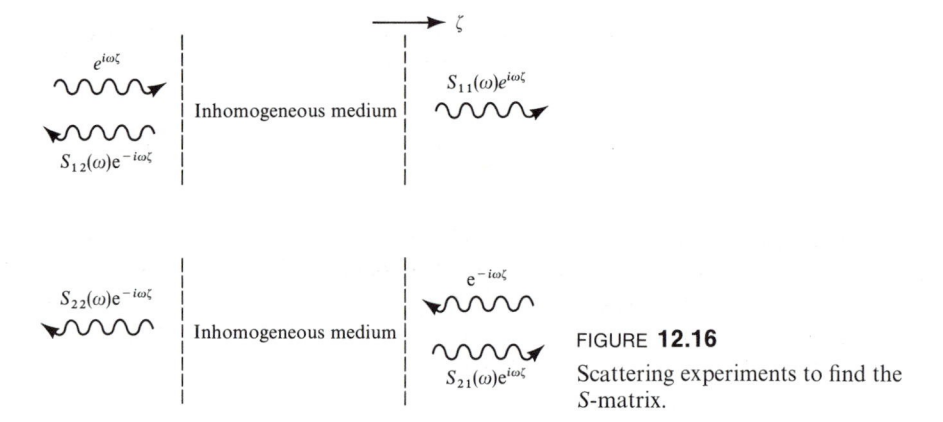

FIGURE **12.16**

Scattering experiments to find the S-matrix.

(1969). Faddeev shows that the matrix $\mathbf{S}(\omega)$ must be unitary as a consequence of general scattering theory. Then, in addition to (12.55), we have $S_{11}S_{21}^* + S_{12}S_{22}^* = 0$. The unitary nature of $\mathbf{S}$ enables us to determine the entire matrix from S_{12} or S_{21} alone. Physically, this additional relation comes from the causality of transmitted waves S_{11}. Because of this causality, the potential $q(x)$ is determined by the specification of just one of the reflection seismograms S_{12} or S_{21}, as shown below. Since $q(x)$ determines the entire matrix, we see that $\mathbf{S}$ is determined from S_{12} or S_{21}.

Let us introduce fundamental solutions $f_1(\zeta, \omega)$ and $f_2(\zeta, \omega)$, which are the solutions of (12.38) for the boundary conditions

$$\lim_{\zeta \to \infty} f_1(\zeta, \omega) = e^{i\omega\zeta} \tag{12.56}$$

and

$$\lim_{\zeta \to -\infty} f_2(\zeta, \omega) = e^{-i\omega\zeta},$$

respectively.

The solution of the scattering problem—e.g., the one depicted at the top of Figure 12.16—can be expressed in terms of the fundamental solutions, either as $S_{11}(\omega)f_1(\zeta, \omega)$ or $f_2(\zeta, -\omega) + S_{12}(\omega)f_2(\zeta, \omega)$. Therefore, we have

$$S_{11}(\omega)f_1(\zeta, \omega) = f_2(\zeta, -\omega) + S_{12}(\omega)f_2(\zeta, \omega). \tag{12.57}$$

Using the analytic properties of f_1 and f_2 under the condition that

$$\int_{-\infty}^{\infty} (1 + |\zeta|)|q(\zeta)| \, d\zeta < \infty,$$

and the causality for the transmitted waves $S_{11}(\omega)$, the above equation leads to the Gel'fand-Levitan integral equation for $A(x, \zeta)$,

$$A(x, \zeta) = -R(\zeta + x) - \int_{-\zeta}^{x} A(x, \tau)R(\zeta + \tau) \, d\tau \qquad \zeta < x, \tag{12.58}$$

from which the potential $q(\zeta)$ is obtained by

$$q(\zeta) = 2\,\frac{dA(\zeta, \zeta)}{d\zeta}, \tag{12.59}$$

where $R(t)$ is the reflection record in the time domain,

$$R(t) = \frac{1}{2\pi} \int_{-\infty}^{\infty} S_{12}(\omega) \exp(-i\omega t) \, d\omega. \tag{12.60}$$

The crux of the above derivation can be easily understood if we follow the analogous steps for the discrete problem. To make the comparison complete, we shall change the field variable from displacement to $\sqrt{\text{impedance}} \times$ displacement. Redefining reflection and transmission coefficients r, t for the new variable, we obtain, in place of (12.33),

$$t_i = \bar{t}_i,$$

$$t_i^2 + r_i^2 = 1,$$

$$\bar{t}_i^2 + \bar{r}_i^2 = 1, \qquad (12.61)$$

$$r_i = -\bar{r}_i = \frac{c_i \rho_i - c_{i+1}\rho_{i+1}}{c_i \rho_i + c_{i+1}\rho_{i+1}} \text{ (unchanged)},$$

which is similar to the relation (12.55) for the S-matrix.

Defining upgoing and downgoing waves U_i and D_i for the new variable in the same way as shown in Figure 12.15, the same propagator matrix (12.45) is obtained with newly defined r_i and t_i.

A discrete analogue of the fundamental solution $f_1(\zeta, \omega)$ can be obtained by solving (12.45) for U_0 and D_0 under the boundary condition

$$\begin{pmatrix} U \\ D \end{pmatrix}_{n+1} = \begin{pmatrix} 0 \\ w^{n+1} \end{pmatrix}, \qquad (12.62)$$

where $w = \exp(i\omega \, \Delta t)$. We then calculate U_k, D_k using U_0, D_0 by the propagator matrix. The result is

$$f_1(k, z) = U_k + D_k$$

$$= \frac{1}{w^k} \{ z^{n+1} G_n(1/z)[F_{k-1}(z) - G_{k-1}(z)]$$

$$+ z^k F_n(z)[F_{k-1}(1/z) - G_{k-1}(1/z)] \}. \qquad (12.63)$$

Likewise, the reversed fundamental solution $\hat{f}_1$ corresponding to $f_1(\zeta, -\omega)$ can be obtained from the boundary condition,

$$\begin{pmatrix} U \\ D \end{pmatrix}_{n+1} = \begin{pmatrix} w^{n+1} \\ 0 \end{pmatrix}, \qquad (12.64)$$

$$\hat{f}_1(k, z) = \frac{1}{w^k} \{ z^{n+1} F_n(1/z)[F_{k-1}(z) - G_{k-1}(z)]$$

$$+ z^k G_n(z)[F_{k-1}(1/z) - G_{k-1}(1/z)] \}. \qquad (12.65)$$

Similarly, the discrete analogue of fundamental solution $f_2(\zeta, \omega)$ can be obtained from the boundary condition

$$\begin{pmatrix} U \\ D \end{pmatrix}_0 = \begin{pmatrix} w^{n+1} \\ 0 \end{pmatrix}, \tag{12.66}$$

and that for $f_2(\zeta, -\omega)$ from

$$\begin{pmatrix} U \\ D \end{pmatrix}_0 = \begin{pmatrix} 0 \\ w^{n+1} \end{pmatrix}. \tag{12.67}$$

The results are

$$f_2(k, z) = w^{n-k+1}[F_{k-1}(z) - G_{k-1}(z)],$$
$$\hat{f}_2(k, z) = w^{n+k+1}[F_{k-1}(1/z) - G_{k-1}(1/z)]. \tag{12.68}$$

Both f_1 and f_2 are related to their reversed functions by

$$f_i(k, z) = z^{n+1}\hat{f}_i(k, 1/z). \tag{12.69}$$

Let us now describe the direct analogue of the scattering experiment defining the reflection seismogram $R(z)$ and transmission seismograms $T(z)$ by

$$\begin{pmatrix} U \\ D \end{pmatrix}_0 = \begin{pmatrix} R(z) \\ 1 \end{pmatrix} \tag{12.70}$$

and

$$\begin{pmatrix} U \\ D \end{pmatrix}_{n+1} = \begin{pmatrix} 0 \\ T(z) \end{pmatrix}. \tag{12.71}$$

The discrete analogue of (12.57) can now be written as

$$\frac{T(z)}{w^{n+1}} f_1(k, z) = \frac{\hat{f}_2(k, z)}{w^{n+1}} + \frac{R(z)}{w^{n+1}} f_2(k, z). \tag{12.72}$$

In the preceding section, we used the fact that the time-shifted transmission seismogram $T(z)/w^n$ has no terms with negative power of z in deriving the inversion formula (12.54). Here we recognize that the incident wave in our scattering problem has not arrived at the interface k before the time $2k\,\Delta t$.

Therefore, (12.72) should be of the form

$$b_k w^k (1 + c_1 z + c_2 z^2 + \cdots),\qquad (12.73)$$

where $b_k = t_0 t_1 \cdots t_{k-1}$. We now introduce function $K_2(k, z)$ to rewrite the left side of (12.72):

$$\frac{\hat{f}_2(k, z)}{w^{n+1}} = b_k [w^k + K_2(k, z)]$$

and

$$\frac{f_2(k, z)}{w^{n+1}} = b_k [w^{-k} + K_2(k, 1/z)].\qquad (12.74)$$

Putting (12.74) into (12.72) and equating (12.73), we have

$$K_2(k, z) + \frac{R(z)}{w^k} + R(z) K_2(k, 1/z) = 0 \qquad (12.75)$$

for all negative powers of w and for positive powers up to and including w^k. Since we know $\hat{f}_2(k, z)$ explicitly, as shown in (12.68), we can write

$$K_2(k, z) = \frac{\hat{f}_2(k, z)}{w^{n+1} b_k} - w^k$$

in terms of $F_k(z)$ and $G_k(z)$ and show that $K_2(k, z)$ is a polynomial of the form

$$K_2(k, z) = K(k, 2 - k)w^{2-k} + K(k, 4 - k)w^{4-k} + \cdots + K(k, k)w^k \qquad k > 2,$$

where

$$K(k, k) = \frac{1}{\displaystyle\prod_{i=0}^{k-2} (1 - r_i)^2 (1 + r_{k-1})} - 1,$$

or

$$r_{k-1} = \frac{1}{\displaystyle\prod_{i=0}^{k-2} (1 - r_i)^2 [1 + K(k, k)]} - 1. \qquad (12.76)$$

Equation (12.75) can be rewritten using $K(k, k')$ as

$$
\left[
\begin{pmatrix}
1 & & & \\
 & 1 & & 0 \\
 & & 1 & \\
 & 0 & & \ddots \\
 & & & & 1
\end{pmatrix}
+
\begin{pmatrix}
 & & & R_0 \\
 & & 0 & R_0 & R_1 \\
 & & & & \ddots & \vdots \\
 & R_0 & & & \\
 R_0 & R_1 & \cdots & R_{k-1}
\end{pmatrix}
\right]
\begin{pmatrix}
K(k, 2-k) \\
K(k, 4-k) \\
\vdots \\
K(k, k)
\end{pmatrix}
=
\begin{pmatrix}
-R_0 \\
-R_1 \\
\vdots \\
-R_{k-1}
\end{pmatrix}
$$

$$(12.77)$$

This is a discrete analogue of the Gel'fand-Levitan integral equation (12.58). $A(x, \zeta)$ corresponds to $K(k, k')$; and the formula (12.59), used to obtain potential $q(\zeta)$ from $A(\zeta, \zeta)$, corresponds to (12.76), which gives the reflection coefficient from $K(k, k)$. The inversion by (12.77) operates on the same reflection seismogram R_k as the inversion by (12.54), but the former is done in two steps, (12.77) and (12.76), and the latter in one step. The former is more general, because R_0 can be different from 1. Koehler and Taner (1977) have applied (12.54) to synthetic data composed of reflection seismograms for a known layered medium and random noise and have shown that computation of the reflection coefficient remains stable down to a signal-noise ratio of 1 to 5.

So far, the inverse problem for reflection seismograms has been solved only for one-dimensional problems. The extension to two- or three-dimensional problems is extremely difficult and has not been accomplished, although the need for such a solution is immense in exploration seismology. Claerbout (1976), however, has made a major effort in approaching the solution of this problem. He starts with an acoustic wave equation involving only velocity perturbation (Section 13.3), neglecting conversions of P to S or S to P. Using the parabolic approximation (Section 13.6), downgoing primary waves are computed for a known source, and upgoing reflected waves are reconstructed from observed records at the surface. His main idea is that a reflector occurs along the surface at which upgoing and downgoing waves are phase coincident. He also formulates the spatial derivatives of reflection coefficients as sources of upgoing waves by a simplified version of the perturbation method described in Section 13.2 and solves the forward and inverse problem by a finite-difference calculation. He has been able to develop a commercially competitive processing scheme based on these ideas.

12.3 The Inverse Problem for a Linearized System

So far, we have considered two basic data sets in seismology—i.e., travel times and reflection seismograms—within the framework of a one-dimensional Earth model. Two other important data sets are (i) the phase and group velocity

of surface waves and (ii) periods of free oscillations. There have been several attempts to find inversion formulas for these data.

For example, Takahashi (1955) was able to reduce the expression for phase velocity of Love waves to the form of Abel's integral equation, and gave the following formula for the depth $z(\beta)$ at which the shear velocity is β from the phase velocity $c(T)$ as a function of period T:

$$z(\beta) \sim 0.87 \frac{1}{\pi^2} \int_0^{T_\beta} \frac{dT}{\sqrt{(c(T))^{-2} - \beta^{-2}}}, \tag{12.78}$$

where T_β is the period at which the phase velocity is β. In deriving the above formula, however, he had to use the WKBJ approximation (valid only for smooth variation of medium properties within a wavelength), and assumed that for a given period there is only one turning point at which the shear velocity is equal to the phase velocity. These assumptions are too restrictive to make the formula practical in most cases.

A general inversion method exists that is applicable to any data within a less restrictive framework for the Earth model if the solution to the direct (or forward) problem is known and if a reasonable initial guess of the model parameters is possible. We perturb the model parameters from the initial guess and calculate the corresponding change in observables. As long as the perturbation is small, we expect that the relation between the change in observable and parameter perturbation is linear. The inverse problem is then reduced to solving a set of linear equations. For surface waves, the change in phase velocity due to perturbation in elastic constants and density at various depths was discussed in Section 7.3 using the variational principle. For example, (7.71) and (7.78) give the linearized formulas for phase velocity of Love waves and Rayleigh waves, respectively. The formulas for toroidal free-oscillation periods were given in (8.64) and (8.65). These formulas can be written in a general form:

$$d_i = \int_0^{r_\oplus} G_i(r)m(r)\, dr \qquad (i = 1, 2, \ldots, N), \tag{12.79}$$

where d_i, representing data, stands for the observed data minus that calculated for the initial model, and $m(r)$, representing model, stands for a profile of some quantity (such as velocity or density) in our desired Earth model, minus that profile for our initial model. Thus $m(r)$ describes the model-parameter perturbation as a function of distance r from the Earth's center. $r_\oplus$ here is the Earth's radius. The kernel $G_i(r)$ is calculated for the initial model.

Our problem is to determine $m(r)$ from a finite set of observed data d_i $(i = 1, \ldots, N)$, each one of which is a known scalar number. The formal solution of equation (12.79) is

$$m(r) = m_p(r) + m_\perp(r), \tag{12.80}$$

where $m_p(r)$ is a particular solution of (12.79) and $m_\perp(r)$ is any solution of the following equation:

$$0 = \int_0^{r_\oplus} G_i(r) m_\perp(r)\, dr \qquad (i = 1, 2, \ldots, N). \tag{12.81}$$

Backus and Gilbert (1967) showed that $m_\perp(r)$ exists for any finite seismological data set. If $m_\perp(r)$ is a solution of (12.81), $m_\perp(r)$ multiplied by any number is also a solution. In other words, the formal solution can take any value from $-\infty$ to $+\infty$. Recognizing this intrinsic nonuniqueness of the solution, Backus and Gilbert (1967, 1968, and 1970) explored the optimal way of making inferences as to the Earth's interior from a given finite data set. They introduced useful concepts of the *spatial resolution* of a particular solution and the *trade-off* between the resolution and error due to noise in the data. We shall describe these concepts in detail, starting with a simpler case of discrete models. Instead of a continuous model $m(r)$, which requires a Hilbert space for its description, we shall consider a discrete model specified by a finite number of parameters m_j ($j = 1, 2, \ldots, M$). The integral equation (12.79) is now reduced to a linear equation

$$d_i = \sum_{j=1}^{M} G_{ij} m_j \qquad (i = 1, 2, \ldots, N), \tag{12.82}$$

which we shall write as

$$\mathbf{d} = \mathbf{Gm}. \tag{12.83}$$

Our inverse problem may be divided into two parts. The first part is to find a particular solution $\mathbf{m} = \mathbf{m}_p$ of (12.83):

$$\mathbf{m}_p = \mathbf{G}_p^{-1} \mathbf{d}. \tag{12.84}$$

$\mathbf{G}_p^{-1}$ operates on data and gives a particular solution. The second part is to find the resolution and error for the particular solution.

Since $\mathbf{d} = \mathbf{Gm}$, we obtain from (12.84) that

$$\mathbf{m}_p = \mathbf{G}_p^{-1} \mathbf{Gm}, \tag{12.85}$$

which expresses the particular solution as a weighted average of the true solution with weights given by row vectors of matrix $\mathbf{G}_p^{-1} \mathbf{G}$. This weight matrix is called the *resolution matrix*. If $\mathbf{G}_p^{-1} \mathbf{G}$ is the identity matrix $\mathbf{I}$, resolution is perfect and the particular solution is equal to the true solution. If the row vectors of $\mathbf{G}_p^{-1} \mathbf{G}$ have components spread around the diagonal (with low values elsewhere), the particular solution represents a smoothed solution over the spread.

The error $\Delta\mathbf{m}_p$ in the estimate of particular solutions due to the error $\Delta\mathbf{d}$ in the data can be described by its covariance matrix $\langle\Delta\mathbf{m}_p\,\Delta\tilde{\mathbf{m}}_p\rangle$, where $\tilde{}$ indicates conjugate transpose and $\langle\ \rangle$ indicates averaging. From (12.84), we find that

$$\langle\Delta\mathbf{m}_p\,\Delta\tilde{\mathbf{m}}_p\rangle = \langle\mathbf{G}_p^{-1}\,\Delta\mathbf{d}\,\Delta\tilde{\mathbf{d}}\tilde{\mathbf{G}}_p^{-1}\rangle$$
$$= \mathbf{G}_p^{-1}\langle\Delta\mathbf{d}\,\Delta\tilde{\mathbf{d}}\rangle\tilde{\mathbf{G}}_p^{-1}. \tag{12.86}$$

The above equation gives the covariance matrix for $\Delta\mathbf{m}_p$ in terms of the covariance matrix of the error in data.

Thus, once the operator $\mathbf{G}_p^{-1}$ for a particular solution is known, the resolution and the error in the solution are easily obtained.

Let us now derive a particular solution called the *generalized inverse*.

12.3.1 *Model space and data space*

Consider the problem of solving the following simultaneous equations:

$$m_1 + m_2 = 1,$$
$$m_3 = 2, \tag{12.87}$$
$$-m_3 = 1.$$

There are three equations for three unknowns, but there are too many solutions for m_1 and m_2 and no solution for m_3.

Our plan for the presentation of linear inverse theory is to describe a thorough way of handling the problems presented by (12.87). This will entail an extensive development of matrix methods that subsequently will allow us to analyze inverse problems in geophysics. Thus we shall describe a natural solution of the above problem, following Lanczos (1961).

In general, our equations are expressed by

$$\mathbf{Gm} = \mathbf{d}, \tag{12.88}$$

where $\mathbf{m} = (m_1, m_2, \ldots, m_M)^T$ is a vector in the model space with M components, $\mathbf{d} = (d_1, d_2, \ldots, d_N)^T$ is a vector in the data space with N components, and $\mathbf{G}$ is an $N \times M$ matrix,

$$\mathbf{G} = \begin{pmatrix} G_{11} & G_{12} & \cdots & G_{1M} \\ G_{21} & G_{22} & \cdots & G_{2M} \\ \vdots & \vdots & & \vdots \\ G_{N1} & G_{N2} & \cdots & G_{NM} \end{pmatrix}.$$

Following Lanczos, we shall construct a Hermitian matrix S, which is composed of G, its complex conjugate, transposed matrix $\tilde{G} \equiv (G^*)^T$, and zeros as shown below:

$$S = \left(\begin{array}{c|c} 0 & G \\ \hline \tilde{G} & 0 \end{array} \right) \begin{array}{l} \} N \\ \} M \end{array}.$$

$$\underbrace{}_{N} \quad \underbrace{}_{M}$$

S is an $(N + M) \times (N + M)$ square matrix, and $\tilde{S} = S$. This assures the existence of an orthogonal set of eigenvectors w_i ($i = 1, 2, \ldots, N + M$) with real eigenvalues λ_i, which satisfy

$$Sw_i = \lambda_i w_i \qquad (i = 1, 2, \ldots, N + M). \qquad (12.89)$$

For nonvanishing w_i to exist for a particular λ, the determinant of $S - \lambda I$ must vanish. Thus the eigenvalues are determined by solving

$$\det(S - \lambda I) = (\lambda_1 - \lambda)(\lambda_2 - \lambda) \cdots (\lambda_{N+M} - \lambda) = 0. \qquad (12.90)$$

The eigenvector w_i has $N + M$ components. We shall divide the components into two parts, u_i and v_i (u_i in the N-dimensional data space and v_i in the M-dimensional model space), and rewrite equation (12.89) as

$$\left(\begin{array}{c|c} 0 & G \\ \hline \tilde{G} & 0 \end{array} \right) \left(\begin{array}{c} u_i \\ v_i \end{array} \right) = \lambda_i \left(\begin{array}{c} u_i \\ v_i \end{array} \right) \begin{array}{l} \} N \\ \} M \end{array}. \qquad (12.91)$$

If λ_i is a nonzero eigenvalue, we get the following coupled equations for the eigenvector pair (u_i, v_i):

$$Gv_i = \lambda_i u_i,$$
$$\tilde{G}u_i = \lambda_i v_i. \qquad (12.92)$$

If we change the sign of λ_i, we find that $(-u_i, v_i)$ is also an eigenvector pair satisfying the above equation. Supposing that there are p pairs of nonzero eigenvalues $\pm \lambda_i$, the corresponding eigenvector pairs are

$$(u_i, v_i) \quad \text{for} \quad \lambda_i \qquad (i = 1, 2, \ldots, p)$$

and

$$(-u_i, v_i) \quad \text{for} \quad -\lambda_i \qquad (i = 1, 2, \ldots, p).$$

For zero eigenvalues, equation (12.92) is decoupled, and $\mathbf{u}_i$ and $\mathbf{v}_i$ become independent:

$$\mathbf{G}\mathbf{v}_i = \mathbf{0} \qquad (i = p + 1, \ldots, M),$$
$$\tilde{\mathbf{G}}\mathbf{u}_i = \mathbf{0} \qquad (i = p + 1, \ldots, N). \tag{12.93}$$

Thus, among $N + M$ eigenvalues of $\mathbf{S}\mathbf{w} = \lambda\mathbf{w}$, $2p$ are nonzero, and the rest $N + M - 2p$ are zero. The data space spanned by $\mathbf{u}_i$ $(i = 1, 2, \ldots, N)$ and the model space spanned by $\mathbf{v}_i$ $(i = 1, 2, \ldots, M)$ are coupled only through nonzero eigenvalues $\pm\lambda_i$ $(i = 1, 2, \ldots, p)$. From (12.92), we find that

$$\tilde{\mathbf{G}}\mathbf{G}\mathbf{v}_i = \lambda_i^2\mathbf{v}_i,$$
$$\mathbf{G}\tilde{\mathbf{G}}\mathbf{u}_i = \lambda_i^2\mathbf{u}_i. \tag{12.94}$$

Since $\tilde{\mathbf{G}}\mathbf{G}$ and $\mathbf{G}\tilde{\mathbf{G}}$ are both Hermitian, each of $\mathbf{v}_i$ and $\mathbf{u}_i$ forms an orthogonal set of eigenvectors with real eigenvalues. After normalization, we can write

$$\tilde{\mathbf{v}}_i\mathbf{v}_j = \delta_{ij} \qquad (i, j = 1, 2, \ldots, M),$$
$$\tilde{\mathbf{u}}_i\mathbf{u}_j = \delta_{ij} \qquad (i, j = 1, 2, \ldots, N). \tag{12.95}$$

We shall define a matrix $\mathbf{V}$ whose column vectors are eigenvectors $\mathbf{v}_i$ with components $(v_{1i}, v_{2i}, \ldots, v_{Mi})^T$:

$$\mathbf{V} = \begin{pmatrix} v_{11} & \cdots & & \cdots & v_{1M} \\ \vdots & & \boxed{\mathbf{v}_i} & & \vdots \\ v_{M1} & \cdots & & \cdots & v_{MM} \end{pmatrix}.$$

Likewise,

$$\mathbf{U} = \begin{pmatrix} u_{11} & \cdots & & \cdots & u_{1N} \\ \vdots & & \boxed{\mathbf{u}_i} & & \vdots \\ u_{N1} & \cdots & & \cdots & u_{NN} \end{pmatrix}.$$

Then equation (12.95) can be written as

$$\tilde{\mathbf{U}}\mathbf{U} = \mathbf{U}\tilde{\mathbf{U}} = \mathbf{I},$$
$$\tilde{\mathbf{V}}\mathbf{V} = \mathbf{V}\tilde{\mathbf{V}} = \mathbf{I}. \tag{12.96}$$

The whole data space is spanned by the columns of $\mathbf{U}$. Let us now divide $\mathbf{U}$ into $\mathbf{U}_p$ and $\mathbf{U}_0$, where $\mathbf{U}_p$ is made up from the eigenvectors with nonzero

eigenvalues and $\mathbf{U}_0$ consists of the eigenvectors with zero eigenvalues. Likewise, we divide $\mathbf{V}$ into $\mathbf{V}_p$ and $\mathbf{V}_0$:

$$\mathbf{U}_p = \begin{pmatrix} u_{11} & \cdots & u_{1p} \\ \vdots & & \vdots \\ u_{N1} & \cdots & u_{Np} \end{pmatrix},$$

$$\mathbf{V}_p = \begin{pmatrix} v_{11} & \cdots & v_{1p} \\ \vdots & & \vdots \\ v_{M1} & \cdots & v_{Mp} \end{pmatrix}.$$

Because of orthogonality we have $\tilde{\mathbf{U}}_p\mathbf{U}_p = \tilde{\mathbf{V}}_p\mathbf{V}_p = \mathbf{I}$, but since $\mathbf{U}_p$ and $\mathbf{V}_p$ are no longer complete, $\mathbf{U}_p\tilde{\mathbf{U}}_p \neq \mathbf{I}$, $\mathbf{V}_p\tilde{\mathbf{V}}_p \neq \mathbf{I}$. $\mathbf{U}_p$ and $\mathbf{V}_p$ are coupled through the nonzero eigenvalues $\lambda_1, \ldots, \lambda_p$. Introducing a diagonal matrix $\mathbf{\Lambda}_p$ whose elements are nonzero eigenvalues, $\lambda_1, \lambda_2, \ldots, \lambda_p$, we can rewrite (12.92) and (12.93) as

$$\mathbf{G}\mathbf{V}_p = \mathbf{U}_p\mathbf{\Lambda}_p,$$
$$\tilde{\mathbf{G}}\mathbf{U}_p = \mathbf{V}_p\mathbf{\Lambda}_p, \qquad (12.97)$$

$$\mathbf{G}\mathbf{V}_0 = \mathbf{0},$$
$$\tilde{\mathbf{G}}\mathbf{U}_0 = \mathbf{0}. \qquad (12.98)$$

Therefore, we can write

$$\mathbf{G}\mathbf{V} = \mathbf{G}(\mathbf{V}_p, \mathbf{V}_0) = (\mathbf{U}_p, \mathbf{U}_0)\begin{pmatrix} \mathbf{\Lambda}_p & \mathbf{0} \\ \mathbf{0} & \mathbf{0} \end{pmatrix},$$

where $(\mathbf{V}_p, \mathbf{V}_0)$ denotes $\mathbf{V}$, partitioned into two matrices, and similarly for $(\mathbf{U}_p, \mathbf{U}_0)$. Since $\mathbf{V}\tilde{\mathbf{V}} = \mathbf{I}$, we have

$$\mathbf{G} = (\mathbf{U}_p, \mathbf{U}_0)\begin{pmatrix} \mathbf{\Lambda}_p & \mathbf{0} \\ \mathbf{0} & \mathbf{0} \end{pmatrix}\begin{pmatrix} \tilde{\mathbf{V}}_p \\ \tilde{\mathbf{V}}_0 \end{pmatrix}$$

$$= \mathbf{U}_p\mathbf{\Lambda}_p\tilde{\mathbf{V}}_p. \qquad (12.99)$$

This is an important factorization of $\mathbf{G}$, and it implies all the results in (12.97) and (12.98). It shows that $\mathbf{G}$ can be constructed by $\mathbf{U}_p$ and $\mathbf{V}_p$ space alone. As Lanczos states, $\mathbf{U}_0$ and $\mathbf{V}_0$ spaces are blind spots not illuminated by the operator $\mathbf{G}$.

At this point, let us explore the $\mathbf{U}_p$, $\mathbf{V}_p$, $\mathbf{U}_0$, and $\mathbf{V}_0$ spaces for the example given in the beginning of this section. From (12.87), we find

$$\mathbf{G} = \begin{pmatrix} 1 & 1 & 0 \\ 0 & 0 & 1 \\ 0 & 0 & -1 \end{pmatrix}, \qquad \tilde{\mathbf{G}} = \begin{pmatrix} 1 & 0 & 0 \\ 1 & 0 & 0 \\ 0 & 1 & -1 \end{pmatrix},$$

$$\mathbf{G}\tilde{\mathbf{G}} = \begin{pmatrix} 2 & 0 & 0 \\ 0 & 1 & -1 \\ 0 & -1 & 1 \end{pmatrix}, \quad \tilde{\mathbf{G}}\mathbf{G} = \begin{pmatrix} 1 & 1 & 0 \\ 1 & 1 & 0 \\ 0 & 0 & 2 \end{pmatrix}.$$

Let us first solve the eigenvalue problem in **U**-space:

$$\mathbf{G}\tilde{\mathbf{G}}u_i = \lambda_i^2 u_i \qquad (i = 1, 2, 3).$$

The eigenvalues are determined by

$$\begin{vmatrix} 2 - \lambda^2 & 0 & 0 \\ 0 & 1 - \lambda^2 & -1 \\ 0 & -1 & 1 - \lambda^2 \end{vmatrix} = (2 - \lambda^2)(\lambda^2 - 2)\lambda^2 = 0.$$

The eigenvectors $\mathbf{u}_1$ and $\mathbf{u}_2$ corresponding to $\lambda_1^2 = \lambda_2^2 = 2$ are found by solving

$$\begin{pmatrix} 2 & 0 & 0 \\ 0 & 1 & -1 \\ 0 & -1 & 1 \end{pmatrix} \begin{pmatrix} u_{11} \\ u_{21} \\ u_{31} \end{pmatrix} = 2 \begin{pmatrix} u_{11} \\ u_{21} \\ u_{31} \end{pmatrix}$$

and normalizing the length as $u_{11}^2 + u_{21}^2 + u_{31}^2 = 1$. We find that

$$\mathbf{u}_1 = \begin{pmatrix} 0 \\ 1/\sqrt{2} \\ -1/\sqrt{2} \end{pmatrix} \quad \text{and} \quad \mathbf{u}_2 = \begin{pmatrix} 1 \\ 0 \\ 0 \end{pmatrix}.$$

The third eigenvalue λ_3^2 is zero. The corresponding eigenvector $\mathbf{u}_3$ is determined by solving

$$\begin{pmatrix} 2 & 0 & 0 \\ 0 & 1 & -1 \\ 0 & -1 & 1 \end{pmatrix} \begin{pmatrix} u_{13} \\ u_{23} \\ u_{33} \end{pmatrix} = 0$$

and normalizing the length. We find that

$$\mathbf{u}_3 = \begin{pmatrix} 0 \\ 1/\sqrt{2} \\ 1/\sqrt{2} \end{pmatrix}.$$

Thus our $\mathbf{U}_p$ and $\mathbf{U}_0$ spaces are

$$\mathbf{U}_p = \begin{pmatrix} 0 & 1 \\ 1/\sqrt{2} & 0 \\ -1/\sqrt{2} & 0 \end{pmatrix} \quad \text{and} \quad \mathbf{U}_0 = \begin{pmatrix} 0 \\ 1/\sqrt{2} \\ 1/\sqrt{2} \end{pmatrix}.$$

Our model $\mathbf{m}$ cannot generate data expressed by $\mathbf{U}_0$ (see (12.87)).
  The corresponding $\mathbf{V}_p$ space can be easily obtained by (12.92) or (12.97):

$$\mathbf{v}_i = \frac{1}{\lambda_i} \tilde{\mathbf{G}} \mathbf{u}_i,$$

and note that division by λ_i is allowed, since this λ_i is not zero. We find that

$$\mathbf{V}_p = (\mathbf{v}_1, \mathbf{v}_2) = \begin{pmatrix} 0 & 1/\sqrt{2} \\ 0 & 1/\sqrt{2} \\ 1 & 0 \end{pmatrix}.$$

To find $\mathbf{V}_0$-space, we have to solve the equation

$$\tilde{\mathbf{G}}\mathbf{G}\mathbf{v}_3 = \mathbf{0},$$

which gives

$$\mathbf{V}_0 = \mathbf{v}_3 = \begin{pmatrix} 1/\sqrt{2} \\ -1/\sqrt{2} \\ 0 \end{pmatrix}.$$

The model $\mathbf{m}$ expressed by $\mathbf{V}_0$ cannot affect any observables (see (12.87)).
  Finally, we can demonstrate the validity of the Lanczos decomposition (12.99),

$$\mathbf{G} = \mathbf{U}_p \mathbf{\Lambda}_p \tilde{\mathbf{V}}_p = \begin{pmatrix} 0 & 1 \\ 1/\sqrt{2} & 0 \\ -1/\sqrt{2} & 0 \end{pmatrix} \begin{pmatrix} \sqrt{2} & 0 \\ 0 & \sqrt{2} \end{pmatrix} \begin{pmatrix} 0 & 0 & 1 \\ 1/\sqrt{2} & 1/\sqrt{2} & 0 \end{pmatrix} = \begin{pmatrix} 1 & 1 & 0 \\ 0 & 0 & 1 \\ 0 & 0 & -1 \end{pmatrix},$$

showing that $\mathbf{G}$ can be constructed using only the $\mathbf{U}_p$ and $\mathbf{V}_p$ spaces. The expression $\mathbf{G} = \mathbf{U}_p \mathbf{\Lambda}_p \tilde{\mathbf{V}}_p$ gives a clear view of the data and model spaces, $\mathbf{U}$ and $\mathbf{V}$, respectively, coupled by the equation $\mathbf{G}\mathbf{m} = \mathbf{d}$.
  Since $\tilde{\mathbf{U}}_0 \mathbf{G}\mathbf{m} = \mathbf{0}$, the prediction $\mathbf{G}\mathbf{m}$ of an observable based on any model $\mathbf{m}$ will have no component in $\mathbf{U}_0$-space, and therefore is restricted to the $\mathbf{U}_p$-space. If there is no $\mathbf{U}_0$-space, one can always find a model that satisfies the equation $\mathbf{G}\mathbf{m} = \mathbf{d}$, because $\mathbf{U}_p$-space is complete in that case. However, if $\mathbf{U}_0$-space exists, and if the data have components in $\mathbf{U}_0$-space, the prediction

$\mathbf{Gm}$ cannot describe the data for any choice of $\mathbf{m}$, because $\mathbf{Gm}$ has no component in $\mathbf{U}_0$. Therefore, $\mathbf{U}_0$-space is the source of discrepancy between the observed data and the prediction by operator $\mathbf{G}$.

On the other hand, $\mathbf{V}_0$-space is the source of nonuniqueness in determining the model from the data. One can add any vectors in $\mathbf{V}_0$-space to the model without contradicting observation, because $\mathbf{GV}_0 = \mathbf{0}$.

12.3.2   Generalized inverse

The exact inverse of $\mathbf{G} = \mathbf{U}\mathbf{\Lambda}\tilde{\mathbf{V}}$, when it exists, can be written as $\mathbf{G}^{-1} = \mathbf{V}\mathbf{\Lambda}^{-1}\tilde{\mathbf{U}}$. Therefore, it is natural to consider the following expression as an inverse operator to the operator $\mathbf{G} = \mathbf{U}_p\mathbf{\Lambda}_p\tilde{\mathbf{V}}_p$:

$$\mathbf{G}_g^{-1} = \mathbf{V}_p\mathbf{\Lambda}_p^{-1}\tilde{\mathbf{U}}_p. \qquad (12.100)$$

We shall call this the *generalized inverse operator* and see how it works.

First, let us consider the case in which no $\mathbf{U}_0$, $\mathbf{V}_0$ spaces exist. In this case, $\mathbf{U}_p\tilde{\mathbf{U}}_p = \mathbf{I}$, $\mathbf{V}_p\tilde{\mathbf{V}}_p = \mathbf{I}$, and therefore $\mathbf{G}_g^{-1} = \mathbf{V}_p\mathbf{\Lambda}_p^{-1}\tilde{\mathbf{U}}_p = (\mathbf{U}_p\mathbf{\Lambda}_p\tilde{\mathbf{V}}_p)^{-1} = \mathbf{G}^{-1}$. The generalized inverse agrees with the exact inverse when there are no $\mathbf{V}_0$, $\mathbf{U}_0$ spaces.

Second, we shall consider the case in which there is no $\mathbf{V}_0$ but $\mathbf{U}_0$-space exists. Then $\tilde{\mathbf{G}}\mathbf{G} = (\mathbf{V}_p\mathbf{\Lambda}_p^2\tilde{\mathbf{V}}_p)$ will have the exact inverse $(\tilde{\mathbf{G}}\mathbf{G})^{-1} = \mathbf{V}_p\mathbf{\Lambda}_p^{-2}\tilde{\mathbf{V}}_p$, and the least-squares method (see p. 693) is applicable. In our notation, the so-called normal equations are written as

$$\tilde{\mathbf{G}}\mathbf{Gm} = \tilde{\mathbf{G}}\mathbf{d},$$

and the solution $\hat{\mathbf{m}}$ is given by

$$\begin{aligned}\hat{\mathbf{m}} &= (\tilde{\mathbf{G}}\mathbf{G})^{-1}\tilde{\mathbf{G}}\mathbf{d} \\ &= \mathbf{V}_p\mathbf{\Lambda}_p^{-2}\tilde{\mathbf{V}}_p \cdot \mathbf{V}_p\mathbf{\Lambda}_p\tilde{\mathbf{U}}_p\mathbf{d} \qquad (12.101) \\ &= \mathbf{V}_p\mathbf{\Lambda}_p^{-1}\tilde{\mathbf{U}}_p\mathbf{d} = \mathbf{G}_g^{-1}\mathbf{d}.\end{aligned}$$

Thus the generalized inverse is nothing but the least-squares solution when $\mathbf{U}_0$-space exists and there is no $\mathbf{V}_0$.

It is easy to show geometrically why the generalized inverse gives the least-squares solution, in which the sum of squares of residuals, $|\mathbf{d} - \mathbf{Gm}|^2$, is minimized.

Putting $\mathbf{m}_g = \mathbf{G}_g^{-1}\mathbf{d}$, we have

$$\begin{aligned}\mathbf{d} - \mathbf{Gm}_g &= \mathbf{d} - \mathbf{U}_p\mathbf{\Lambda}_p\tilde{\mathbf{V}}_p\mathbf{V}_p\mathbf{\Lambda}_p^{-1}\tilde{\mathbf{U}}_p\mathbf{d} \\ &= \mathbf{d} - \mathbf{U}_p\tilde{\mathbf{U}}_p\mathbf{d}.\end{aligned}$$

Since $\tilde{U}_p U_p = I$, we find that

$$\tilde{U}_p(d - Gm_g) = \tilde{U}_p d - \tilde{U}_p U_p \tilde{U}_p d$$
$$= 0.$$

The residual for the generalized inverse, $d - Gm_g$, has no components in U_p-space. On the other hand, since $\tilde{U}_0 Gm_g = 0$, Gm_g has no components in U_0-space. Therefore, the vector Gm_g is perpendicular to the residual vector $d - Gm_g$, as shown in Figure 12.17. As mentioned before, the vector Gm for any m is restricted to U_p-space. The distance between the data vector d and Gm, or $|d - Gm|^2$, is obviously minimized when $d - Gm$ is perpendicular to Gm.

Third, let us consider the case in which there is no U_0 but V_0-space exists. In this case, one can show immediately that the generalized inverse m_g satisfies the equation $Gm_g = d$. Since $U_p \tilde{U}_p = I$,

$$Gm_g = GG_g^{-1}d$$
$$= U_p \Lambda_p \tilde{V}_p V_p \Lambda_p^{-1} \tilde{U}_p d$$
$$= d.$$

The generalized inverse gives a model that satisfies the equation $Gm = d$, and is restricted to V_p-space. The solution of $Gm = d$ can be expressed in general as

$$m = m_g + \sum_{i=p+1}^{M} a_i v_i,$$

where v_i are the eigenvectors of the V_0-space. Since $\tilde{v}_i v_j = \delta_{ij}$, we get

$$|m|^2 = |m_g|^2 + \Sigma a_i^2$$
$$\geqq |m_g|^2.$$

The generalized inverse gives the minimum of all possible solutions.

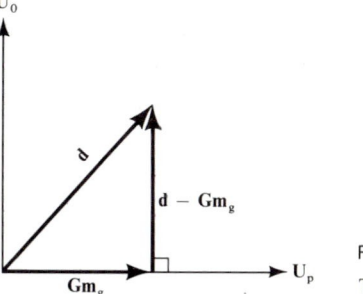

FIGURE **12.17**

The residual $d-Gm_g$ is perpendicular to Gm_g.

An arbitrary model vector restricted in V_p-space can be generated from a data vector, say $\mathbf{f}$, by the operator $\tilde{\mathbf{G}} = V_p \Lambda_p \tilde{U}_p$. Putting such a model vector $\tilde{\mathbf{G}}\mathbf{f}$ into the equation $\mathbf{Gm} = \mathbf{d}$, we find $\mathbf{f} = (\mathbf{G}\tilde{\mathbf{G}})^{-1}\mathbf{d}$. The model $\mathbf{m} = \tilde{\mathbf{G}}(\mathbf{G}\tilde{\mathbf{G}})^{-1}\mathbf{d}$ satisfies $\mathbf{Gm} = \mathbf{d}$ and is restricted to V_p-space. Therefore, the operator $\tilde{\mathbf{G}}(\mathbf{G}\tilde{\mathbf{G}})^{-1}$ should be identical to $\mathbf{G}_g^{-1}$:

$$\mathbf{G}_g^{-1} = \tilde{\mathbf{G}}(\mathbf{G}\tilde{\mathbf{G}})^{-1}. \tag{12.102}$$

Thus, when there is no U_0-space, we can calculate the generalized inverse directly from $\mathbf{G}$, without the eigenvector analysis, as in the case of the least-squares solution (12.101) when there was no V_0-space.

Finally, when there are both V_0- and U_0-spaces, the generalized inverse $\mathbf{G}_g^{-1} = V_p \Lambda_p^{-1} \tilde{U}_p$ will simultaneously minimize $|\mathbf{d} - \mathbf{Gm}|^2$ in the data space and $|\mathbf{m}|^2$ in the model space. This is the case for our example (12.87), and we find

$$\mathbf{G}_g^{-1} = \begin{pmatrix} 0 & 1/\sqrt{2} \\ 0 & 1/\sqrt{2} \\ 1 & 0 \end{pmatrix} \begin{pmatrix} 1/\sqrt{2} & 0 \\ 0 & 1/\sqrt{2} \end{pmatrix} \begin{pmatrix} 0 & 1/\sqrt{2} & -1/\sqrt{2} \\ 1 & 0 & 0 \end{pmatrix}$$

$$= \begin{pmatrix} 1/2 & 0 & 0 \\ 1/2 & 0 & 0 \\ 0 & 1/2 & -1/2 \end{pmatrix},$$

which gives the generalized inverse solution

$$\mathbf{m}_g = \mathbf{G}_g^{-1}\mathbf{d} = \begin{pmatrix} 1/2 \\ 1/2 \\ 1/2 \end{pmatrix}.$$

12.3.3   Resolution and error of the generalized inverse solution

The solution of the generalized inverse problem given in the previous section, however natural, is meaningless without some description of its uniqueness and reliability.

Let us first find the relation between the generalized inverse solution $\mathbf{m}_g$ and the true Earth model $\mathbf{m}$. Since $\mathbf{m}_g = \mathbf{G}_g^{-1}\mathbf{d}$ and $\mathbf{d} = \mathbf{Gm}$, we find

$$\mathbf{m}_g = \mathbf{G}_g^{-1}\mathbf{Gm}. \tag{12.103}$$

When the data vector $\mathbf{d}$ has a component in U_0 space, the equation $\mathbf{d} = \mathbf{Gm}$ does not hold. The above relation between $\mathbf{m}_g$ and $\mathbf{m}$ is valid even in that case, because $\tilde{U}_p U_0 = \mathbf{0}$ and the operator $\mathbf{G}_g^{-1} = V_p \Lambda_p^{-1} \tilde{U}_p$ annihilates U_0-space anyway (i.e., any component from U_0 is removed by the operator $\mathbf{G}_g^{-1}$).

Using equations (12.99) and (12.100), the above relation becomes

$$\mathbf{m}_g = \mathbf{V}_p \mathbf{\Lambda}_p^{-1} \tilde{\mathbf{U}}_p \mathbf{U}_p \mathbf{\Lambda}_p \tilde{\mathbf{V}}_p \mathbf{m}$$
$$= \mathbf{V}_p \tilde{\mathbf{V}}_p \mathbf{m}. \qquad (12.104)$$

If there is no $\mathbf{V}_0$-space, $\mathbf{V}_p \tilde{\mathbf{V}}_p = \mathbf{I}$ and $\mathbf{m}_g = \mathbf{m}$. Thus, when $\mathbf{V}_0 = \mathbf{0}$, the solution is unique whether $\mathbf{U}_0$-space exists or not. When $\mathbf{V}_0$-space does exist, (12.104) shows that the row vectors of $\mathbf{V}_p \tilde{\mathbf{V}}_p$ form weighting coefficients with which $\mathbf{m}_g$ is expressed as a weighted average of $\mathbf{m}$. This is similar to the representation of the beamforming wavenumber spectrum by a weighted average of the true spectrum, as was discussed in Chapter 11. The matrix $\mathbf{V}_p \tilde{\mathbf{V}}_p$ is the *resolution matrix*. As we shall see in Section 12.3.6, the row vector of $\mathbf{V}_p \tilde{\mathbf{V}}_p$ is the closest to the delta function (unit diagonal element and zero otherwise) in a least-squares sense.

The diagonal elements of $\mathbf{V}_p \tilde{\mathbf{V}}_p$ are useful measures of resolution. The total sum of diagonal elements (trace of $\mathbf{V}_p \tilde{\mathbf{V}}_p$) is equal to p, i.e., the number of eigenvectors $\mathbf{v}_i$ that form $\mathbf{V}_p$. The reason is that $\mathbf{v}_i$ $(i = 1, 2, \ldots, p)$ are eigenvectors of $\mathbf{V}_p \tilde{\mathbf{V}}_p$ with unit eigenvalues, and the trace is equal to the sum of the eigenvalues.

Suppose that the components m_i $(i = 1, 2, \ldots, M)$ of model vector $\mathbf{m}$ are arranged in a physically related fashion, such as the case of density at depth h_i where h_i increases monotonically with i. If $\mathbf{V}_p \tilde{\mathbf{V}}_p = \mathbf{I}$, all the values of m_i at M depths are uniquely determined. If $\mathbf{V}_p \tilde{\mathbf{V}}_p \neq \mathbf{I}$, our estimate of m_i is expressed as a weighted average of true values at the neighbor points around h_i. In the extreme case of $p = 1$, the total sum of M diagonal elements is 1, and the average value of each diagonal element will be around $1/M$. Thus the M components of each row vector of $\mathbf{V}_p \tilde{\mathbf{V}}_p$ will have roughly uniform amplitudes of $1/M$, and the estimates at all the depths will be indistinguishable. As a matter of fact, when $p = 1$, only one independent parameter estimate can be obtained for the whole range of depth. Since $\tilde{\mathbf{V}}_0 \mathbf{m}_g = \mathbf{0}$, $M - p$ constraints apply to M components of $\mathbf{m}_g$, and only p components of $\mathbf{m}_g$ are independent.

Wiggins (1972) illustrated these results, using the distribution of diagonal elements of $\mathbf{V}_p \tilde{\mathbf{V}}_p$ as a rough measure of resolution in model space. He divided the total depth range into p portions, within each of which the sum of the diagonal elements is equal to 1. One independent average estimate of model parameter can be assigned to each of such portions.

The resolution matrix of our example (12.87) is

$$\mathbf{V}_p \tilde{\mathbf{V}}_p = \begin{pmatrix} 0 & 1/\sqrt{2} \\ 0 & 1/\sqrt{2} \\ 1 & 0 \end{pmatrix} \begin{pmatrix} 0 & 0 & 1 \\ 1/\sqrt{2} & 1/\sqrt{2} & 0 \end{pmatrix} = \begin{pmatrix} 1/2 & 1/2 & 0 \\ 1/2 & 1/2 & 0 \\ 0 & 0 & 1 \end{pmatrix}.$$

This shows that data can determine only one average over m_1 and m_2, but can uniquely determine m_3.

Let us now turn our attention to the resolution in data space. The observed data vector $\mathbf{d}$ can be related to the predicted $\mathbf{d}_g$ by the generalized inverse as follows:

$$\mathbf{d}_g = \mathbf{G}\mathbf{G}_g^{-1}\mathbf{d} \tag{12.105}$$

$$= \mathbf{U}_p\tilde{\mathbf{U}}_p\mathbf{d}. \tag{12.106}$$

Thus, when there is no $\mathbf{U}_0$, $\mathbf{U}_p\tilde{\mathbf{U}}_p = \mathbf{I}$ and a perfect fit is obtained between the observed and the predicted. If $\mathbf{U}_0$ exists, a discrepancy between them occurs, and the predicted is expressed as a weighted average of the observed. The weighting coefficients are given by the row vectors of $\mathbf{U}_p\tilde{\mathbf{U}}_p$.

Since $\tilde{\mathbf{U}}_0\mathbf{d}_g = \mathbf{0}$, $N - p$ constraints exist among N components of $\mathbf{d}_g$. Therefore, only p components of the predicted data vector $\mathbf{d}_g$ are independent. Since the trace of $\mathbf{U}_p\tilde{\mathbf{U}}_p$ is also p, the diagonal elements of $\mathbf{U}_p\tilde{\mathbf{U}}_p$ can be used to divide the data into p portions, to each of which one independent prediction can be assigned. To make more predictions is meaningless, because they give only redundant information.

In our example,

$$\mathbf{U}_p\tilde{\mathbf{U}}_p = \begin{pmatrix} 0 & 1 \\ 1/\sqrt{2} & 0 \\ -1/\sqrt{2} & 0 \end{pmatrix} \begin{pmatrix} 0 & 1/\sqrt{2} & -1/\sqrt{2} \\ 1 & 0 & 0 \end{pmatrix}$$

$$= \begin{pmatrix} 1 & 0 & 0 \\ 0 & 1/2 & -1/2 \\ 0 & -1/2 & 1/2 \end{pmatrix}.$$

This shows that the prediction will fit the observed for d_1, but can give only one independent average value over d_2 and d_3. (In this case, the weight for the neighbor point is negative. The two averages are related as $d_{g2} = -d_{g3}$.)

Finally, we shall measure the reliability of the solution by its covariance matrix. The error $\Delta\mathbf{m}_g$ in the solution due to error $\Delta\mathbf{d}$ in data can be written as

$$\Delta\mathbf{m}_g = \mathbf{G}_g^{-1}\,\Delta\mathbf{d}.$$

Therefore, their covariance matrices are related by

$$\langle\Delta\mathbf{m}_g\Delta\tilde{\mathbf{m}}_g\rangle = \mathbf{G}_g^{-1}\langle\Delta\mathbf{d}\,\Delta\tilde{\mathbf{d}}\rangle\tilde{\mathbf{G}}_g^{-1}.$$

Assuming that all the components of the data vector are statistically independent and share the same variance σ_d^2, we have

$$\langle\Delta\mathbf{m}_g\Delta\tilde{\mathbf{m}}_g\rangle = \sigma_d^2\mathbf{G}_g^{-1}\tilde{\mathbf{G}}_g^{-1}. \tag{12.107}$$

For $\mathbf{U}_0 = \mathbf{V}_0 = \mathbf{0}$, putting $\mathbf{G}_g^{-1} = \mathbf{G}^{-1}$ gives

$$\langle \Delta \mathbf{m}_g \Delta \tilde{\mathbf{m}}_g \rangle = \sigma_d^2 (\tilde{\mathbf{G}} \mathbf{G})^{-1}. \tag{12.108}$$

For $\mathbf{U}_0 \neq \mathbf{0}$ and $\mathbf{V}_0 = \mathbf{0}$, putting $\mathbf{G}_g^{-1} = (\tilde{\mathbf{G}} \mathbf{G})^{-1} \tilde{\mathbf{G}}$ (see (12.101)) gives

$$\begin{aligned}
\langle \Delta \mathbf{m}_g \Delta \tilde{\mathbf{m}}_g \rangle &= \sigma_d^2 (\tilde{\mathbf{G}} \mathbf{G})^{-1} \tilde{\mathbf{G}} \mathbf{G} (\tilde{\mathbf{G}} \mathbf{G})^{-1} \\
&= \sigma_d^2 (\tilde{\mathbf{G}} \mathbf{G})^{-1}.
\end{aligned} \tag{12.109}$$

This is the familiar formula used in the usual least-squares method for obtaining the variance of the solution.

For $\mathbf{U}_0 = \mathbf{0}$ and $\mathbf{V}_0 \neq \mathbf{0}$, putting $\mathbf{G}_g^{-1} = \tilde{\mathbf{G}}(\mathbf{G} \tilde{\mathbf{G}})^{-1}$ (see (12.102)) gives

$$\langle \Delta \mathbf{m}_g \Delta \tilde{\mathbf{m}}_g \rangle = \sigma_d^2 \cdot \tilde{\mathbf{G}} (\mathbf{G} \tilde{\mathbf{G}})^{-1} (\mathbf{G} \tilde{\mathbf{G}})^{-1} \mathbf{G}. \tag{12.110}$$

In general, putting $\mathbf{G}_g^{-1} = \mathbf{V}_p \mathbf{\Lambda}_p^{-1} \tilde{\mathbf{U}}_p$ gives

$$\begin{aligned}
\langle \Delta \mathbf{m}_g \Delta \tilde{\mathbf{m}}_g \rangle &= \sigma_d^2 \mathbf{V}_p \mathbf{\Lambda}_p^{-1} \tilde{\mathbf{U}}_p \mathbf{U}_p \mathbf{\Lambda}_p^{-1} \tilde{\mathbf{V}}_p \\
&= \sigma_d^2 \mathbf{V}_p \mathbf{\Lambda}_p^{-2} \tilde{\mathbf{V}}_p,
\end{aligned} \tag{12.111}$$

where $\mathbf{\Lambda}_p^{-2}$ is a diagonal matrix with elements λ_i^{-2} $(i = 1, 2, \ldots, p)$.

Obviously, the covariance of the solution becomes large when λ_i is small. In a strategy used by Wiggins (1972), eigenvectors with small eigenvalues were eliminated in order to keep $\langle \Delta \mathbf{m}_g \Delta \tilde{\mathbf{m}}_g \rangle$ below a certain level. This, however, reduces the number p of nonzero eigenvectors, degrading the resolution in both model and data spaces. An appropriate p can be selected by studying the trade-off between the resolution and the variance of the solution due to errors in data.

In our example,

$$\begin{aligned}
\langle \Delta \mathbf{m}_g \, \Delta \tilde{\mathbf{m}}_g \rangle &= \sigma_d^2 \begin{pmatrix} 0 & 1/\sqrt{2} \\ 0 & 1/\sqrt{2} \\ 1 & 0 \end{pmatrix} \begin{pmatrix} 1/2 & 0 \\ 0 & 1/2 \end{pmatrix} \begin{pmatrix} 0 & 0 & 1 \\ 1/\sqrt{2} & 1/\sqrt{2} & 0 \end{pmatrix} \\
&= \sigma_d^2 \begin{pmatrix} 1/4 & 1/4 & 1/2 \\ 1/4 & 1/4 & 0 \\ 0 & 0 & 1/2 \end{pmatrix}.
\end{aligned}$$

This shows that the variances of m_{g_1} and m_{g_2} are one-half that of m_{g_3} and that m_{g_1} and m_{g_2} are completely correlated with the correlation coefficient $+1$. This correlation results from the form of our particular solution $\mathbf{G}_g^{-1} \mathbf{d}$, which gives $m_1 = m_2 = d_1/2$. The original observation $m_1 + m_2 = 1$ does not imply this particular relation between m_1 and m_2. There are some examples in the geophysical literature in which such an apparent correlation among estimated model parameters is taken seriously as a real physical relation.

12.3.4   *Maximum-likelihood inverse*

In the preceding section, we have calculated the covariance of the solution on the assumption that the data covariance has a particularly simple form. If the data covariance is not $\sigma_d^2 \mathbf{I}$, but of the general form $\mathbf{R}_{dd}$, not only does the calculation of the solution variance become complex, but a fundamental problem of statistical estimation also arises. As shown in Figure 12.17, the generalized inverse was found to minimize $|\mathbf{d} - \mathbf{Gm}|^2$. This minimization gives the maximum-likelihood estimation when d_i are statistically independent and share the same variance σ_d^2, but not when $\mathbf{R}_{dd}$ is more general.

As we have shown in discussing the "maximum-likelihood filter" in Section 11.4.2, the probability density function for the multivariate Gaussian distribution with covariance matrix $\mathbf{R}_{dd}$ is written as

$$f(\mathbf{d}) = \frac{|\mathbf{R}_{dd}^{-1}|^{1/2}}{(2\pi)^{N/2}} \exp\left[-\tfrac{1}{2}\widetilde{(\mathbf{d} - \mathbf{Gm})}\mathbf{R}_{dd}^{-1}(\mathbf{d} - \mathbf{Gm})\right]. \qquad (12.112)$$

This shows that for an estimation made by maximizing the likelihood function, we must minimize $\widetilde{(\mathbf{d} - \mathbf{Gm})}\mathbf{R}_{dd}^{-1}(\mathbf{d} - \mathbf{Gm})$ instead of $|\mathbf{d} - \mathbf{Gm}|^2$. In other words, one has to minimize the weighted sum of the squared residual, with the weight matrix being the inverse of the data-covariance matrix.

The same line of reasoning suggests a similar modification in the minimization in model space. The minimization of $|\mathbf{m}|^2$ associated with the generalized inverse should be replaced by the minimization of $\tilde{\mathbf{m}}\mathbf{R}_{mm}^{-1}\mathbf{m}$ if $\mathbf{R}_{mm}$ is not proportional to the identity matrix.

Unlike $\mathbf{R}_{dd}$, however, $\mathbf{R}_{mm}$ cannot be measured directly. It is somewhat artificial and brings a subjective element into the inverse problem. One obvious case that naturally requires nonidentity $\mathbf{R}_{mm}$ occurs when the parameter of an Earth model is discretized with different layer thicknesses. When the model $m(r)$ is a continuous function of distance r from the center of the Earth the generalized inverse requires the minimization of $\int_0^1 |m(r)|^2 \, dr$, where the upper limit of the integral corresponds to the Earth's surface.

If we approximate the integral by a finite sum as

$$\int_0^1 |m(r)|^2 \, dr \sim \sum_i |m(r_i)|^2 \, \Delta r_i$$

and the layer thickness Δr_i is chosen to be nonuniform, then obviously Δr_i should be taken as the weight in the minimization. In this case, $\mathbf{R}_{mm}$ is a diagonal matrix with elements $1/\Delta r_i$.

Another simple example is the case in which model parameters are of different physical dimensions or measured in different units. In this case, they should be properly normalized before minimizing $|\mathbf{m}|^2$, and $\mathbf{R}_{mm}$ can be chosen

as a diagonal matrix consisting of the expected variance of each parameter in the Earth.

Furthermore, we may have some *a priori* idea about the smoothness of fluctuation of a physical property in the Earth's interior. In that case, we can use such *a priori* knowledge in forming appropriate $\mathbf{R}_{mm}$.

Let us now find the solution that minimizes $(\widetilde{\mathbf{d} - \mathbf{Gm}})\mathbf{R}_{dd}^{-1}(\mathbf{d} - \mathbf{Gm})$ in data space and $\tilde{\mathbf{m}}\mathbf{R}_{mm}^{-1}\mathbf{m}$ in model space. Analogous to the naming of a filter designed on a similar basis, we may call this the *maximum-likelihood inverse*. The solution can be given by a generalized inverse in transformed coordinates, as shown by Wiggins (1972) and Jackson (1972).

Since $\mathbf{R}_{dd}$ and $\mathbf{R}_{mm}$ are positive definite matrices, we can always find orthogonal eigenvector matrices $\mathbf{D}$ and $\mathbf{M}$ and express them as

$$\mathbf{R}_{dd} = \mathbf{D}\mathbf{\Lambda}_d\tilde{\mathbf{D}},$$
$$\mathbf{R}_{mm} = \mathbf{M}\mathbf{\Lambda}_m\tilde{\mathbf{M}},$$

(12.113)

where $\mathbf{\Lambda}_d$ and $\mathbf{\Lambda}_m$ are diagonal matrices whose elements are the eigenvalues of $\mathbf{R}_{dd}$ and $\mathbf{R}_{mm}$, respectively. None of the eigenvalues are zero, and $\mathbf{D}$ and $\mathbf{M}$ are complete (i.e., the column vectors of the matrix span the whole space). $\mathbf{D}\tilde{\mathbf{D}} = \tilde{\mathbf{D}}\mathbf{D} = \mathbf{I}$, $\mathbf{M}\tilde{\mathbf{M}} = \tilde{\mathbf{M}}\mathbf{M} = \mathbf{I}$.

We shall introduce a new set of variables by the following relations:

$$\mathbf{d}' = \mathbf{\Lambda}_d^{-1/2}\tilde{\mathbf{D}}\mathbf{d},$$
$$\mathbf{m}' = \mathbf{\Lambda}_m^{-1/2}\tilde{\mathbf{M}}\mathbf{m},$$
$$\mathbf{G}' = \mathbf{\Lambda}_d^{-1/2}\tilde{\mathbf{D}}\mathbf{G}\mathbf{M}\mathbf{\Lambda}_m^{1/2},$$

(12.114)

or

$$\mathbf{d} = \mathbf{D}\mathbf{\Lambda}_d^{1/2}\mathbf{d}',$$
$$\mathbf{m} = \mathbf{M}\mathbf{\Lambda}_m^{1/2}\mathbf{m}',$$
$$\mathbf{G} = \mathbf{D}\mathbf{\Lambda}_d^{1/2}\mathbf{G}'\mathbf{\Lambda}_m^{-1/2}\tilde{\mathbf{M}}.$$

(12.115)

By these transformations, we have

$$(\widetilde{\mathbf{d} - \mathbf{Gm}})\mathbf{R}_{dd}^{-1}(\mathbf{d} - \mathbf{Gm}) = |\mathbf{d}' - \mathbf{G}'\mathbf{m}'|^2$$

and

$$\tilde{\mathbf{m}}\mathbf{R}_{mm}^{-1}\mathbf{m} = |\mathbf{m}'|^2.$$

The generalized inverse for $\mathbf{G}'$ will minimize the above two quantities. We can write the result with new variables as

$$\mathbf{m}'_g = \mathbf{G}'^{-1}_g \mathbf{d}'.$$

Transforming the above equation back to old variables, we can write the maximum-likelihood inverse solution $\mathbf{m}_{\text{max}}$ as

$$\mathbf{m}_{\text{max}} = \mathbf{M}\Lambda^{1/2}_m \mathbf{m}'_g. \tag{12.116}$$

Both $\mathbf{R}_{dd}$ and $\mathbf{R}_{mm}$ are assumed to be diagonal in the analysis by Wiggins (1972), and the above transformation of variables was a trivial problem because $\mathbf{M}$ and $\mathbf{D}$ were assumed to be identity matrices.

BOX **12.3**

Earthquake location

By far the oldest inverse problem studied in seismology is that of using arrival-time data at several different stations for body waves originating from some earthquake or explosion and inverting for the coordinates of the source in space and time. The context of this problem makes it appropriate to consider only a point source and to consider the travel time as being given by an integral along a geometrical ray.

The Earth structure and identification of the ray path (for a particular observed arrival) are usually assumed to be known and fixed, so that there are only four model parameters: the depth, two horizontal coordinates, and origin time.

If t_i is the ith datum (some arrival time at a particular station, observed presumably with some error) and $T_i = T_i(\mathbf{a})$ is the theoretical value (where $\mathbf{a} = (a_1, a_2, a_3, a_4)$ are the model parameters), then clearly if the errors are unbiassed we wish in some way to minimize all the residuals $t_i - T_i(\mathbf{a})$. Supposing that σ_i is the standard deviation of t_i, we shall define

$$\chi^2 = \sum_{i=1}^{n} [t_i - T_i(\mathbf{a})]^2/\sigma_i^2, \tag{1}$$

where n is the number of observations and we choose to estimate the model parameters by that value $\tilde{\mathbf{a}}$ that minimizes χ^2. The idea of minimizing a sum of weighted squared residuals to estimate hypocentral coordinates apparently originated with L. Geiger in about 1910.

If we make a trial guess $\mathbf{a}^0$ at $\hat{\mathbf{a}}$, which is sufficiently close to $\hat{\mathbf{a}}$ so that $\partial T_i/\partial a_j$ is virtually unchanged between $\mathbf{a} = \hat{\mathbf{a}}$ and $\mathbf{a} = \mathbf{a}^0$, then (using the summation convention)

$$0 = \left.\frac{\partial \chi^2}{\partial a_j}\right|_{\hat{\mathbf{a}}} \sim 2\left(d_i(\mathbf{a}^0) + (\hat{a}_k - a_k^0)\frac{\partial d_i}{\partial a_k}(\mathbf{a}^0)\right)\frac{\partial d_i}{\partial a_j}(\mathbf{a}^0), \tag{2}$$

where $d_i(\mathbf{a}) = [t_i - T_i(\mathbf{a})]/\sigma_i$. It follows that the model correction $\mathbf{m} = \mathbf{a}^0 - \hat{\mathbf{a}}$ can be obtained from residuals $\mathbf{d}$ by writing (2) as

$$\tilde{\mathbf{G}}\mathbf{d} = \tilde{\mathbf{G}}\mathbf{G}\mathbf{m}, \tag{3}$$

where

$$G_{ij} = \frac{\partial d_i}{\partial a_j}(\mathbf{a}^0) \quad \text{and} \quad \tilde{G}_{ij} = G_{ji}.$$

In the theory of parameter estimation by least squares, equations (3) are known as the *normal equations*. If partial derivatives are significantly different at $\hat{\mathbf{a}}$ and $\mathbf{a}^0$, it may be necessary to iterate toward $\tilde{\mathbf{a}}$. Note that (3) is solved by $\mathbf{d} = \mathbf{G}\mathbf{m}$, which may be obtained by linearizing the equations $t_i - T_i(\tilde{\mathbf{a}})/\sigma_i = 0$, i.e., requiring that each residual is made small. In the usual application, there are more observations than model parameters, so that (in the terminology of Section 12.3.1) there is no $\mathbf{V}_0$, but $\mathbf{U}_0$ exists. From (12.101) we see that the generalized inverse of (4) is merely the usual solution of (3), $\tilde{\mathbf{m}} = (\tilde{\mathbf{G}}\mathbf{G})^{-1}\tilde{\mathbf{G}}\mathbf{d}$.

Although the problem of earthquake location can thus be set up as an example of generalized linear-inverse theory, the progress made in this theory since the late 1960's has not yet been fully brought to bear upon location problems. It is perhaps more important that the many decades of experience gained in special studies of earthquake location can shed some light on problems that are likely to arise in other inverse studies, particularly where $\mathbf{U}_0$ exists. In the remainder of this box, we therefore describe the findings of three special analyses of earthquake location.

(i) A numerical difficulty can arise in the usual solution of (3), since $(\tilde{\mathbf{G}}\mathbf{G})$ can be a nearly singular matrix, and its computed inverse may be inaccurate. Lee and Lahr (1972) associate the near singularity with model parameters that are poorly constrained by the data. They choose, therefore, to hold such ill-constrained components of $\mathbf{a}$ (often, the earthquake depth) fixed at each interation step, eliminating them from the normal equations, and do not proceed to a local minimum of χ^2. Buland (1976) quantifies the numerical difficulty by working with the *condition number* of $(\tilde{\mathbf{G}}\mathbf{G})$, which is the ratio of its largest and smallest eigenvalue, and finds that for the inversion of just P-arrival times, from a source that may be a few array diameters outside the array of stations used to obtain the data, the condition number may exceed 10^{29}. He advocates inversion of $\mathbf{d} = \mathbf{G}\mathbf{m}$ (using Householder's QR algorithm), since this involves a condition number that is the square root of the condition number for (3).

The condition number can be drastically improved if S arrivals, depth phases, or readings from a station near the source become available. Moreover, Smith (1976) notes that a proper choice of units for the model parameter can lead to a scaling of the columns of $\mathbf{G}$ which will improve the condition number.

(ii) Note that χ^2 in (1) has a so-called χ^2 distribution with n degrees of freedom in an experiment where repeated earthquakes occur at the same known hypocenter with known origin time and where observed suites of travel times are obtained, one suite for each event. Our situation is different, as only one suite of data is available for each unknown hypocenter. The equation $\tilde{\mathbf{G}}(\mathbf{d} - \mathbf{G}\mathbf{m}) = \mathbf{0}$ imposes (in our case) four linear constraints on the model parameters, so that there are only $n - 4$ degrees of freedom. For a known number of

degrees of freedom, χ^2 should, 95% of the time (say), take on a certain range of values that can be found by consulting standard statistical tables. By finding the minimum value $\chi^2(\hat{\mathbf{a}})$ of (1), we can see if it falls into the range of values for 95% confidence, and in this sense check if the solution must be rejected (perhaps because a poor Earth model was assumed).

We gave the solution variance for model parameters in (12.109), and here this reduces to

$$\langle \Delta\mathbf{m}\Delta\tilde{\mathbf{m}} \rangle = (\tilde{\mathbf{G}}\mathbf{G})^{-1}, \tag{4}$$

since we normalized the residuals to give $\langle \Delta\mathbf{d}\Delta\tilde{\mathbf{d}} \rangle = \mathbf{I}$. If the arrival-time errors are *not* known *a priori*, then from a knowledge of the *relative* errors r_i between the t_i, it is possible to estimate absolute errors σ_i and hence apply (5). This can be done by forming the sum in (1) with r_i in place of σ_i, again minimizing to find $\hat{\mathbf{a}}$, and assuming that $\sigma_i^2 = \lambda r_i^2$. Here, λ is some overall constant for the data, which, once known, will give absolute errors, and hence the model variance. Since the expected value of χ^2 for $n - 4$ degrees of freedom is $n - 4$, the value of λ is estimated (Flinn, 1965) by

$$\hat{\lambda}(n - 4) = \sum_{i=1}^{n} [t_i - T_i(\hat{\mathbf{a}})]^2/r_i^2.$$

In practice, the resulting estimate for absolute errors σ_i may be very poor, e.g., because of model errors (see (iii) below). Then special statistical trials, involving random errors added to computed travel times, may be necessary to assess the location errors associated with a given array (Evernden, 1969).

(iii) The effect of using an inaccurate Earth model in computing the observable $T_i(\mathbf{a})$ can seriously upset any attempt to estimate hypocentral coordinates that have any scientific relevance. There is a warning here for those who tackle other inverse problems.

If Earth structure in the region containing source and receivers is laterally homogeneous but depth dependent, then structural parameters may be added to a_1, a_2, a_3, a_4, and the Earth model itself is estimated from suites of arrival-time data, along with the hypocenters (Crosson, 1976a,b). With lateral heterogeneity too, the approach of Section 12.3.9 can be taken, but here the effort to locate earthquakes is effectively subordinated to the problem of determining the large number of model parameters for a three-dimensional Earth structure. Instead, an approach often adopted is to fix the Earth structure (choosing one for which the calculation of $T_i(\mathbf{a})$ is relatively simple) and to assume that part of the residual $t_i - T_i(\mathbf{a})$ is due to a station correction. For example, this could be a constant delay due to sediments under the ith station, plus a term depending on the azimuth of the epicenter from the station to compensate for local dipping structures. Thus, in addition to the model parameters originally of interest, $\mathbf{a} = (a_1, a_2, a_3, a_4)$, there are new unknowns $\mathbf{b}$ that model the data residual in a particular fashion. The effort to parameterize the geophysical problem has turned into an effort to parameterize the data more directly. Since the $\mathbf{b}$ vector should be the same for different hypocenters, sets of arrival-time data for different earthquakes can jointly be inverted to estimate station corrections along with all the hypocenters. Such methods of joint hypocentral determination were developed by Douglas (1967) and Freedman (1967). It is necessary to have a wide distribution of sources, so that the azimuthally dependent station corrections can be identified. Methods of joint hypocentral determination have also been used to locate many earthquakes relative to one particular earthquake (often called the *master event*, or the *calibration event*). A wide distribution is undesirable: instead, the earthquakes should all be fairly close to the master event. Dewey

(1972) has given details of this method, and applications have now been made in many seismically active regions of the world.

E. Smith (1978) has examined the consequences of using the wrong Earth model for calculation of $T_i(\mathbf{a})$, and points out that the problem of model error is not due to its possible departure from a Gaussian distribution. (Errors need not be normally distributed for a statistical interpretation of the least-squares solution.) Rather, in the context of joint hypocentral determination, the problem is that model error introduces a correlation between the residual for the ith event at the jth station, and the residual for the kth event at the lth station. Where the correlation cannot be estimated accurately, confidence regions for earthquake location cannot be constructed on a sound basis.

12.3.5 The stochastic inverse

The calculation of the generalized inverse requires the eigenvector analysis described in previous sections. The stochastic inverse introduced by Franklin (1970) requires only matrix multiplication and inversion, and gives an alternative solution to the generalized inverse and the maximum-likelihood inverse solution.

We consider that the data consist of signal and noise,

$$\mathbf{d} = \mathbf{Gm} + \mathbf{n} \tag{12.117}$$

and that both $\mathbf{m}$ and $\mathbf{n}$ are stochastic processes. We shall assume that their means are zero,

$$\langle \mathbf{m} \rangle = \langle \mathbf{n} \rangle = \mathbf{0} \tag{12.118}$$

(if not, we can subtract the mean from the original process), and that their covariance matrices are given by

$$\langle \mathbf{m\tilde{m}} \rangle = \mathbf{R}_{mm},$$
$$\langle \mathbf{n\tilde{n}} \rangle = \mathbf{R}_{nn}. \tag{12.119}$$

The *stochastic inverse operator* $\mathbf{L}$ is determined by minimizing the statistical average of the discrepancy between $\mathbf{m}$ and $\mathbf{Ld}$. Consider repeated experiments in which $\mathbf{m}$ and $\mathbf{n}$ are generated. Suppose their sample values at the kth experiment are $\mathbf{m}^{(k)}$ and $\mathbf{n}^{(k)}$. For each experiment, we compute $\mathbf{Ld}$, and seek $\mathbf{L}$ which minimizes

$$\frac{1}{n} \sum_{k=1}^{n} \left(m_i^{(k)} - \sum_{j=1}^{N} L_{ij} d_j^{(k)} \right)^2.$$

The minimization of this with respect to the operator $\mathbf{L}$ can be easily made by differentiating with respect to L_{il} and equating to zero:

$$\frac{1}{n} \sum_{k=1}^{n} \left(m_i^{(k)} - \sum_{j=1}^{N} L_{ij} d_j^{(k)} \right) d_l^{(k)} = 0,$$

which will reduce to

$$\langle \mathbf{m}\tilde{\mathbf{d}} \rangle = \mathbf{L} \langle \mathbf{d}\tilde{\mathbf{d}} \rangle$$

or

$$\mathbf{L} = \mathbf{R}_{md} \mathbf{R}_{dd}^{-1}. \tag{12.120}$$

On the other hand, if $\mathbf{m}$ and $\mathbf{n}$ are uncorrelated ($\langle \mathbf{m}\tilde{\mathbf{n}} \rangle = \mathbf{0}$), we obtain

$$\begin{aligned}
\mathbf{R}_{dd} &= \langle \mathbf{d}\tilde{\mathbf{d}} \rangle \\
&= \langle (\mathbf{Gm} + \mathbf{n})(\widetilde{\mathbf{Gm} + \mathbf{n}}) \rangle \\
&= \mathbf{G}\mathbf{R}_{mm}\tilde{\mathbf{G}} + \mathbf{R}_{nn}.
\end{aligned} \tag{12.121}$$

Likewise, we find

$$\mathbf{R}_{md} = \mathbf{R}_{mm}\tilde{\mathbf{G}}. \tag{12.122}$$

Putting (12.121) and (12.122) into (12.120), we obtain

$$\mathbf{L} = \mathbf{R}_{mm}\tilde{\mathbf{G}}(\mathbf{G}\mathbf{R}_{mm}\tilde{\mathbf{G}} + \mathbf{R}_{nn})^{-1}. \tag{12.123}$$

It is interesting to note that the transformation of variables used in the preceding section simplifies the stochastic inverse. Expressing the stochastic inverse solution as $\hat{\mathbf{m}}$, we find from (12.120) and (12.122)

$$\begin{aligned}
\hat{\mathbf{m}} &= \mathbf{R}_{md}\mathbf{R}_{dd}^{-1}\mathbf{d} \\
&= \mathbf{R}_{mm}\tilde{\mathbf{G}}\mathbf{R}_{dd}^{-1}\mathbf{d}.
\end{aligned}$$

Using the transformations given in (12.113) and (12.115), we get

$$\begin{aligned}
\mathbf{R}_{mm} &= \mathbf{M}\mathbf{\Lambda}_m\tilde{\mathbf{M}}, \\
\mathbf{R}_{dd}^{-1} &= \mathbf{D}\mathbf{\Lambda}_d^{-1}\tilde{\mathbf{D}}, \\
\mathbf{m} &= \mathbf{M}\mathbf{\Lambda}_m^{1/2}\mathbf{m}', \\
\mathbf{d} &= \mathbf{D}\mathbf{\Lambda}_d^{1/2}\mathbf{d}'.
\end{aligned} \tag{12.124}$$

Therefore,

$$\mathbf{M}\Lambda_m^{1/2}\hat{\mathbf{m}}' = \mathbf{M}\Lambda_m\tilde{\mathbf{M}}\tilde{\mathbf{G}}\mathbf{D}\Lambda_d^{-1}\tilde{\mathbf{D}}\mathbf{d},$$

which can be written, using (12.114), as

$$\hat{\mathbf{m}}' = \tilde{\mathbf{G}}'\mathbf{d}'. \tag{12.125}$$

The stochastic inverse operator is nothing but the transpose of $\mathbf{G}$ in the transformed coordinates. A similar simplified inversion process is discussed by Gilbert (1971b), as we shall see in the next section.

A special case of the stochastic inverse, in which

$$\mathbf{R}_{mm} = \sigma_m^2\mathbf{I},$$
$$\mathbf{R}_{nn} = \sigma_n^2\mathbf{I}, \tag{12.126}$$

gives a good approximation to the generalized inverse. Putting (12.126) into (12.123), we have the stochastic inverse operator

$$\mathbf{L}_0 = \tilde{\mathbf{G}}(\mathbf{G}\tilde{\mathbf{G}} + \varepsilon^2\mathbf{I})^{-1} \tag{12.127}$$

where

$$\varepsilon^2 = \sigma_n^2/\sigma_m^2.$$

To find the above solution in terms of eigenvectors, we use the complete set of eigenvectors $\mathbf{U} = [\mathbf{U}_p, \mathbf{U}_0]$ and obtain

$$(\mathbf{G}\tilde{\mathbf{G}} + \varepsilon^2\mathbf{I})^{-1} = [\mathbf{U}_p, \mathbf{U}_0]\begin{pmatrix}(\Lambda_p^2 + \varepsilon^2\mathbf{I})^{-1} & \mathbf{0} \\ \mathbf{0} & \varepsilon^{-2}\mathbf{I}\end{pmatrix}\begin{pmatrix}\tilde{\mathbf{U}}_p \\ \tilde{\mathbf{U}}_0\end{pmatrix}$$
$$= \mathbf{U}_p(\Lambda_p^2 + \varepsilon^2\mathbf{I})^{-1}\mathbf{U}_p + \mathbf{U}_0\varepsilon^{-2}\tilde{\mathbf{U}}_0.$$

Since $\tilde{\mathbf{G}} = \mathbf{V}_p\Lambda_p\tilde{\mathbf{U}}_p$ and $\tilde{\mathbf{U}}_p\mathbf{U}_0 = \mathbf{0}$, we find

$$\mathbf{L}_0 = \tilde{\mathbf{G}}(\mathbf{G}\tilde{\mathbf{G}} + \varepsilon^2\mathbf{I})^{-1}$$
$$= \mathbf{V}_p\frac{\Lambda_p}{\Lambda_p^2 + \varepsilon^2\mathbf{I}}\tilde{\mathbf{U}}_p. \tag{12.128}$$

Comparing this equation with the generalized inverse (12.100), we see that $\mathbf{L}_0$ is an approximation to the latter. The contributions of eigenvectors with eigenvalues smaller than ε^2 are suppressed in the stochastic inverse.

The operator $\mathbf{L}_0$ can be written also as

$$\mathbf{L}_0 = (\tilde{\mathbf{G}}\mathbf{G} + \varepsilon^2\mathbf{I})^{-1}\tilde{\mathbf{G}} \tag{12.129}$$

because $(\tilde{\mathbf{G}}\mathbf{G} + \varepsilon^2\mathbf{I})^{-1} = \mathbf{V}_p(\Lambda_p^2 + \varepsilon^2\mathbf{I})^{-1}\mathbf{V}_p + \mathbf{V}_0\varepsilon^{-2}\tilde{\mathbf{V}}_0$, and $\tilde{\mathbf{V}}_0\mathbf{V}_p = \mathbf{0}$. The inverse solution given by (12.129) has been known as *damped least squares.* Levenberg (1944) obtained it by minimizing the sum of the squares of data residual and model parameter with weights inversely proportional to their variances; i.e., $\sigma_n^{-2}|\mathbf{d} - \mathbf{Gm}|^2 + \sigma_m^{-2}|\mathbf{m}|^2$, where again $\varepsilon^2 = \sigma_n^2/\sigma_m^2$.

The resolution matrix for $\mathbf{L}_0$ is given by

$$\mathbf{L}_0\mathbf{G} = \mathbf{V}_p \frac{\Lambda_p^2}{\Lambda_p^2 + \varepsilon^2\mathbf{I}} \tilde{\mathbf{V}}_p. \tag{12.130}$$

The trace of $\mathbf{L}_0\mathbf{G}$, which is a measure of resolution in model space, as discussed in Section 12.3.3, can be written as

$$\text{trace of } \mathbf{L}_0\mathbf{G} = \sum_{i=1}^{p} \frac{\lambda_i^2}{\lambda_i^2 + \varepsilon^2}, \tag{12.131}$$

which is clearly smaller than p. Thus the introduction of ε^2 will degrade resolution, but will stabilize the solution by reducing the covariance. The covariance matrix is given by (12.86) as

$$\langle \Delta\hat{\mathbf{m}} \, \Delta\tilde{\hat{\mathbf{m}}} \rangle = \sigma_d^2\mathbf{L}_0\tilde{\mathbf{L}}_0$$
$$= \sigma_d^2\mathbf{V}_p \frac{\Lambda_p^2}{(\Lambda_p^2 + \varepsilon^2\mathbf{I})^2} \tilde{\mathbf{V}}_p, \tag{12.132}$$

where σ_d^2 is the variance of the error $\Delta\mathbf{d}$ in data $\mathbf{d}$, assuming a uniform and independent error for each individual measurement. In our stochastic model, $\Delta\mathbf{d}$ corresponds to $\mathbf{n} = \mathbf{d} - \mathbf{Gm}$, and $\sigma_d^2 = \sigma_n^2$. Thus the increase in ε^2 reduces the error of model-parameter estimates, thereby sacrificing the resolution. In the stochastic inverse scheme, the best choice of ε^2 is σ_n^2/σ_m^2 (the ratio of noise variance to model variance), as shown in (12.127).

Let us find $\mathbf{L}_0$ for our example problem and compare it with the generalized inverse $\mathbf{G}_g^{-1}$ obtained earlier:

Since

$$\mathbf{G}\tilde{\mathbf{G}} = \begin{pmatrix} 2 & 0 & 0 \\ 0 & 1 & -1 \\ 0 & -1 & 1 \end{pmatrix}$$

and

$$(G\tilde{G} + \varepsilon^2 I)^{-1} = \begin{pmatrix} \dfrac{1}{2 + \varepsilon^2} & 0 & 0 \\[2ex] 0 & \dfrac{1 + \varepsilon^2}{(1 + \varepsilon^2)^2 - 1} & \dfrac{1}{(1 + \varepsilon^2)^2 - 1} \\[2ex] 0 & \dfrac{1}{(1 + \varepsilon^2)^2 - 1} & \dfrac{1 + \varepsilon^2}{(1 + \varepsilon^2)^2 - 1} \end{pmatrix}$$

we find that

$$L_0 = \tilde{G}(G\tilde{G} + \varepsilon^2 I)^{-1}$$

$$= \begin{pmatrix} \dfrac{1}{2 + \varepsilon^2} & 0 & 0 \\[2ex] \dfrac{1}{2 + \varepsilon^2} & 0 & 0 \\[2ex] 0 & \dfrac{1}{2 + \varepsilon^2} & \dfrac{-1}{2 + \varepsilon^2} \end{pmatrix}$$

In the limit as ε^2 approaches zero—i.e., as the noise variance vanishes—the stochastic inverse converges to the generalized inverse. In our example,

$$\operatorname*{Lim}_{\varepsilon \to 0} L_0 = \begin{pmatrix} \tfrac{1}{2} & 0 & 0 \\ \tfrac{1}{2} & 0 & 0 \\ 0 & \tfrac{1}{2} & -\tfrac{1}{2} \end{pmatrix} = G_g^{-1}.$$

12.3.6   Methods of Backus and Gilbert

In previous sections, we solved the inverse problem of a general linear system by Lanczos's elegant method of eigenvector analysis, extended it to the case of a general covariance matrix for both data and model vectors, and introduced Franklin's stochastic inverse. We showed that a special case of the stochastic inverse gives a good approximation to the generalized inverse.

In a series of papers published since 1967, Backus and Gilbert not only attracted the attention of seismologists to the inverse problem of a general linear system, but also made several unique contributions that are distinct from the approaches covered in previous sections. We shall briefly outline their methods here.

In their first paper (1967), they showed the high degree of nonuniqueness in geophysical inverse problems and offered a practical solution that satisfies the observed data and minimizes the departure from the initial guess. The process of finding the minimum solution, taking into account the errors in data, was later described by Gilbert (1971b).

Their minimum solution is identical to the generalized inverse for $U_0 = 0$. In this case, the minimum solution $\hat{\mathbf{m}}$ is given by (12.102):

$$\hat{\mathbf{m}} = \tilde{\mathbf{G}}(\mathbf{G}\tilde{\mathbf{G}})^{-1}\mathbf{d}.$$

Introducing a vector $\mathbf{v}$ in the data space, we can rewrite this as

$$\hat{\mathbf{m}} = \tilde{\mathbf{G}}\mathbf{v}, \quad \text{where} \quad (\mathbf{G}\tilde{\mathbf{G}})\mathbf{v} = \mathbf{d}. \tag{12.133}$$

Backus and Gilbert consider the model as a continuous function of distance r from the Earth's center. The data are expressed by (12.79),

$$d_i = \int_0^{r_\oplus} G_i(r)m(r)\,dr.$$

In order to translate the result for discrete cases into the solution of a continuous problem, we put

$$
\begin{aligned}
d_i &= \sum_{k=1}^{M} G_i(k\,\Delta r)m(k\,\Delta r)\,\Delta r, \\
m_k &= m(k\,\Delta r)(\Delta r)^{1/2}, \\
G_{ik} &= G_i(k\,\Delta r)(\Delta r)^{1/2}.
\end{aligned}
\tag{12.134}
$$

Then $\hat{m}(r)$, which minimizes

$$\sum_{k=1}^{M} |m(k\,\Delta r)|^2\,\Delta r,$$

is given by equation (12.133) as

$$\hat{m}(r) = \sum_{j=1}^{N} G_j(r)v_j, \quad \text{with } v_j \text{ given by}$$

$$\sum_{j=1}^{N} v_j \int_0^{r_\oplus} G_i(r)G_j(r)\,dr = d_i. \tag{12.135}$$

When the data have errors, the above equation may be replaced by

$$d_i - \sigma_i \leqq \sum_{j=1}^{N} v_j \int_0^{r_\oplus} G_i(r)G_j(r)\, dr \leqq d_i + \sigma_i, \qquad (12.136)$$

where σ_i is the standard error of d_i. Now we want to find v that satisfies the above inequality and minimizes

$$\int |\mathbf{m}|^2\, dr = \int |\Sigma G_j(r)v_j|^2\, dr. \qquad (12.137)$$

We rewrite (12.136) and (12.137) as

$$\mathbf{d} - \boldsymbol{\sigma} \leqq \mathbf{G}\tilde{\mathbf{G}}v \leqq \mathbf{d} + \boldsymbol{\sigma} \qquad (12.138)$$

and

$$|\mathbf{m}|^2 = \tilde{v}\mathbf{G}\tilde{\mathbf{G}}v, \qquad (12.139)$$

respectively.

To simplify the above process of finding v, we introduce a transformation $\mathbf{T}$ that diagonalizes $\mathbf{G}\tilde{\mathbf{G}}$ and $\mathbf{R}_{dd} = \langle \Delta\mathbf{d}\Delta\tilde{\mathbf{d}} \rangle$ simultaneously. From (12.113), we have

$$\mathbf{R}_{dd} = \mathbf{D}\boldsymbol{\Lambda}_d\tilde{\mathbf{D}}.$$

Defining a positive definite matrix

$$\mathbf{A} = \boldsymbol{\Lambda}_d^{-1/2}\tilde{\mathbf{D}}\mathbf{G}\tilde{\mathbf{G}}\mathbf{D}\boldsymbol{\Lambda}_d^{-1/2}$$

and expressing its eigenvector matrix as $\mathbf{R}$, we write

$$\mathbf{A} = \mathbf{R}\boldsymbol{\Lambda}\tilde{\mathbf{R}},$$

where $\boldsymbol{\Lambda}$ is a diagonal matrix. $\mathbf{T}$ is then defined as

$$\mathbf{T} = \boldsymbol{\Lambda}^{-1/2}\tilde{\mathbf{R}}\boldsymbol{\Lambda}_d^{-1/2}\tilde{\mathbf{D}} \qquad (12.140)$$

It is easily shown that $\mathbf{T}\mathbf{G}\tilde{\mathbf{G}}\tilde{\mathbf{T}} = \mathbf{I}$ and $\mathbf{T}\mathbf{R}_{dd}\tilde{\mathbf{T}} = \boldsymbol{\Lambda}^{-1}$.

Introducing a new set of variables by

$$\mathbf{d}' = \mathbf{T}\mathbf{d},$$

$$v = \tilde{\mathbf{T}}v', \qquad (12.141)$$

(12.138) and (12.139) can be simplified as

$$\mathbf{d}' - \boldsymbol{\sigma}' \leq \mathbf{v}' \leq \mathbf{d}' + \boldsymbol{\sigma}' \tag{12.142}$$

and

$$|\mathbf{m}|^2 = |\mathbf{v}'|^2. \tag{12.143}$$

With this transformation, the process for finding $\mathbf{v}'$ that minimizes $|\mathbf{m}|^2$ becomes very simple. We first arrange the data components d_i' in increasing order of error σ_i'. If $|d_1'| < |\sigma_1'|$, we set $v_1' = 0$ and proceed to d_2'. If not, v_1' is determined by comparing $|d_1' + \sigma_1'|$ and $|d_1' - \sigma_1'|$. If $|d_1' + \sigma_1'|$ is larger than $|d_1' - \sigma_1'|$, we set $v_1' = d_1' - \sigma_1'$. If $|d_1' - \sigma_1'|$ is larger than $|d_1' + \sigma_1'|$, we set $v_1' = d_1' + \sigma_1'$ and then proceed to d_2'. This process assures the minimization of $|\mathbf{m}|^2 = \sum_i v_i'^2$.

In their second paper (1968), Backus and Gilbert focused their attention on the resolution in model space and introduced the *deltaness criterion*.

An averaging kernel $A(r_0, r)$ is defined as the weight function with which an estimate $\hat{m}(r_0)$ of the model at r_0 is expressed as a weighted average of the "true" model $m(r)$:

$$\hat{m}(r_0) = \int_0^{r_\oplus} A(r_0, r) m(r) \, dr. \tag{12.144}$$

$A(r_0, r)$ is equivalent to the resolution matrix of discrete problems. Since, in the linear inversion, the model estimate $\hat{m}(r_0)$ is a linear combination of data, we can write

$$\hat{m}(r_0) = \sum_{i=1}^{N} a_i d_i = \tilde{\mathbf{a}}\mathbf{d}, \tag{12.145}$$

where $a_i = a_i(r_0)$ is a function of r_0. From (12.144), (12.145), and (12.79), we find

$$A(r_0, r) = \sum_{i=1}^{N} a_i G_i(r) = \tilde{\mathbf{a}}\mathbf{G}(r). \tag{12.146}$$

We now want to determine a_i by imposing the deltaness criterion on $A(r_0, r)$—i.e., by making $A(r_0, r)$ as much like a Dirac delta function as possible—so that $\hat{m}(r_0)$ in (12.144) is close to $m(r_0)$.

Backus and Gilbert considered the following two criteria:

$$J = \int_0^{r_\oplus} [A(r_0, r) - \delta(r - r_0)]^2 \, dr,$$

$$K = 12 \int_0^{r_\oplus} (r - r_0)^2 [A(r_0, r) - \delta(r - r_0)]^2 \, dr.$$

Minimizing J will make $A(r_0, r)$ the closest to a δ-function in a least-squares sense. It is easy to show that the J-criterion gives the same result as the minimum solution. Putting

$$(G\tilde{G})_{ij} = \int_0^{r_\oplus} G_i(r)G_j(r) \, dr, \tag{12.147}$$

we can write

$$J = \sum_i \sum_j a_i a_j (G\tilde{G})_{ij} - 2 \sum_i a_i G_i(r_0) + \int_0^{r_\oplus} \{\delta(r - r_0)\}^2 \, dr.$$

Minimizing J with respect to a_i, we obtain

$$\mathbf{a} = (G\tilde{G})^{-1} G(r_0),$$

and then

$$A(r_0, r) = \tilde{G}(r_0)(G\tilde{G})^{-1} G(r),$$

$$\hat{m}(r_0) = \int A(r_0, r)m(r) \, dr$$

$$= \tilde{G}(r_0)(G\tilde{G})^{-1} \int G(r)m(r) \, dr$$

$$= G(r_0)(G\tilde{G})^{-1} \mathbf{d}. \tag{12.148}$$

These results are the same as for the minimum solution (12.102) or the generalized inverse for $\mathbf{U}_0 = \mathbf{0}$.

The K-criterion, on the other hand, leads to a new result. This criterion avoids the squared δ-function in J and also suppresses the side lobes of $A(r_0, r)$ more efficiently than J. K has the dimension of r, and is a measure of spread of the average kernel along the r-axis around r_0. In the K-criterion, K is minimized under the constraint that

$$\int_0^{r_\oplus} A(r_0, r) \, dr = 1. \tag{12.149}$$

The factor 12 appearing above in the definition of K was chosen so that K is a measure of the width (spread) of the peak in $A(r_0, r)$ when $A(r_0, r)$ resembles $\delta(r - r_0)$.

Introducing the spread matrix $\mathbf{S}$ with components given by

$$S_{ij} = 12 \int_0^{r_\oplus} dr(r - r_0)^2 G_i(r)G_j(r),$$

we can write

$$K = \sum_i \sum_j a_i a_j S_{ij}$$
$$= \tilde{\mathbf{a}} \mathbf{S} \mathbf{a}. \tag{12.150}$$

The constraint given by (12.149) can be written as

$$\sum_i \int_0^{r_\oplus} a_i G_i(r) \, dr = 1$$

or

$$\tilde{\mathbf{a}} \mathbf{u} = 1, \tag{12.151}$$

where

$$u_i = \int_0^{r_\oplus} G_i(r) \, dr. \tag{12.152}$$

The minimization of $\tilde{\mathbf{a}} \mathbf{S} \mathbf{a}$ under the constraint $\tilde{\mathbf{a}} \mathbf{u} = 1$ is most easily made by inspecting the geometry. Since $\mathbf{S}$ is positive definite,

$$\tilde{\mathbf{a}} \mathbf{S} \mathbf{a} = \text{const.}$$

expresses a spheroidal surface in the multidimensional space spanned by the vector $\mathbf{a}$. The constraint $\tilde{\mathbf{a}} \mathbf{u} = 1$ represents a fixed plane. As shown in Figure 12.18, the minimum of $\tilde{\mathbf{a}} \mathbf{S} \mathbf{a}$ occurs when the spheroidal surface is tangent to

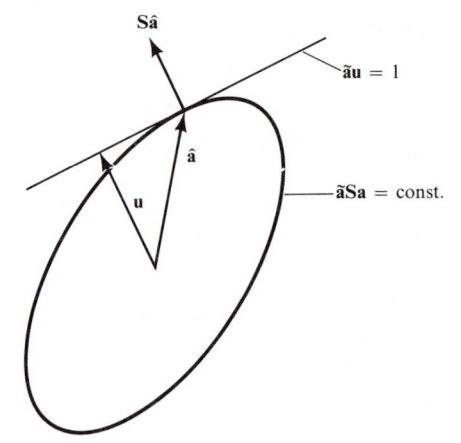

FIGURE **12.18**
Geometric interpretation of estimate $\hat{\mathbf{a}}$.

the plane. If the spheroid is smaller than this, there is no solution; if it is greater, $\tilde{\mathbf{a}}\mathbf{S}\mathbf{a}$ is also greater.

Since the normal to the spheroidal surface is $\mathbf{S}\mathbf{a}$, and since it should be parallel to the normal to the plane at the point of contact, $\mathbf{S}\hat{\mathbf{a}}$ must be parallel to $\mathbf{u}$ (where $\hat{\mathbf{a}}$ is the desired solution). Then

$$\hat{\mathbf{a}} = \lambda \mathbf{S}^{-1}\mathbf{u},$$

where λ is a constant scalar. Since $\tilde{\hat{\mathbf{a}}}\mathbf{u} = 1$, $\lambda = 1/(\tilde{\mathbf{u}}\mathbf{S}^{-1}\mathbf{u})$. We have, therefore,

$$\hat{\mathbf{a}} = \frac{\mathbf{S}^{-1}\mathbf{u}}{\tilde{\mathbf{u}}\mathbf{S}^{-1}\mathbf{u}}.$$

The corresponding model estimate is given by (12.145) as

$$\hat{m}(r_0) = \tilde{\hat{\mathbf{a}}}\mathbf{d}$$
$$= \frac{\tilde{\mathbf{u}}\mathbf{S}^{-1}\mathbf{d}}{\tilde{\mathbf{u}}\mathbf{S}^{-1}\mathbf{u}}, \tag{12.153}$$

and the averaging kernel is determined as

$$A(r_0, r) = \tilde{\hat{\mathbf{a}}}\mathbf{G}(r)$$
$$= \frac{\tilde{\mathbf{u}}\mathbf{S}^{-1}\mathbf{G}(r)}{\tilde{\mathbf{u}}\mathbf{S}^{-1}\mathbf{u}}. \tag{12.154}$$

Although the solution $\hat{m}(r_0)$ does not exactly satisfy the original equation (12.79), it is useful because it represents a weighted average of the "true" model $m(r)$ with the known and well-shaped weight-function $A(r_0, r)$.

Finally, in the third paper by Backus and Gilbert (1970), an extensive investigation was made of the trade-off between the ability to resolve detail of a model and the reliability of the estimate of model parameters.

The error in the model estimate $\hat{m}(r_0)$ is related to the error $\Delta\mathbf{d}$ in data as

$$\Delta\hat{m}(r_0) = \tilde{\hat{\mathbf{a}}}\,\Delta\mathbf{d}.$$

The variance of $\hat{m}(r_0)$, therefore, is

$$\langle|\Delta\hat{m}(r_0)|^2\rangle = \tilde{\hat{\mathbf{a}}}\langle\Delta\mathbf{d}\,\Delta\tilde{\mathbf{d}}\rangle\mathbf{a}$$
$$= \tilde{\hat{\mathbf{a}}}\mathbf{R}_{dd}\mathbf{a}. \tag{12.155}$$

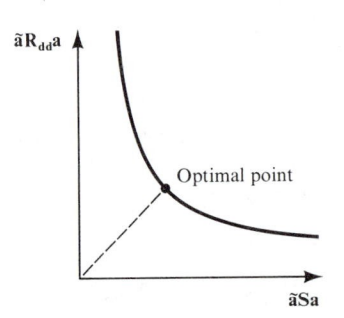

FIGURE **12.19**

Trade-off between resolution and error.

The resolution was again measured by $\tilde{\mathbf{a}}\mathbf{S}\mathbf{a}$. It was shown that the reduction of the error measure increased the measure of spread, and vice versa. In order to find a compromise, a linear combination $\tilde{\mathbf{a}}\mathbf{W}\mathbf{a}$ of the two measures was minimized, where

$$\mathbf{W} = (1 - \alpha)\mathbf{S} + \alpha\mathbf{R}_{dd} \qquad 0 \leqq \alpha \leqq 1, \qquad (12.156)$$

under the constraint $\tilde{\mathbf{a}}\mathbf{u} = 1$. The solution is given by

$$\mathbf{a}(\alpha) = \frac{\mathbf{W}^{-1}\mathbf{u}}{\tilde{\mathbf{u}}\mathbf{W}^{-1}\mathbf{u}}. \qquad (12.157)$$

Calculating $\tilde{\mathbf{a}}\mathbf{S}\mathbf{a}$ and $\tilde{\mathbf{a}}\mathbf{R}_{dd}\mathbf{a}$ for $\mathbf{a} = \mathbf{a}(\alpha)$, and plotting them in a diagram like Figure 12.19 for trial values of α, one can obtain an optimal value of α that corresponds to the point on the curve closest to the origin. This curve is called the *trade-off curve*, and may be constructed independently for different r_0.

Similar but somewhat different measures of resolution and error have been proposed by others. For example, Der et al. (1970) proposed the minimization of the off-diagonal elements of the resolution matrix under the constraint that its diagonal element is unity. Jordan (1972) used a slightly different error estimate from the one used by Backus and Gilbert to show that the stochastic inverse lies on the optimal point on the trade-off curve.

We have now described all the main features and methods of linearized inversion of data with Gaussian errors. These methods have been quite extensively applied in seismology. For example, surface-wave dispersion data were first inverted (using least squares) by Dorman and Ewing (1962) and Brune and Dorman (1963). A thorough study of source parameters and Earth structure, using normal-mode data extracted from WWSSN records, was given by Gilbert and Dziewonski (1975). Langston (1976) has used WWSSN waveforms, maximizing a correlation function suggested by Mellman and Burdick and defined jointly on data and synthetic seismograms, to infer source parameters from teleseismically observed body waves.

12.3.7   Limitation due to nonlinearity

All the results obtained so far in Section 12.3 are valid only for the range of variables in which the perturbation in the model is linearly related to the corresponding change in the observable. If the nonlinearity is severe, the correction to the starting model may lead to a poorer agreement with data, resulting in divergence of the linearization-iteration process.

We shall describe a remedy due to Marquardt (1963), who suggested the repeated use of the damped least-squares solution (12.128) with proper choice of damping constant ε^2.

Suppose that we want to minimize

$$\Phi = |\mathbf{D} - \mathbf{f}(\mathbf{M})|^2$$

$$= \sum_{i=1}^{N} (D_i - f_i(\mathbf{M}))^2, \qquad (12.158)$$

where D_i $(i = 1, 2, \ldots, N)$ and M_j $(j = 1, 2, \ldots, M)$ are data and model, respectively. Having guessed an initial model $\mathbf{M}^0$, we linearize the relation between data and model by putting

$$\mathbf{f}(\mathbf{M}) \sim \mathbf{f}(\mathbf{M}^0) + \mathbf{Gm},$$

where

$$G_{ij} = \left(\frac{\partial f_i}{\partial M_j}\right)_{\mathbf{M}=\mathbf{M}^0},$$

$$\mathbf{m} = \mathbf{M} - \mathbf{M}^0.$$

Putting $\mathbf{d} = \mathbf{D} - \mathbf{f}(\mathbf{M}^0)$, we get

$$\Phi \sim |\mathbf{d} - \mathbf{Gm}|^2.$$

Minimizing Φ, we find the familiar least-squares solution (12.101),

$$\mathbf{m}^L = (\tilde{\mathbf{G}}\mathbf{G})^{-1}\tilde{\mathbf{G}}\mathbf{d}. \qquad (12.159)$$

The standard linearization-iteration proceeds by constructing the second model $\mathbf{M}^1 = \mathbf{M}^0 + \mathbf{m}^L$, revising the linear relation by putting

$$\mathbf{f}(\mathbf{M}) \sim \mathbf{f}(\mathbf{M}^1) + \mathbf{Gm},$$

where now

$$G_{ij} = \left(\frac{\partial f_i}{\partial M_j}\right)_{\mathbf{M}=\mathbf{M}^1},$$

and repeating the minimization process.

In practice, the above process may not converge (Hartley, 1961). A remedy is to make the second model like

$$\mathbf{M}^1 = \mathbf{M}^0 + K\mathbf{m}^L,$$

where $0 < K \leq 1$. Even with small K, failure to converge is not uncommon.

An entirely different approach is the gradient method, in which the direction of most rapid change of Φ is obtained in the model space. Since the equation of a plane tangent to the surface $\Phi = $ const. at $M = M^0$ is given by

$$\sum_j \frac{\partial \Phi}{\partial M_j}(M_j - M_j^0) = 0,$$

the direction of most rapid change, which is normal to the plane, is given by the vector $\mathbf{m}^G$, with components $(-\partial\Phi/\partial M_1, -\partial\Phi/\partial M_2, \ldots, -\partial\Phi/\partial M_M)$ evaluated at $\mathbf{M} = \mathbf{M}^0$.

From (12.158),

$$\begin{aligned}
\frac{\partial \Phi}{\partial M_j} &= 2 \sum_i (D_i - f_i(M)) \frac{\partial f_i}{\partial M_j} \\
&= 2 \sum_i d_i \frac{\partial f_i}{\partial M_j} \\
&= 2 \sum_i G_{ij} d_i,
\end{aligned} \tag{12.160}$$

and we find that $\mathbf{m}^G$ is parallel to $\tilde{\mathbf{G}}\mathbf{d}$.

The correction of a model by the gradient method always can be made to converge. The convergence, however, may be very slow. On the other hand, the least-squares iteration converges very rapidly when it does converge. The vectors $\mathbf{m}^G$ and $\mathbf{m}^L$ are often nearly 90° apart from each other.

It therefore seems reasonable to take an intermediate direction between $\mathbf{m}^G$ and $\mathbf{m}^L$. Marquardt shows that the damped least-squares solution (12.128),

$$\hat{\mathbf{m}} = (\tilde{\mathbf{G}}\mathbf{G} + \varepsilon^2\mathbf{I})^{-1}\tilde{\mathbf{G}}\mathbf{d},$$

points in such an intermediate direction. In fact, when $\varepsilon^2 = 0$, $\hat{\mathbf{m}} = \mathbf{m}^L$; and when $\varepsilon^2 \to \infty$, the direction of $\hat{\mathbf{m}}$ approaches that of $\tilde{\mathbf{G}}\mathbf{d}$, and therefore of $\mathbf{m}^G$.

For a small ε^2, the process may diverge, and for a large ε^2, the convergence may be too slow. After a few trials, an optimal ε^2 may be found for a fast convergence.

In the U.S.S.R., according to Keilis-Borok and Yanovskaya (1967), the gradient method and the Monte Carlo method have been widely used in the search for minima of Φ. They particularly recommended the *hedgehog method* developed by V. Valus, in which the Monte Carlo method first found a point in the minimum region, after which a systematic search was made to define the extent of acceptably low values of Φ in the neighborhood of the minimum.

The Monte Carlo method was also extensively used by Press (1968, 1970). This method gives bounds of successful models, but is an extremely slow process. If the system is completely linear, there is a very quick way of finding such bounds. This is the method of linear programming used by Johnson (1972) to find the maximum and minimum shear velocities and densities at various depths in the Earth which satisfy the observed data.

Consider a model space with the coordinates m_i $(i = 1, 2, \ldots, M)$. The observed data d_i, with the error σ_i, will define a space sandwiched by N pairs of planes:

$$d_i + \sigma_i \geq \sum_{j=1}^{M} G_{ij}m_j \geq d_i - \sigma_i \qquad (i = 1, 2, \ldots, N). \qquad (12.161)$$

Such a space is a convex polyhedron, and any linear combination Z of m_i will take the maximum or minimum value at a vertex. The SIMPLEX algorithm (Dantzig, 1963) allows us to find the neighboring vertex from a starting vertex. The value of Z is found at each vertex, and the search is made from vertex to vertex until the minimum of Z is found. The process is extremely fast because the search is done over a series of points, as compared to the M-dimensional search of the Monte Carlo or hedgehog methods.

12.3.8 Non-Gaussian errors

In addition to the linearity assumption, the basic assumption of Gaussian statistics underlying the linear-inversion process can pose a serious limitation to its application to geophysical data.

If the data contain errors that do not obey the Gaussian distribution, the least-squares criterion (the most basic one in the linear inversion) may not work.

Suppose we want to estimate the mean value of data d_i $(i = 1, \ldots, N)$. The least-squares criterion finds the estimate by minimizing the so-called L_2-norm:

$$\sum_{i=1}^{n} (d_i - m)^2.$$

Taking the derivative with respect to m and setting it equal to zero, we find the mean as an arithmetic average:

$$m = \frac{\sum\limits_{i=1}^{n} d_i}{n}. \tag{12.162}$$

We have also shown in Section 11.4.2 that the weighted least-squares criterion that minimizes $\sum_i \sum_j W_{ij}(d_i - m)(d_j - m)$ corresponds to the maximum-likelihood estimate for a multivariate Gaussian distribution with the covariance matrix $\mathbf{W}^{-1}$. In this case, the estimate of the mean is given by a weighted average:

$$m = \frac{\sum\limits_i \sum\limits_j W_{ij}d_i}{\sum\limits_i \sum\limits_j W_{ij}}. \tag{12.163}$$

A classic example of non-Gaussian seismological data is the distribution of travel-time residuals encountered by Jeffreys during the process of constructing the Jeffreys-Bullen tables. As described in Problem 11.5, he used a weighted average with a weight depending on the deviation of each observed value of d_i from the mean estimated for the sample. Another example is the signal-generated noise, as mentioned but not treated in Chapter 11.

An extreme case of a long-tailed distribution is Cauchy's distribution, referred to by Tukey (1965) in his discussion of geophysical data. The probability density function is $f(x) = (1/\pi)/(1 + x^2)$. It has the remarkable property that its arithmetic average

$$\bar{x} = \frac{\sum\limits_{i=1}^{n} x_i}{n}$$

obeys the same distribution as x. In this case, repeating experiments to find an average is merely wasting time.

For such a long-tailed distribution, criteria based on the L_1 norm may be more effective. Instead of the sum of squares in the L_2 norm, the sum of absolute values is minimized in the L_1 norm criterion. Differentiating $\sum_i |x_i - m|$ with respect to m and setting it equal to zero, we find

$$\sum\limits_{i=1}^{n} \text{sign}(x_i - m) = 0. \tag{12.164}$$

This gives the median of x_i for m if n is odd. If n is even, m lies between the two median x_i. In any case, this estimate of the mean is practically unaffected by

such a blunder as putting one of the x_i equal to infinity. Such a blunder will destroy the arithmetic average of the L_2 norm. Thus the L_1 norm estimate is more robust. An evaluation of the L_1 norm in geophysical data analysis is given by Claerbout and Muir (1973).

12.3.9   Determination of a three-dimensional seismic-velocity structure: an example of linear inversion

Consider an array of seismograph stations covering an area of the Earth's surface, say 100×100 km^2. We shall measure the time of arrival of the first P-waves from a seismic event at distances greater than, say, 30°. Thus our data set consists of arrival times t_{ij}^{OBS} observed at the ith station for the jth event. A typical array of 30 stations and 100 events around the world will give 3000 measurements.

To invert this data set, we must first set up an initial model and calculate the arrival time t_{ij}^{CAL} expected for the ith station and jth event. As an initial model for the structure under the array, we shall use a stack of homogeneous layers with horizontal boundaries overlying a standard spherically symmetric Earth. Such a layered medium with laterally averaged velocities is usually known under an array from a refraction survey. Then, using published epicenter, focal depth, and origin time of the jth event, we can calculate t_{ij}^{CAL} for the initial model.

We then form the difference between the observed and calculated, or the residual, which contains new information about the Earth beyond our initial model. We shall try to explain these residuals by perturbing our initial model. Following Aki, Christoffersson, and Husebye (1976), we divide the layered medium into many blocks, as shown in Figure 12.20, and assign each block a parameter that describes the fractional perturbation of slowness (reciprocal of

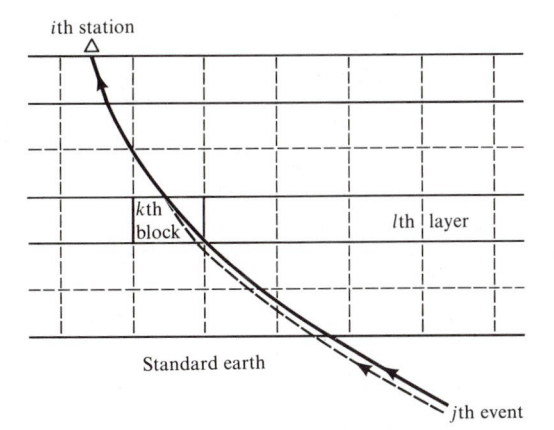

FIGURE **12.20**

Block model for three-dimensional inversion of teleseismic travel-time data observed at a two-dimensional array.

velocity) relative to the average for the layer to which the block belongs. Expressing the parameter for the kth block as m_k, we can write the residual as

$$\Delta t_{ij} = t_{ij}^{OBS} - t_{ij}^{CAL} = \sum_k g_{ijk} m_k + t_j, \qquad (12.165)$$

where

$$g_{ijk} = \frac{d_l}{v_l \cos \theta_l},$$

d_l and v_l are the thickness and velocity of the lth layer, θ_l is the angle between the vertical and the unperturbed ray path in the lth layer, and the summation with respect to k is made over one block from each layer that contains most of the unperturbed path in the layer. Since, from Fermat's principle, the arrival time is stationary with respect to the perturbation of ray path around the real one, it can be calculated along the unperturbed path instead of the perturbed one (see Problem 13.3). The term t_j is introduced as a source parameter to absorb errors common to all the stations, such as the error in origin time of the seismic event. Note that the sum over k in (12.165) is a sum over all blocks from all layers. The selection from each layer of only one contributing block in $\sum_k$ is achieved by setting $g_{ijk} = 0$ for most (ijk).

There are two sources of nonuniqueness in this problem. One is due to insufficient data, and occurs when two blocks always share common ray paths. This nonuniqueness can be removed by adding ray paths that sample the two blocks separately. The other nonuniqueness is intrinsic to our problem, and cannot be removed by increasing data. This is due to the fact that a uniform perturbation over a layer will change the arrival times at all stations by the same amount and therefore cannot be distinguished from the effect of source parameter t_j.

The determination of m_k and t_j can be separated, because we can eliminate t_j from (12.168) by first taking an average over i for each j and subtracting the average from the individual equation. We have

$$\Delta t_{ij} - \overline{\Delta t_{ij}}^{(j)} = \sum_k (g_{ijk} - \overline{g_{ijk}}^{(j)}) m_k, \qquad (12.166)$$

where $\overline{}^{(j)}$ stands for the average over stations for the jth event. This is the equation (12.83), $\mathbf{d} = \mathbf{Gm}$, of our problem. Because of the intrinsic nonuniqueness mentioned above, the $\mathbf{V}$-space contains as many eigenvectors with zero eigenvalues as the number of layers in the initial model. If there are no additional zero eigenvalues due to insufficient data, we obtain a very simple resolution matrix for the generalized inverse of our problem. Consider the true solution m_k for all the blocks ($k = 1, 2, \ldots, N$) in a layer. Because of the nonuniqueness, we cannot determine the mean value of m_k. Since the generalized inverse gives the minimum solution in the least-squares sense, and since the

least-squares estimate of a mean value is the arithmetic average, the generalized inverse $\hat{m}_k$ can be written as

$$\hat{m}_k = m_k - \frac{\sum\limits_{k=1}^{N} m_k}{N}. \tag{12.167}$$

Thus we find that the resolution matrix $\mathbf{R}$ for the generalized inverse solution of our problem has elements

$$R_{kk} = 1 - \frac{1}{N},$$

$$R_{km} = -\frac{1}{N} \quad \text{(if block numbers } k, m \text{ are in the same layer)}, \tag{12.168}$$

$$R_{km} = 0 \quad \text{(for } k, m \text{ in different layers).}$$

The interlayer resolution is perfect if there are sufficient data. In practice, there are usually several blocks that share common rays with one or two other blocks and cause additional zero eigenvalues. The generalized inverse solution is constructed by the use of eigenvectors with nonzero eigenvalues, as discussed in Section 12.3.2.

The generalized inverse solution is given by

$$\hat{\mathbf{m}} = \mathbf{V}_p \mathbf{\Lambda}_p^{-1} \tilde{\mathbf{U}}_p \mathbf{d}, \tag{12.100 again}$$

showing that $\hat{\mathbf{m}}$ is a linear combination of column vectors of $\mathbf{V}_p$, i.e., eigenvectors with nonzero eigenvalues. The linear coefficient of an eigenvector is proportional to the reciprocal of its eigenvalue. If some of the nonzero eigenvalues are very small, random errors in data can be magnified in the contribution of the corresponding eigenvectors. For example, the damped least-squares solution (12.128) was introduced to suppress the contribution of eigenvectors with eigenvalues smaller than the damping factor ε^2.

It is instructive to find how the eigenvectors look in the model space for large and small eigenvalues. Let us take an example from the work of Husebye et al. (1976) done with the central California array. They used arrival time data from about 100 events recorded at about 30 stations distributed over an area 100×200 km^2. The Earth under the array was divided into five layers of equal thickness 25 km. The top two layers are divided into blocks with side length 25 km, and the bottom three layers into blocks with side length 30 km.

For the above data set and model configuration, they found 205 nonzero eigenvalues. Figures 12.21 and 12.22 show schematically (the actual block size is greater for the bottom three layers) the eigenvectors for the largest and smallest

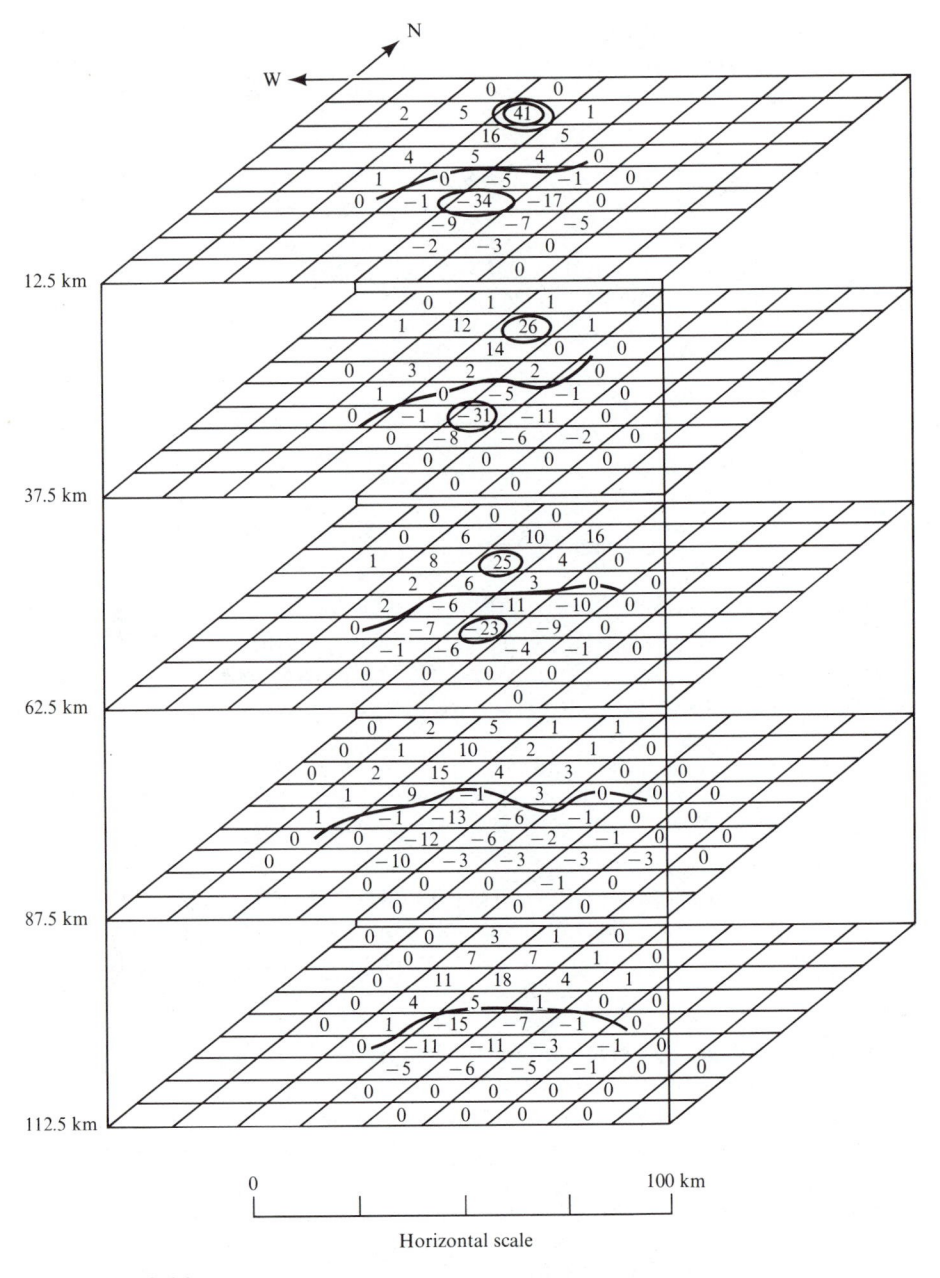

FIGURE **12.21**

Eigenvector for the largest eigenvalue in the case of central California studied by Husebye et al., 1976. Each sheet represents a layer in the initial model, placed at its median depth. The number in each square is the component of the eigenvector for the block, in percent of the initial velocity model. Contours are drawn at 20% intervals. [From Aki, 1977.]

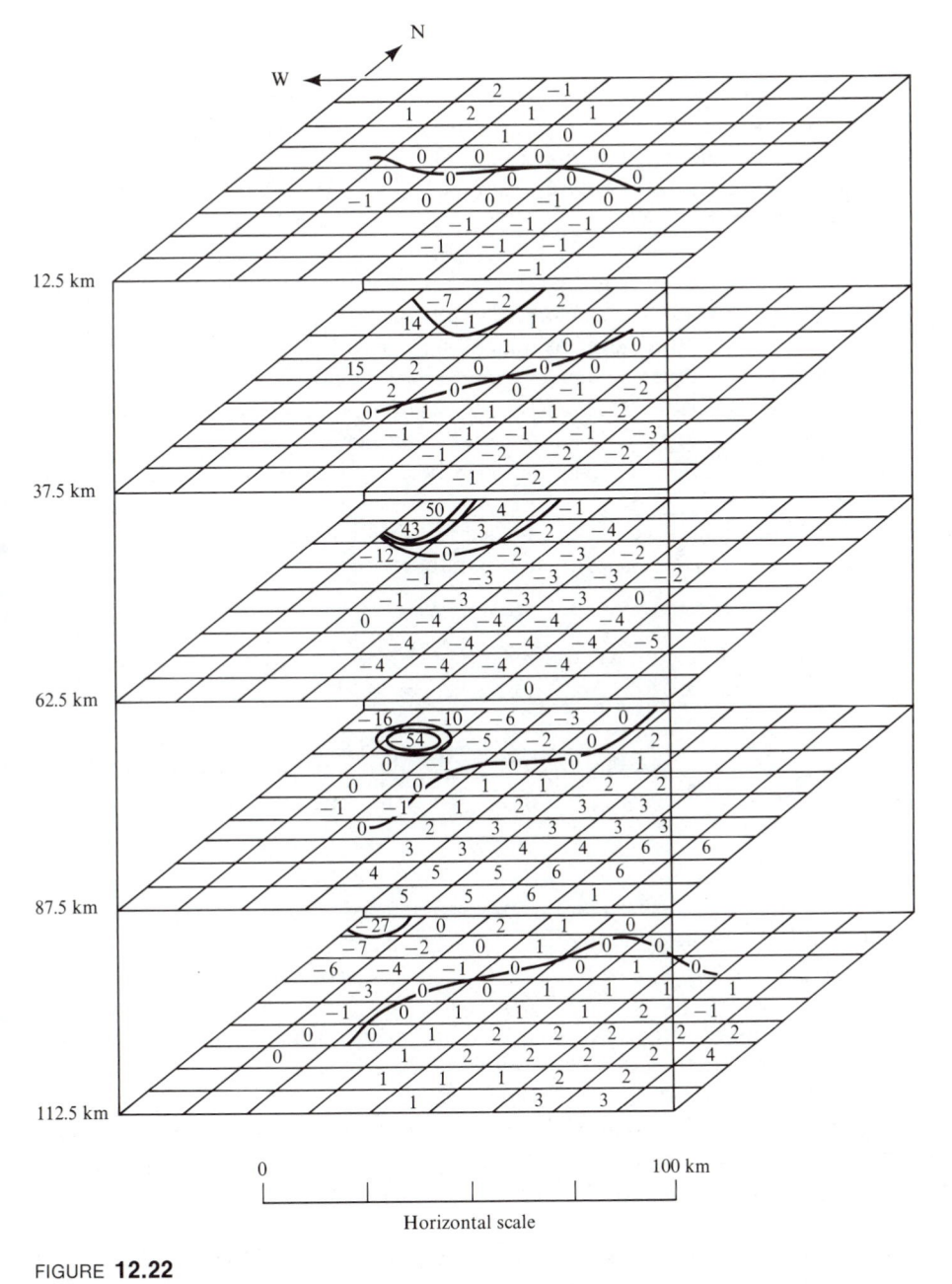

FIGURE **12.22**

Eigenvector for the smallest eigenvalue in the case of central California studied by
Husebye et al., 1976. Otherwise, the legend of Figure 12.21 applies. [From Aki, 1977.]

eigenvalues, respectively. A linear combination of these and 203 more patterns constitute our solution.

The eigenvector corresponding to the largest eigenvalue shows one large positive anomaly and one large negative anomaly in the central area, and the anomaly pattern is roughly the same in all layers. This is the pattern of slowness distribution that can be determined most reliably from our data. On the other hand, the eigenvector corresponding to the smallest eigenvalue shows large values only at a corner, where the sign of the anomaly alternates between neighboring layers. We need this eigenvector if we want to resolve the difference in anomaly pattern at this corner between neighboring layers. Thus we lose vertical resolution in our three-dimensional seismic image if we suppress eigenvectors with small eigenvalues. The effect of random error, however, will cause large fluctuations in the peripheral area if we do not suppress them. To find a compromise solution, we use our *a priori* knowledge about what is a reasonable magnitude of model fluctuation in assigning a proper value of the damping constant as given by (12.127).

Since the generalized inverse $\mathbf{m}_g$ is the minimum solution, the magnitude of true fluctuation $\mathbf{m}$ is always greater than that of $\mathbf{m}_g$:

$$\tilde{\mathbf{m}}\mathbf{m} > \tilde{\mathbf{m}}_g\mathbf{m}_g. \tag{12.169}$$

Our estimate $\hat{\mathbf{m}}$ of the generalized inverse $\mathbf{m}_g$ contains the error $\Delta\mathbf{m}_g$, whose covariance matrix $\langle \Delta\mathbf{m}_g\, \Delta\tilde{\mathbf{m}}_g \rangle$ was given by (12.111). Thus

$$
\begin{aligned}
\langle \tilde{\hat{\mathbf{m}}}\hat{\mathbf{m}} \rangle &= \langle \widetilde{(\mathbf{m}_g + \Delta\mathbf{m}_g)}(\mathbf{m}_g + \Delta\mathbf{m}_g) \rangle \\
&= \langle \tilde{\mathbf{m}}_g\mathbf{m}_g \rangle + \langle \Delta\tilde{\mathbf{m}}_g\, \Delta\mathbf{m}_g \rangle \\
&= \langle \tilde{\mathbf{m}}_g\mathbf{m}_g \rangle + \sigma_d^2\, \mathrm{tr}(\mathbf{V}_p\boldsymbol{\Lambda}_p^{-2}\tilde{\boldsymbol{\Lambda}}_p),
\end{aligned}
$$

where we have used the relation $\tilde{\mathbf{x}}\mathbf{x} = \mathrm{trace}\ \mathbf{x}\tilde{\mathbf{x}}$. Thus we can obtain the lower limit of true slowness fluctuation $\langle \tilde{\mathbf{m}}_g\mathbf{m}_g \rangle$ from the fluctuation of our estimate of the generalized inverse and the standard error σ_d in data. For example, a RMS fluctuation of 3.1% was obtained for $\langle \tilde{\mathbf{m}}_g\mathbf{m}_g \rangle$ in central California.

This method has been applied to many arrays around the world, and has revealed small-scale (20–50 km) velocity anomalies with contrast at least 5% to the depth of 100 km or more under most of the arrays. The velocity anomalies in the crust correlate well with geology and other geophysical data in young active areas, such as central California, Yellowstone, and Hawaii. The correlation was poor in old stable areas, such as eastern Montana and southeastern Norway.

The method is not yet complete, because we have not closed the cycle of our linearization-iteration process. We must modify the initial model according to our solution and then set up a new linearized system of equations on the basis of the revised model. This requires an efficient method of ray tracing in a three-dimensionally inhomogeneous medium. In practice, the iteration may

not be needed in most cases, because the actual slowness variation may be small.

The method can be extended to use data from local earthquakes occurring under the array. In this case, although the number of unknowns increases with the number of earthquakes (four unknowns per single event), the intrinsic nonuniqueness associated with teleseismic data is eliminated. If there are sufficient data, we can uniquely determine both the source and medium parameters. A preliminary study by Aki and Lee (1976) based on a homogeneous initial model demonstrated that part of the low-velocity zone a few kilometers in width between the San Andreas and Calaveras faults can be resolved by the use of microearthquake P-time data recorded by a dense network of stations.

SUGGESTIONS FOR FURTHER READING

Claerbout, J. F., and S. M. Doherty. Downward continuation of moveout corrected seismograms. *Geophysics*, **37**, 741–768, 1972.

Claerbout, J. F., and A. G. Johnson. Extrapolation of time-dependent waveforms along their path of propagation. *Geophysical Journal of the Royal Astronomical Society*, **26**, 285–293, 1971.

Claerbout, J. F. *Fundamentals of Geophysical Data Processing With Applications to Petroleum Prospecting*. New York: McGraw-Hill, 1976.

Colin, L. (editor), *Mathematics of Profile Inversion*. Proceedings of a workshop, NASA Technical Memorandum C-62, 150, August 1972. (Contains many reports of inverse theory as applied in seismology and other branches of physical science.)

Faddeev, L. D. Properties of the S-matrix of the one-dimensional Schrödinger equation. *American Mathematical Society, translations*, Series 2, **65**, 139–166, 1967.

Gilbert, F. Inverse problems for the earth's normal modes. *In* E. C. Robinson (editor), *The Nature of the Solid Earth*. New York: McGraw-Hill, 1972.

Gilbert, F., A. Dziewonski, and J. Brune. An informative solution to a seismological inverse problem. *Proceedings of the National Academy of Sciences*, **70**, 1410–1413, 1973.

Jordan, T. H., and J. Franklin. Optimal solutions to a linear inverse problem in geophysics. *Proceedings of the National Academy of Science*, **68**, 291–293, 1971.

Lanczos, C. *Linear Differential Operators* (Chap. 3). New York: Van Nostrand, 1961.

Minster, J. B., T. H. Jordan, P. Molnar, and E. Haines. Numerical modeling of instantaneous plate tectonics. *Geophysical Journal of the Royal Astronomical Society*, **36**, 541–576, 1974.

Parker, R. L. The inverse problem of electrical conductivity in the mantle. *Geophysical Journal of the Royal Astronomical Society*, **22**, 121–138, 1970.

Wiggins, R. A. General linear inverse problem—Implication of surface waves and free oscillations for earth structure. *Reviews of Geophysics and Space Physics*, **10**, 251–285, 1972.

Wilkinson, J. H. *The Algebraic Eigenvalue Problem.* Oxford: Clarendon Press, 1965.
Zelen, M. Linear estimation and related topics. *In* J. Todd (editor), *A Survey of Numerical Analysis.* New York: McGraw-Hill, 1962.

PROBLEMS

12.1   Using (12.10), show that the velocity distribution is given by $c(z) = c_0 \cosh \pi z / X_c$ when all the rays emerge at the same distance X_c from the source point.

12.2   For $c(z) = c_0 + gz$, show that

$$X(p) = 2 \frac{\sqrt{1 - c_0^2 p^2}}{gp},$$

$$Z(p) = \frac{1}{g}(p^{-1} - c_0).$$

Next consider a low-velocity channel with a constant velocity c_1 with its top at z_1 and bottom at $\bar{z}_1$ in the above medium. Find $X(p)$ for this case. Consider $X(p)$ as observed, and discuss the application of the Gerver-Markushevich method to find bounds for the solution.

12.3   The constraint that $Z(p)$ is a nonincreasing function of p gives the upper bound for the thickness of a low-velocity layer. Show that the condition $dZ(p)/dp = 0$ imposed on equation (12.23) and evaluated at $p \leq p_1$ leads to Slichter's equation (12.16) for maximum thickness of the LVL.

12.4   Consider the case of a homogeneous medium for $x < 0$ and a linearly increasing impedance for $x > 0$. The density $\rho(x)$ and velocity $c(x)$ are expressed as

$$\rho(x) = c(x) = 1 + [(1 + 2x)^{1/2} - 1]H(x),$$

where $H(x)$ is a unit step function. The elements of the scattering matrix for this medium are

$$S_{11}(\omega) = \frac{2i\omega}{2i\omega + 1},$$

$$S_{12}(\omega) = \frac{-1}{2i\omega + 1},$$

and from (12.60) $R(t) = -\frac{1}{2}e^{-t/2}H(t)$. Apply the Gel'fand Levitan equation (12.58) to this reflected wave $R(t)$, and recover the impedance as a function of vertical travel time.

12.5   Using the formulas obtained in Section 12.2.2 for upgoing and downgoing waves, derive the two-point filter (11.24) for removing the effect of the surface layer from the seismogram.

12.6 Construct a synthetic reflection seismogram for a layered medium, and mix with various types of noise in the manner of Koehler and Taner (1977). Apply the inversion algorithms (12.54) and (12.77), and compare the results.

12.7 Suppose that travel time T is observed to depend on distance Δ like $T \propto \sin \lambda \Delta$ for some constant λ. If $r_\oplus =$ Earth's radius, we can introduce constants a and b by the relation

$$T = \frac{2r_\oplus^{1-b}}{a(1-b)} \sin\left[(1-b)\frac{\Delta}{2}\right].$$

Deduce from an analytic inverse theory that the associated velocity profile for this travel time curve is $c = ar^b$. Hint: You will probably need the result

$$\int_0^X \cosh^{-1}\left(\frac{\cos \lambda x}{\cos \lambda X}\right) dx = \frac{\pi}{2\lambda} \log(\sec \lambda X).$$

12.8 Define a linear inverse problem with three unknowns similar to (12.87). Find eigenvalues and eigenvectors of $\mathbf{V}_0$, $\mathbf{V}_p$, $\mathbf{U}_0$, and $\mathbf{U}_p$ spaces. Calculate the generalized inverse by (12.100), resolution matrix by (12.104), and covariance of solution by (12.111).

12.9 For the damped least-squares solution, the resolution matrix and the covariance matrix are given by (12.130) and (12.132), respectively. Putting the diagonal elements of resolution and covariance matrices as R_{ii} and C_{ii}, respectively, show that

$$C_{ii} \le \frac{\sigma_d^2}{\varepsilon^2}(R_{ii} - R_{ii}^2).$$

This relation is due to Ellsworth (1977), and offers a criterion for choosing ε^2 on the basis of variance of solution.

CHAPTER 13

Seismic Waves in
Three-dimensionally Inhomogeneous Media

So far, we have studied body waves, surface waves, and free oscillations in a medium in which the elastic constants and density vary only with depth or with distance from the center of the Earth, except for a ray-theoretical treatment of the forward problem of a general inhomogeneous medium (Section 4.4) and the corresponding inverse problem (Section 12.3.9). With the introduction of the WWSSN and modern array stations with well-calibrated sensors (see Chapter 11 for a historical description of improvement in seismic data), it became immediately clear that the classical Earth model with lateral homogeneity cannot satisfactorily explain many new observations.

For example, the methods described in Chapter 7 as applicable to long-period surface waves break down not only when the travel path includes such an obvious lateral inhomogeneity as the ocean-continent transition zone, but also when the travel distance approaches roughly 50 times (as a rule of thumb) the wavelength, even for paths totally contained in the same ocean or continent. Another example is the apparent variation in free-oscillation period of a given mode with different locations of station or earthquake, reflecting the width of splitting due to lateral heterogeneity. The variation of period for fundamental-mode toroidal oscillations is of the order of 1 to 1.5% for angular orders l between 20 and 40. The variation for fundamental-mode spheroidal oscillations is 0.5 to 1% for $20 < l < 50$. At a modern array station with aperture around

100 km, we find that both the amplitude and the arrival times of teleseismic *P*-waves show a strong fluctuation within the array (an order of magnitude difference in amplitude, and as much as 1 sec variation in arrival time), which cannot be explained by a laterally homogeneous Earth model.

Of course, anybody who has glanced at a geologic map knows that the Earth is strongly inhomogeneous laterally. It was indeed a surprising result that seismologists were able to explain so many observations so well within the framework of a laterally homogeneous Earth model. It must be that the lateral inhomogeneity is weaker than the vertical inhomogeneity in our Earth.

There are many ways to deal with the lateral inhomogeneity. If the inhomogeneity is smooth and the medium properties change little within a wavelength, then ray theory will be most effective.

If the linear dimension of the Earth volume in which seismic scattering occurs is not too large as compared to wavelength, numerical methods such as the finite-difference and finite-element methods will be useful.

The perturbation method is applicable to weak inhomogeneities, depending also on the scale length *a* of the inhomogeneity, the wavelength *λ*, and the travel distance *L*. In a class of problems involving weakly heterogeneous media, we shall find that the eigenfunctions for unperturbed media are effectively use to construct an *ansatz* (trial solution). Rayleigh's method for optical scattering from gratings has been used to study the seismic scattering in a layered medium with irregular interfaces. First-order perturbation theory is able to explain the fluctuation of amplitude and phase of teleseismic *P*-waves for periods longer than 1 sec observed at a large-aperture seismic array.

The problem becomes more complex as the linear dimension *L* of inhomogeneous Earth volume becomes greater as compared to wavelength *λ* and inhomogeneity scale length *a*. We shall describe the problem more explicitly in this chapter in terms of *L*, *λ*, and *a*. The greater L/λ and L/a, the more difficult the problem is to deal with deterministically. The statistical approach, then, may be used to find some average properties of the Earth.

In this chapter, we shall describe the ray-theoretical approach, the perturbation theory, the numerical methods and, the statistical approach, which have all been useful in interpreting observed seismograms.

13.1   Ray-tracing in Inhomogeneous Media

In Chapter 4, we considered motions near a wavefront in a general inhomogeneous isotropic medium and used a trial solution of the form (4.36):

$$\mathbf{u}(\mathbf{x}, t) = \mathbf{U}[t - T(\mathbf{x})]f(\mathbf{x}). \tag{13.1}$$

Putting this *ansatz* into the equation of motion and recognizing that the matrix

of coefficients $\ddot{U}_k f$ must be singular, we derived the eikonal equation (4.41)

$$(\nabla T)^2 = \frac{1}{c^2} = n^2, \tag{13.2}$$

where c is either the local P-wave speed, $\sqrt{[\lambda(\mathbf{x}) + 2\mu(\mathbf{x})]/\rho(\mathbf{x})} = \alpha$, or the local S-wave speed, $\sqrt{\mu(\mathbf{x})/\rho(\mathbf{x})} = \beta$, and n is the corresponding slowness. In optics, n would be the *refractive index*, in units for which $n = 1$ is a vacuum.

The quantity T plays two roles. First, we can say that the equation $T = T(\mathbf{x})$ with T held constant is an implicit equation for a surface, the position of the wavefront at a particular time. As values of the "constant" are increased, the wavefront moves ("propagates") to new positions. On the other hand, rays are fixed in space, and our present interest is in finding their position for a given source in a medium with given $n = n(\mathbf{x})$. The second use of T is then via the equation $\mathbf{x} = \mathbf{x}(T)$, which is the explicit equation for a particular ray as T varies, with T here meaning the travel time along the ray from some particular reference point (usually the source).

From the definition (valid only in isotropic media) of rays as normals to the system of wavefronts, it follows that

$$\mathbf{s} \equiv \nabla T \tag{13.3}$$

is a vector directed along a ray. It is natural to call $\mathbf{s}$ the *slowness vector*, and clearly $|\mathbf{s}| = n$ (from 13.2). Then from (4.42)–(4.47), using $\mathbf{x} = \mathbf{x}(T)$ as the ray equation,

$$\mathbf{s} = n^2 \frac{d\mathbf{x}}{dT} \quad \text{and} \quad n \frac{d}{dT}\left[n^2 \frac{d\mathbf{x}}{dT}\right] = \nabla n. \tag{13.4}$$

Writing these as

$$\frac{dx_i}{dT} = \frac{1}{n^2} s_i, \qquad \frac{ds_i}{dT} = \frac{1}{n}\frac{\partial n}{\partial x_i}, \tag{13.5}$$

we have derived a set of first-order ordinary differential equations for s_i and x_i, which can be solved by standard numerical methods. So far, we have used Cartesian coordinates, which are convenient for source-receiver distances up to about 100 km. But for global problems, we need to work in spherical geometry.

Let us consider spherical coordinates (r, θ, ϕ) and define unit vectors $\hat{\mathbf{r}}$, $\hat{\boldsymbol{\theta}}$, and $\hat{\boldsymbol{\phi}}$ in the directions r, θ, and ϕ, respectively, as shown in Figure 13.1. We

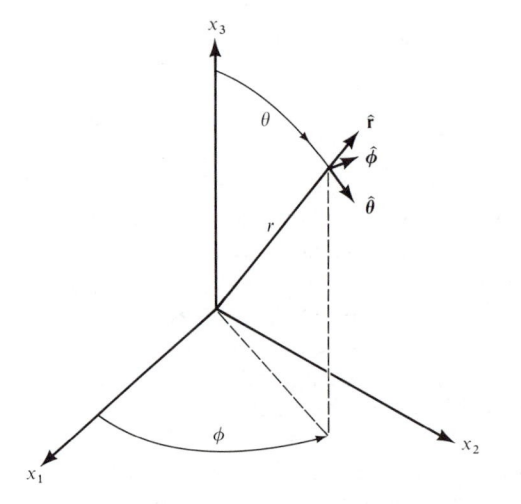

FIGURE **13.1**

Orthogonal unit vectors $(\hat{\mathbf{r}}, \hat{\boldsymbol{\theta}}, \hat{\boldsymbol{\phi}})$ in (r, θ, ϕ) directions of the spherical coordinate.

shall define the direction of a ray path by two angles i and ζ: i is the angle between the ray direction and $\hat{\mathbf{r}}$; ζ is the angle between $\hat{\boldsymbol{\theta}}$ and the projection of the ray path into the $\theta\phi$-plane. Then the first equation of (13.4) is transformed into

$$\frac{dr}{dT} = \frac{1}{n}\cos i,$$

$$r\frac{d\theta}{dT} = \frac{1}{n}\sin i \cos \zeta, \qquad (13.6)$$

$$r\sin\theta\frac{d\phi}{dT} = \frac{1}{n}\sin i \sin \zeta.$$

The transformation of $d\mathbf{s}/dT = n^{-1}\nabla n$ is not so simple, because the spherical coordinate frame is rotated as the point moves along the ray. We must formulate the result in fixed coordinates, such as (x_1, x_2, x_3), using the parameters of spherical coordinates. The components of $\mathbf{s}$ in (x_1, x_2, x_3) and (r, θ, ϕ) coordinates are related by

$$\begin{pmatrix} s_1 \\ s_2 \\ s_3 \end{pmatrix} = \begin{pmatrix} \sin\theta\cos\phi & \cos\theta\cos\phi & -\sin\phi \\ \sin\theta\sin\phi & \cos\theta\sin\phi & \cos\phi \\ \cos\theta & -\sin\theta & 0 \end{pmatrix} \begin{pmatrix} s_r \\ s_\theta \\ s_\phi \end{pmatrix}. \qquad (13.7)$$

Applying the above transformation to both $\mathbf{s}$ and ∇n, we get, for example, the following equation for the x_3-component:

$$\frac{ds_3}{dT} = \frac{dn}{dT}(\cos i \cos \theta - \sin i \cos \zeta \sin \theta)$$

$$+ n(-\sin i \cos \theta - \cos i \cos \zeta \sin \theta)\frac{di}{dT}$$

$$+ n(-\cos i \sin \theta - \sin i \cos \zeta \cos \theta)\frac{d\theta}{dT}$$

$$+ n \sin i \sin \zeta \sin \theta \frac{d\zeta}{dT}$$

$$= \frac{1}{n}\left(\frac{\partial n}{\partial r}\cos \theta - \frac{1}{r}\frac{\partial n}{\partial \theta}\sin \theta\right).$$

From this equation and those for x_1- and x_2-components, and using (13.6) to eliminate dr/dT, $d\theta/dT$, and $d\phi/dT$, we get

$$\frac{di}{dT} = -\sin i\left(\frac{1}{nr} + \frac{1}{n^2}\frac{\partial n}{\partial r}\right)$$

$$+ \frac{\cos i}{r}\left(\cos \zeta \frac{1}{n^2}\frac{\partial n}{\partial \theta} + \frac{\sin \zeta}{\sin \theta}\frac{1}{n^2}\frac{\partial n}{\partial \phi}\right),$$

$$\frac{d\zeta}{dT} = -\frac{\sin \zeta}{r \sin i}\frac{1}{n^2}\frac{\partial n}{\partial \theta}$$

$$+ \frac{\cos \zeta}{\sin i}\frac{1}{r \sin \theta}\frac{1}{n^2}\frac{\partial n}{\partial \phi} - \frac{1}{nr}\sin i \sin \zeta \cot \theta.$$

(13.8)

The above equations were first obtained by Julian (1970), and have been used in determining the effect of three-dimensional structure and also locating earthquake foci in such structure. Figure 13.2 gives an example of ray-tracing results for earthquakes in the mid-Atlantic ridge and in the Aleutian arc.

The ray-tracing equations for surface waves traveling along the surface of a sphere can be obtained as a special case of the equations for body waves. Setting r to be a constant R and restricting $i = \pi/2$ forces a ray to lie on the surface of a sphere with radius R. Then equations (13.6) and (13.8) reduce to

$$\frac{d\theta}{dT} = \frac{1}{nR}\cos \zeta,$$

$$\frac{d\phi}{dT} = \frac{1}{nR \sin \theta}\sin \zeta,$$

(13.9)

$$\frac{d\zeta}{dT} = -\frac{\sin \zeta}{n^2 R}\frac{\partial n}{\partial \theta} + \frac{\cos \zeta}{n^2 R \sin \theta}\frac{\partial n}{\partial \phi} - \frac{1}{nR}\sin \zeta \cot \theta.$$

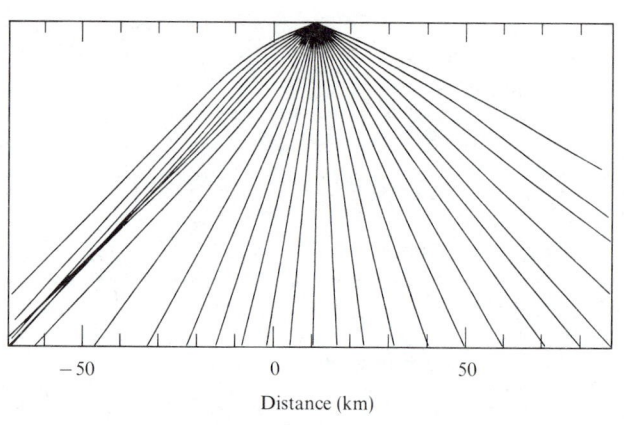

(a)

Distance (km)

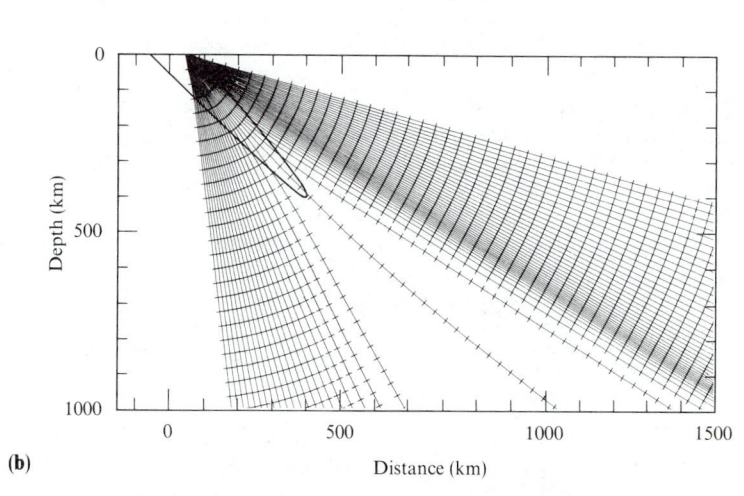

(b)

Depth (km)

Distance (km)

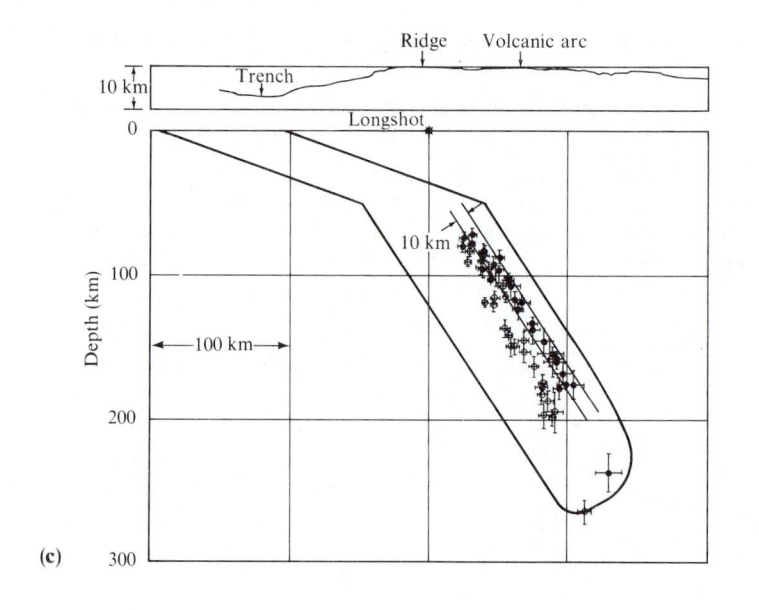

(c)

Depth (km)

Equations (13.9) give the ray path for a given slowness distribution $n(\theta, \phi)$ on the spherical surface.

The calculation of ray paths by equations (13.6) and (13.8) is straightforward when the starting point and initial direction are given. The problem becomes complicated for cases in which the starting point and the ending point are given. There are basically two approaches to this problem. One, called the "shooting method," uses trial values for the initial direction and adjusts them to make the calculated ray paths go as near as may be desired to the given ending point. The other approach is to use trial ray paths with the right starting and ending points and then adjust the path to satisfy the differential equation for the ray trajectory. Since the first approach needs no further explanation, we turn now to a description of the second method, sometimes called the "bending method."

For simplicity, we shall use Cartesian coordinates and specify the ray $\mathbf{x} = \mathbf{x}(\sigma)$, where σ is defined by

$$\frac{dx_i}{d\sigma} = \frac{\partial T}{\partial x_i}. \tag{13.10}$$

For this choice of independent variable, the ray equation (following 4.46) becomes

$$\frac{d^2 x_i}{d\sigma^2} = n \frac{\partial n}{\partial x_i} = \frac{1}{2} \frac{\partial n^2}{\partial x_i}. \tag{13.11}$$

Let $x_i^{(0)}(\sigma)$ be the initial trial path, which starts at a point x_i^s and ends at a point x_i^e. Corresponding values of σ are zero and σ_e, so that

$$x_i^{(0)}(0) = x_i^s \quad \text{and} \quad x_i^{(0)}(\sigma_e) = x_i^e \quad (i = 1, 2, 3). \tag{13.12}$$

FIGURE **13.2**

Examples of the effect (on rays) of laterally varying structure. **(a)** Ray paths from a surface earthquake occurring 10 km away from the axis of a model of a mid-ocean ridge. [From Solomon and Julian, 1974.] **(b)** Ray paths from an earthquake occurring within and near the top edge of a high-velocity slab that dips 45° into a homogeneous half-space. In general, rays within heterogeneous media do not lie in a vertical plane. However, for the two-dimensional structures underlying (a) and (b), rays propagating normal to the strike of ridge and slab do lie in a vertical plane, as shown. [From Toksöz et al., 1971; copyrighted by the American Geophysical Union.] **(c)** Relocation of earthquake hypocenters in the Central Aleutians. Hypocenters have been projected onto a vertical section across a high-velocity slab that extends down to a depth of 250 km. Open circles are earthquake locations, assuming a spherically symmetric Earth model, and solid circles show relocations once the lateral inhomogeneity (the slabs, assumed to have velocity 7% higher than that outside at the same depth) has been taken into account. A shift of about 30 km occurs, and the relocations define a narrow belt of activity near the upper edge of the slab. [From Engdahl, 1973.]

From (13.10) and (13.2) we obtain $|dx| = n\, d\sigma$, so that the value of σ along the trial path is

$$\sigma = \int_0^{x_1^{(0)}} \frac{1}{n} \left[1 + \left(\frac{dx_2^{(0)}}{dx_1^{(0)}} \right)^2 + \left(\frac{dx_3^{(0)}}{dx_1^{(0)}} \right)^2 \right]^{1/2} dx_1^{(0)}. \tag{13.13}$$

We now add a correction term $\delta x_i(\sigma)$ to the trial path so that the sum will satisfy equation (13.11):

$$\frac{d^2}{d\sigma^2} \left[x_i^{(0)} + \delta x_i \right] = \frac{1}{2} \frac{\partial n^2}{\partial x_i} \bigg|_{x_i = x_i^{(0)} + \delta x_i} \qquad (i = 1, 2, 3). \tag{13.14}$$

Expanding the right-hand side into a Taylor series in δx_i and neglecting terms of order higher than the first, we get a linear differential equation for δx_i:

$$\frac{d^2 \delta x_i}{d\sigma^2} - \frac{1}{2} \frac{\partial^2 n^2}{\partial x_j \partial x_i} \bigg|_{x_i = x_i^{(0)}} \delta x_j = \frac{1}{2} \frac{\partial n^2}{\partial x_i} \bigg|_{x_i = x_i^{(0)}} - \frac{d^2 x_i^{(0)}}{d\sigma^2} \qquad (i = 1, 2, 3). \tag{13.15}$$

The coefficients of δx_j and the right-hand side are calculated for the initial path. The above equation for the boundary conditions $\delta x_i(0) = \delta x_i(\sigma_e) = 0$ can be easily solved by a finite-difference method. The revised path can then be used as the second trial path, and the process can be iterated until it converges to a final solution. A version of this method was suggested by Wesson (1971), and Julian and Gubbins (1977) have found it a much faster way of getting the travel time between source and receiver than the shooting method mentioned earlier.

Yang and Lee (1976) reduced the second-order ray equations (13.11) to a set of first-order equations and then solved them using an adaptive finite-difference program written by Lentini and Pereyra (1977), with improved accuracy and less computer time as compared to earlier methods.

In this section, we have emphasized ray tracing for smoothly varying media. The effect of dipping interfaces has been described by Langston (1977), using ray theory, and by Hong and Helmberger (1977), using generalized ray theory. This topic will be discussed in Section 13.4.2.

13.2 Elastic Waves in Weakly Inhomogeneous Media

In the ray-tracing discussed in the preceding section, we confined our attention to the time of arrival of a wavefront. We shall now look at the entire seismogram and consider it to consist of two parts: namely, *primary waves* and

scattered waves. The primary waves are assumed to be known, and would constitute the whole seismogram if the inhomogeneities were absent. The scattered waves are generated by the interaction between primary waves and the inhomogeneities. We shall refer to media with and without inhomogeneities as "perturbed" and "unperturbed."

In this section, the unperturbed medium is a homogeneous, isotropic, unbounded body, and we shall consider plane *P*- and *S*-waves as primary waves. We shall assume a weak inhomogeneity, so that the scattered waves obey wave equations for the unperturbed medium with the source terms determined by the interaction of inhomogeneity and primary waves. The body-force equivalent to a particular scattering source can be easily obtained. As discussed in the introduction to this chapter, the power carried by the scattered waves will be explicitly given in terms of three important lengths involved in any scattering phenomena: the linear dimension L of the inhomogeneous region, the scale length a of the inhomogeneity, and the wavelength. Finally, we shall consider the amplitude and phase fluctuation caused by the scattered waves, with special reference to the teleseismic *P*-wave observations at a large-aperture seismic array.

Let us start with the equation of motion for displacement **u** in a general inhomogeneous, isotropic, elastic body:

$$\rho \ddot{u}_i = (\lambda \nabla \cdot \mathbf{u})_{,i} + [\mu(u_{i,j} + u_{j,i})]_{,j}. \tag{13.16}$$

We shall use ρ_0, λ_0, and μ_0 for the unperturbed homogeneous medium and express those for the perturbed medium as

$$\rho = \rho_0 + \delta\rho,$$

$$\lambda = \lambda_0 + \delta\lambda, \tag{13.17}$$

$$\mu = \mu_0 + \delta\mu.$$

The terms $\delta\rho$, $\delta\lambda$, and $\delta\mu$ are functions of space, but their magnitudes are assumed to be much smaller than the corresponding unperturbed values.

Putting (13.17) into (13.16) and arranging the unperturbed terms on one side and the perturbed one on the other, we get

$$\rho_0 \ddot{u}_i - (\lambda_0 + \mu_0)(\nabla \cdot \mathbf{u})_{,i} - \mu_0 \nabla^2 u_i = -\delta\rho \ddot{u}_i + (\delta\lambda + \delta\mu)(\nabla \cdot \mathbf{u})_{,i} + \delta\mu \nabla^2 u_i + (\delta\lambda)_{,i} \nabla \cdot \mathbf{u} + (\delta\mu)_{,j}(u_{i,j} + u_{j,i}). \tag{13.18}$$

We now write the solution **u** as the sum of "primary waves" $\mathbf{u}^0$ and "scattered waves" $\mathbf{u}^1$:

$$\mathbf{u} = \mathbf{u}^0 + \mathbf{u}^1. \tag{13.19}$$

As defined earlier, $\mathbf{u}^0$ is a solution for the unperturbed medium and satisfies the following equation:

$$\rho_0 \ddot{u}_i^0 - (\lambda_0 + \mu_0)(\nabla \cdot \mathbf{u}^0)_{,i} - \mu_0 \nabla^2 u_i^0 = 0. \tag{13.20}$$

Putting (13.19) into (13.18), subtracting (13.20) from (13.18), assuming $|u_i^1| \ll |u_i^0|$, and neglecting second- and higher-order terms, we obtain the following equation for $\mathbf{u}^1$:

$$\rho_0 \ddot{u}_i^1 - (\lambda_0 + \mu_0)(\nabla \cdot \mathbf{u}^1)_{,i} - \mu_0 \nabla^2 u_i^1 = Q_i \tag{13.21}$$

where

$$Q_i = -\delta\rho \ddot{u}_i^0 + (\delta\lambda + \delta\mu)(\nabla \cdot \mathbf{u}^0)_{,i} + \delta\mu \nabla^2 u_i^0$$
$$+ (\delta\lambda)_{,i}(\nabla \cdot \mathbf{u}^0) + (\delta\mu)_{,j}[u_{i,j}^0 + u_{j,i}^0]. \tag{13.22}$$

Thus (13.21) is nothing but the equation of motion (4.1) in a homogeneous, unbounded, isotropic, elastic medium with body force $\mathbf{Q}$. Let us find which body force corresponds to $\delta\rho$, $\delta\lambda$, $\delta\mu$, $(\delta\lambda)_{,i}$ and $(\delta\mu)_{,j}$ for both primary plane P-waves and S-waves.

13.2.1   Primary plane P-waves

Taking the x_1-direction as the propagation direction of primary plane P-waves, we write

$$u_i^0 = \delta_{1i} \exp[-i\omega(t - x_1/\alpha_0)], \tag{13.23}$$

where $\alpha_0 = \sqrt{(\lambda_0 + 2\mu_0)/\rho_0}$ is the P-wave speed in the unperturbed medium. Putting (13.23) into (13.22), we obtain the corresponding body-force components as

$$Q_1 = \left[\delta\rho\omega^2 - \frac{(\delta\lambda + 2\,\delta\mu)\omega^2}{\alpha_0^2} + i\frac{\omega}{\alpha_0}(\delta\lambda)_{,1} \right.$$
$$\left. + 2i\frac{\omega}{\alpha_0}(\delta\mu)_{,1} \right] \exp[-i\omega(t - x_1/\alpha_0)],$$

$$Q_2 = i\frac{\omega}{\alpha_0}(\delta\lambda)_{,2} \exp[-i\omega(t - x_1/\alpha_0)], \tag{13.24}$$

$$Q_3 = i\frac{\omega}{\alpha_0}(\delta\lambda)_{,3} \exp[-i\omega(t - x_1/\alpha_0)].$$

The first two terms in the right side of the equation for Q_1 can be lumped together as a perturbation in P-wave speed. We have

$$\delta\rho\omega^2 - \frac{(\delta\lambda + 2\,\delta\mu)\omega^2}{\alpha_0^2} = -\omega^2\rho_0\left(-\frac{\delta\rho}{\rho_0} + \frac{\delta\lambda + 2\,\delta\mu}{\lambda_0 + 2\mu_0}\right)$$

$$= -2\omega^2\rho_0\frac{\delta\alpha}{\alpha_0}. \tag{13.25}$$

This term is present only in the x_1-component and acts as a single force in the x_1-direction. We shall refer to this term as "velocity perturbation," and later describe in detail the scattered waves due to this term. A small element of this source will generate scattered far-field P-waves and S-waves, as shown at the top of Figure 13.3.

The term due to $\nabla(\delta\lambda)$, on the other hand, corresponds to three mutually perpendicular dipoles if $\delta\lambda$ is localized in a small region V and vanishes outside of V. Since, in that case,

$$\int_V x_i \frac{\partial(\delta\lambda)}{\partial x_k}\,dV = -\delta_{ik}\int_V \delta\lambda\,dV,$$

the moment tensor will be diagonal, with equal elements proportional to $-\int_V \delta\lambda\,dV$. Thus an element of this source will generate uniform P-waves in all directions, but no S-waves, as shown in the middle of Figure 13.3.

The final term due to $(\delta\mu)_{,1}$ corresponds to the dipole $(1, 1)$, shown in Figure 3.7, if $\delta\mu$ is localized in a small region V and vanishes outside of V. The moment tensor will have only an M_{11} element, which is proportional to $\int_V \delta\mu\,dV$. The radiation pattern of P- and S-waves from an element of this source is depicted at the bottom of Figure 13.3.

Note that there are no scattered far-field S-waves due to primary P-waves in the direction of primary-wave propagation.

13.2.2 Primary plane S-waves

For primary plane S-waves with particle motion in the x_2-direction and propagation in the x_1-direction, we write

$$u_i^0 = \delta_{2i}\exp[-i\omega(t - x_1/\beta_0)], \tag{13.26}$$

where $\beta_0 = \sqrt{\mu_0/\rho_0}$ is the S-wave speed in the unperturbed medium. Putting

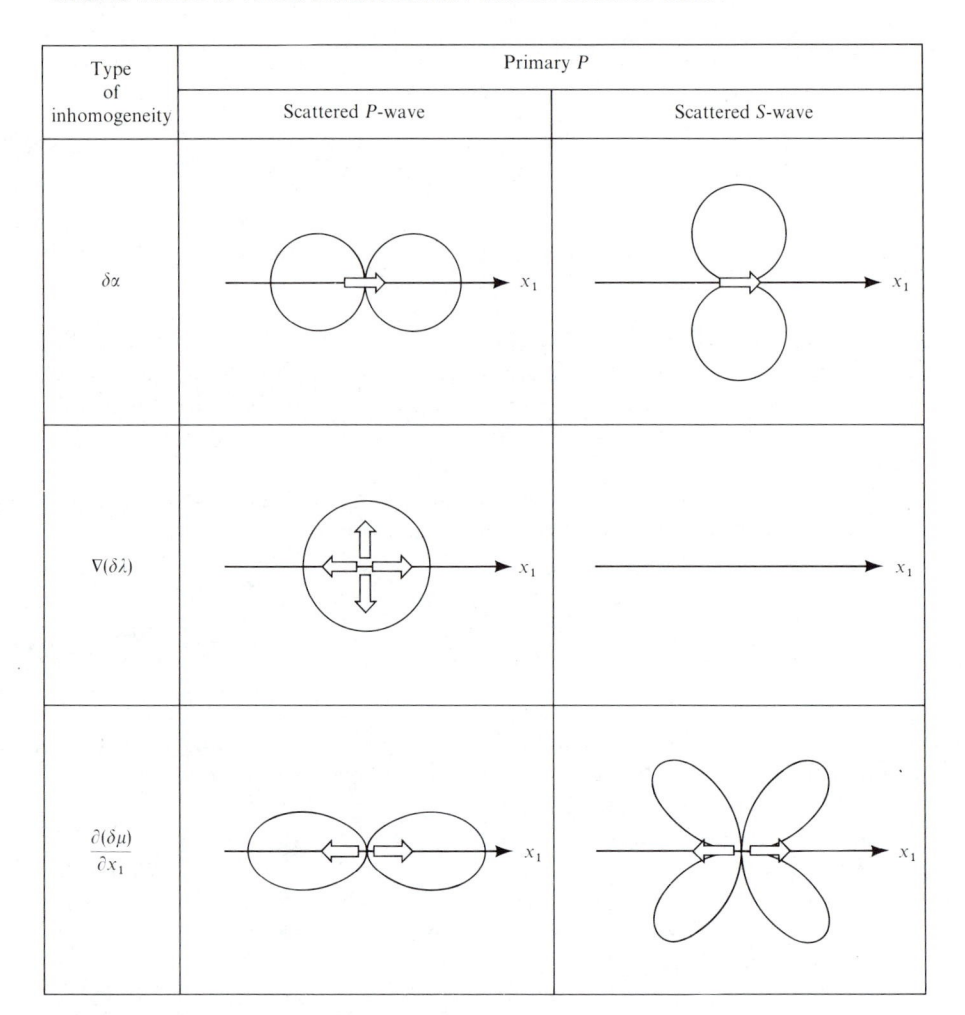

FIGURE **13.3**

The radiation pattern of far-field P- and S-waves scattered from three types of localized inhomogeneity when plane P-waves are incident in the x_1-direction.

(13.26) into (13.22), we obtain the body-force components in the form

$$Q_1 = i \frac{\omega}{\beta_0} (\delta\mu)_{,2} \exp[-i\omega(t - x_1/\beta_0)],$$

$$Q_2 = \left[\delta\rho\omega^2 - \delta\mu \frac{\omega^2}{\beta^2} + i \frac{\omega}{\beta_0} (\delta\mu)_{,1} \right] \exp[-i\omega(t - x_1/\beta_0)], \quad (13.27)$$

$$Q_3 = 0.$$

Again, we can lump the first two terms in the equation for Q_2 as a "velocity perturbation,"

$$\delta\rho\omega^2 - \delta\mu\frac{\omega^2}{\beta^2} = -2\omega^2\rho_0\frac{\delta\beta}{\beta_0}. \tag{13.28}$$

This inhomogeneity corresponds to a single force in the x_2-direction (the direction of particle motion for the primary waves).

The inhomogeneity in λ does not interact with S-waves, and there are no scattering sources involving $\delta\lambda$ or $\nabla\,\delta\lambda$.

The term due to the spatial derivatives of $\delta\mu$, on the other hand, corresponds to a double couple if $\delta\mu$ is localized in a small region V and vanishes outside V. The moment tensor will have nonvanishing elements $M_{12} = M_{21}$ proportional to $\int_V \delta\mu\,dV$.

The radiation patterns for scattered far-field P- and S-waves in the case of the primary plane S-wave are shown in Figure 13.4. Note again that there are no scattered P-waves due to primary S-waves in the direction of primary-wave propagation.

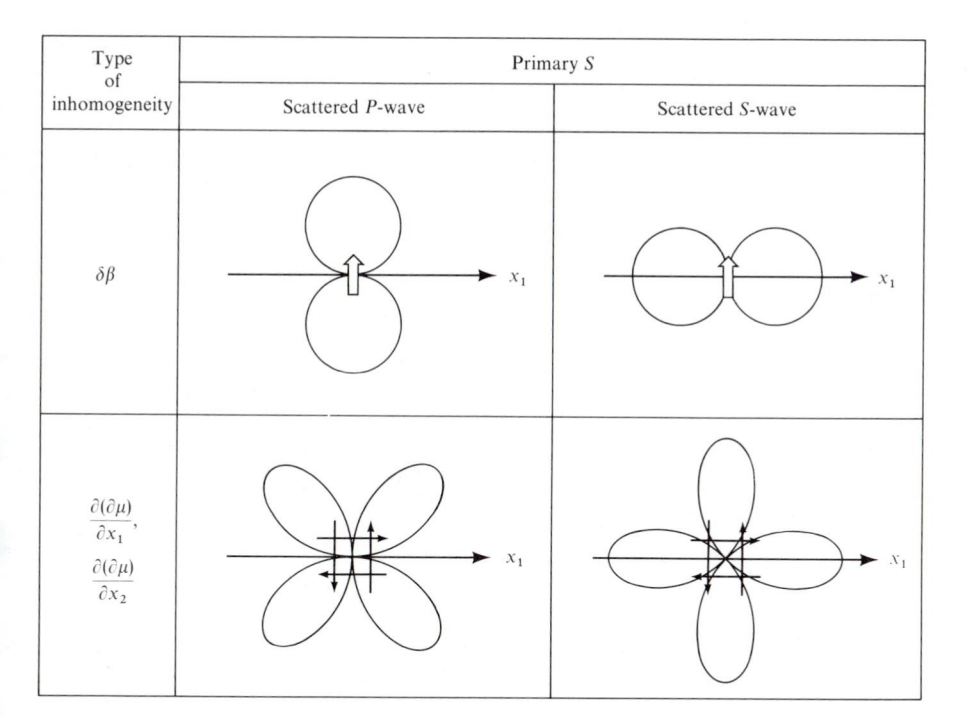

Type of inhomogeneity	Primary S	
	Scattered P-wave	Scattered S-wave
$\delta\beta$		
$\dfrac{\partial(\partial\mu)}{\partial x_1}$, $\dfrac{\partial(\partial\mu)}{\partial x_2}$		

FIGURE **13.4**

The radiation pattern of far-field P- and S-waves scattered from two types of localized inhomogeneity when plane S-waves are incident in the x_1-direction

13.2.3   Solution for scattered waves

Now that we have a good idea about the nature of the body force **Q** responsible for scattered waves, we shall obtain the solution of (13.21). Equation (13.21) can be rewritten as a vector equation,

$$\rho_0 \ddot{\mathbf{u}}^1 - (\lambda_0 + 2\mu_0)\nabla(\nabla \cdot \mathbf{u}^1) + \mu_0\nabla \times (\nabla \times \mathbf{u}^1) = \mathbf{Q}.$$

Taking divergence and curl of this equation, we get a pair of inhomogeneous wave equations, for which we know that the solution is given by the retarded potential (4.6). The wave equations are

$$\nabla \cdot \ddot{\mathbf{u}}^1 - \alpha_0^2\nabla^2(\nabla \cdot \mathbf{u}^1) = \nabla \cdot \mathbf{Q}/\rho_0 \tag{13.29}$$

and

$$\nabla \times \ddot{\mathbf{u}}^1 - \beta_0^2\nabla^2(\nabla \times \mathbf{u}^1) = \nabla \times \mathbf{Q}/\rho_0. \tag{13.30}$$

Their solutions are

$$\nabla \cdot \mathbf{u}^1(\mathbf{x}, t) = \frac{1}{4\pi\alpha_0^2\rho_0} \int_V \frac{1}{|\mathbf{x} - \boldsymbol{\xi}|} \nabla \cdot \mathbf{Q}\left(\boldsymbol{\xi}, t - \frac{|\mathbf{x} - \boldsymbol{\xi}|}{\alpha_0}\right) dV(\boldsymbol{\xi}) \tag{13.31}$$

and

$$\nabla \times \mathbf{u}^1(\mathbf{x}, t) = \frac{1}{4\pi\beta_0^2\rho_0} \int_V \frac{1}{|\mathbf{x} - \boldsymbol{\xi}|} \nabla \times \mathbf{Q}\left(\boldsymbol{\xi}, t - \frac{|\mathbf{x} - \boldsymbol{\xi}|}{\beta_0}\right) dV(\boldsymbol{\xi}), \tag{13.32}$$

where V is the inhomogeneous region in which **Q** is nonzero.

   In order to check the applicability of the above formula, we shall consider a simple case of known solution. Suppose that plane P-waves are incident normally upon an infinite slab of constant thickness d and uniform velocity perturbation $\delta\alpha$ (see Fig. 13.5). We expect that, after passage through the slab, the total waves (primary plus scattered) will show the time delay given by

$$\frac{d}{\alpha_0 + \delta\alpha} - \frac{d}{\alpha_0} \sim -\frac{\delta\alpha d}{\alpha_0^2}$$

relative to primary waves in the absence of a slab.

   Since the body-force equivalents due to the gradients of $\delta\lambda$ and $\delta\mu$ at one face of the slab have opposite signs from those at the other, their effects approximately cancel in the forward direction, and may be neglected in the present problem. Then, for primary waves $u_i^0 = \delta_{1i}\exp[-i\omega(t - x_1/\alpha_0)]$, and from

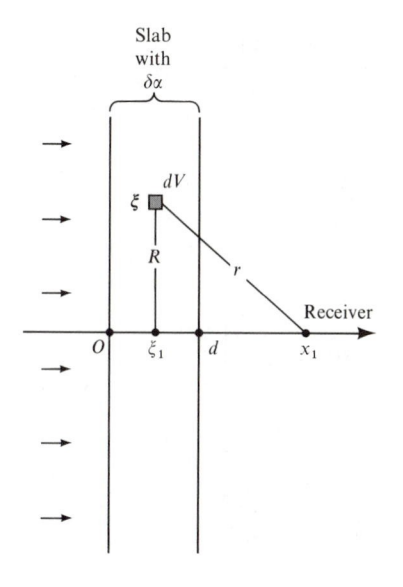

FIGURE **13.5**

Plane P-waves are incident upon a slab with uniform
velocity perturbation $\delta\alpha$.

(13.24) and (13.25), we have

$$
\nabla \cdot \mathbf{Q} = \frac{\partial}{\partial x_1}\left[-2\omega^2 \rho_0 \frac{\delta\alpha}{\alpha_0} \exp[-i\omega(t - x_1/\alpha_0)] \right]
$$

$$
\sim -2\omega^2 \rho_0 \frac{\delta\alpha}{\alpha_0} i \frac{\omega}{\alpha_0} \exp[-i\omega(t - x_1/\alpha_0)], \tag{13.33}
$$

where we again omit the term involving the gradient of velocity perturbation.
Putting (13.33) into (13.31), we obtain the dilatation for scattered waves as

$$
\nabla \cdot \mathbf{u}^1(\mathbf{x}, t) \sim \frac{-1}{4\pi\alpha_0^2} \int_V \frac{1}{r} 2\omega^2 \frac{\delta\alpha}{\alpha_0} i \frac{\omega}{\alpha_0} \exp\left[-i\omega\left(t - \frac{r}{\alpha_0} - \frac{\xi_1}{\alpha_0} \right) \right] dV(\boldsymbol{\xi}). \tag{13.34}
$$

Using variables defined in Figure 13.5, we note that

$$
R^2 = r^2 - (x_1 - \xi_1)^2, \text{ and hence } dV(\boldsymbol{\xi}) = 2\pi R\, dR\, d\xi_1 = 2\pi r\, dr\, d\xi_1.
$$

Thus

$$
\nabla \cdot \mathbf{u}^1(\mathbf{x}, t) \sim \frac{-1}{\alpha_0^2} \int_0^d \int_{x_1 - \xi_1}^\infty \omega^2 i \frac{\delta\alpha \cdot \omega}{\alpha_0^2} \exp\left[-i\omega\left(t - \frac{r}{\alpha_0} - \frac{\xi_1}{\alpha_0} \right) \right] dr\, d\xi_1
$$

$$
= -i \frac{\omega^3}{\alpha_0^2} \frac{\delta\alpha}{\alpha_0^2} \int_0^d \exp\left[-i\omega\left(t - \frac{\xi_1}{\alpha_0} \right) \right] d\xi_1 \int_{x_1 - \xi_1}^\infty \exp(i\omega r/\alpha_0)\, dr.
$$

Since the contribution from $r = \infty$ should be negligible (α_0^{-1} has a small positive imaginary part),

$$\int_{x_1 - \xi_1}^{\infty} e^{i\omega r/\alpha_0} \, dr = -\frac{\alpha_0}{i\omega} \exp\left[i\omega \frac{(x_1 - \xi_1)}{\alpha_0} \right].$$

Then we have

$$\nabla \cdot \mathbf{u}^1(\mathbf{x}, t) = \omega^2 \frac{\delta\alpha}{\alpha_0^3} d \exp\left[-i\omega\left(t - \frac{x_1}{\alpha_0} \right) \right]. \tag{13.35}$$

The dilatation for primary waves was

$$\nabla \cdot \mathbf{u}^0(\mathbf{x}, t) = i\frac{\omega}{\alpha_0} \exp\left[-i\omega\left(t - \frac{x_1}{\alpha_0} \right) \right].$$

Therefore, the total wave is

$$
\begin{aligned}
\nabla \cdot (\mathbf{u}^1 + \mathbf{u}^0) &= i\frac{\omega}{\alpha_0}\left[1 - i\omega \frac{\delta\alpha d}{\alpha_0^2} \right] \exp\left[-i\omega\left(t - \frac{x_1}{\alpha_0} \right) \right] \\
&\sim i\frac{\omega}{\alpha_0} \exp\left[-i\omega\left(t - \frac{x_1}{\alpha_0} + \frac{\delta\alpha \cdot d}{\alpha_0^2} \right) \right],
\end{aligned}
\tag{13.36}
$$

showing that an infinite slab of thickness d and uniform velocity perturbation $\delta\alpha$ gives rise to a time delay of

$$-\frac{\delta\alpha d}{\alpha_0^2},$$

as expected.

The exact solution of the above problem is easy to obtain, and the dilatation of transmitted waves is given by

$$\nabla \cdot \mathbf{u} = \frac{i\omega}{\alpha_0} \frac{4\alpha_0\rho_0\alpha_1\rho_1 \exp\left[-i\omega\left(t - \frac{x_1}{\alpha_0} + \frac{d}{\alpha_0} \right) \right]}{(\alpha_0\rho_0 + \alpha_1\rho_1)^2 e^{-i\omega d/\alpha_1} - (\alpha_0\rho_0 - \alpha_1\rho_1)^2 e^{i\omega d/\alpha_1}},$$

where α_1 and ρ_1 are, respectively, the wave velocity and density of the slab. The above exact solution reduces to (13.36) if we neglect the second-order terms in

impedance contrast, $\alpha_0 \rho_0 - \alpha_1 \rho_1$, and velocity contrast. Thus, to the first order in parameter perturbation, the phase of transmitted waves is affected by the slab, but the amplitude is not.

13.3 Scattering due to Velocity Perturbation

As shown in the preceding section, an infinite slab (of finite thickness) with constant velocity perturbation produced a change in the phase delay of normally incident plane waves but did not affect their amplitude in the forward direction. We shall now proceed to the case in which the velocity varies in all directions with a finite scale length, and consider only a scalar wave Φ (e.g., $\nabla \cdot \mathbf{u}$). For simplicity, we shall again neglect the effect of spatial gradients in medium parameters and concentrate on the effect of velocity perturbation alone. This is justified if the inhomogeneity is smooth within the wavelength. In this case, for primary waves of the form

$$\Phi^0 = A \exp[-i\omega(t - x_1/c_0)], \tag{13.37}$$

the scattered waves are given by (see (13.34))

$$\Phi^1(\mathbf{x}, t) = \frac{A\omega^2}{2\pi c_0^2} \int_V \left(-\frac{\delta c}{c_0}\right) \frac{\exp\left[-i\omega\left(t - \frac{r}{c_0} - \frac{\xi_1}{c_0}\right)\right]}{r} dV(\xi), \tag{13.38}$$

where c is the wave velocity, α or β, V is the inhomogeneous region in which $\delta c \neq 0$, and $r = |\mathbf{x} - \xi|$.

In order to define the scale length of inhomogeneity clearly, we consider a random medium and describe the spatial fluctuation $-\delta c/c_0$ of slowness by its autocorrelation function. Putting

$$\mu = -\delta c/c_0, \tag{13.39}$$

we shall assume that the fluctuation of μ is isotropic and stationary in space. We then define the normalized autocorrelation function by

$$N(\mathbf{r}) = \frac{\langle \mu(\mathbf{r}')\mu(\mathbf{r}' + \mathbf{r})\rangle}{\langle \mu^2 \rangle}. \tag{13.40}$$

Following Chernov (1960), we shall consider the cases $N(\mathbf{r}) = e^{-|\mathbf{r}|/a}$ and $N(\mathbf{r}) = e^{-|\mathbf{r}|^2/a^2}$, where a is the measure of scale length of inhomogeneity, called the *correlation distance*.

BOX **13.1**

Statistical model of inhomogeneities in a medium

The correlation function $\langle \mu(\mathbf{r}_1)\mu(\mathbf{r}_2) \rangle$ can be easily understood if $\mu(\mathbf{r})$ is a random function of time, such as the density of air when temperature fluctuates. In that case, the averaging operation $\langle \quad \rangle$ may be taken over many statistically independent time samples. In the solid Earth, however, $\mu(\mathbf{r}_1)$ is a fixed function of space at least for the durations of seismic experiments. How, then, do we define a statistical average when the variable is actually fixed in space?

To answer this question, let us consider measurements of densities of rock samples collected from an area. Since the density will vary from sample to sample, one can construct a distribution function $f(\rho)$ for density, i.e., the probability of finding density in the range $(\rho - \frac{1}{2}\Delta\rho, \rho + \frac{1}{2}\Delta\rho)$ is $f(\rho)\,\Delta\rho$. From such a distribution function we can get the average value for the density of rocks in the area, their variance, etc. These statistical properties give some useful information about the density of rocks in the area. The density of a rock sample becomes a statistical variable because we choose not to be bothered by the specific location from which the sample came. We are interested only in a rough description of the density of rocks in the area.

The same argument may be applied to the slowness fluctuation $\mu(\mathbf{r})$. We assume that the fluctuation of $\mu(\mathbf{r})$ has stationary statistical properties within the volume considered. Then we can define the correlation function as an average of $\mu(\mathbf{r}_1)\mu(\mathbf{r}_1 + \mathbf{r})$ over the independent sample points $\mathbf{r}_1$. This function will describe the magnitude and smoothness of the fluctuation of $\mu(\mathbf{r})$, which may be used in estimating the energy of waves scattered from the volume or the statistical properties of the fluctuation in amplitudes and arrival times of observed waves.

13.3.1 Power spectra of scattered waves

Let us first study the scattered waves observed at a distance far away from an inhomogeneous region confined in a small volume V with linear dimension L, as shown in Figure 13.6. In order to evaluate the integral in (13.38), we shall use the same approximation as used for the calculation of the far-field term from a finite seismic source (14.11), namely,

$$r = (|\mathbf{x}|^2 + |\boldsymbol{\xi}|^2 - 2\mathbf{x} \cdot \boldsymbol{\xi})^{1/2}$$
$$\sim |\mathbf{x}| - \mathbf{n} \cdot \boldsymbol{\xi}, \tag{13.41}$$

where $\mathbf{n}$ is the unit vector in the direction of $\mathbf{x}$. This approximation is valid if $kL^2/(2|\mathbf{x}|) \ll \pi/2$. Putting (13.41) into the integrand exponent in (13.38) and replacing $1/r$ by $1/|\mathbf{x}|$, we get

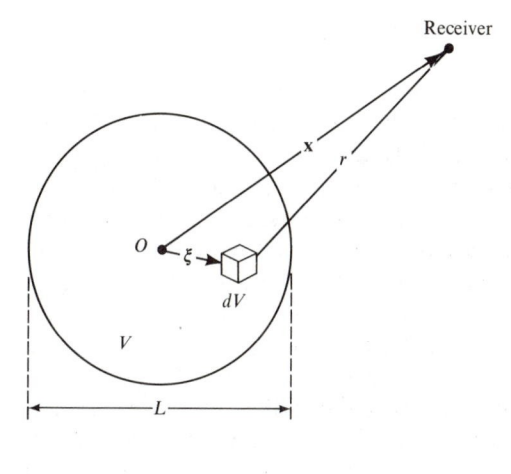

FIGURE **13.6**

A volume element $dV(\xi)$ in the inhomogeneity V is located at ξ from origin O. The distance between volume element and station is $r = |\mathbf{x} - \xi|$.

$$\Phi^1(\mathbf{x}, t) = \frac{Ak^2 \exp[-i(\omega t - k|\mathbf{x}|)]}{2\pi|\mathbf{x}|} \int_V \mu(\xi) \exp[ik(\xi_1 - \mathbf{n} \cdot \xi)] \, dV(\xi), \quad (13.42)$$

where we put $k = \omega/c_0$.

Let us now estimate the power carried by scattered waves. Since $|\Phi^1|^2$ is equal to the product of Φ^1 and its complex conjugate, we have

$$|\Phi^1|^2 = \frac{A^2 k^4}{(2\pi)^2 |\mathbf{x}|^2} \int_V \int_V \mu(\xi')\mu(\xi)$$

$$\times \exp\{ik[\xi_1 - \xi_1' - \mathbf{n} \cdot (\xi - \xi')]\} \, dV(\xi) \, dV(\xi'). \quad (13.43)$$

Expressing the unit vector in the ξ_1-direction as $\mathbf{e}_1$, we define $\mathbf{K} = \mathbf{e}_1 - \mathbf{n}$. From Figure 13.7, it is easy to see that $|\mathbf{K}|^2 = 1 + 1 - 2\cos\theta = 4\sin^2(\theta/2)$, hence $|\mathbf{K}| = 2\sin(\theta/2)$. Taking the statistical average of (13.43) using (13.40), we get

$$\langle |\Phi^1|^2 \rangle = \frac{A^2 k^4 \langle \mu^2 \rangle}{(2\pi)^2 |\mathbf{x}|^2} \int_V \int_V N(\xi - \xi')$$

$$\times \exp[ik\mathbf{K} \cdot (\xi - \xi')] \, dV(\xi) \, dV(\xi'). \quad (13.44)$$

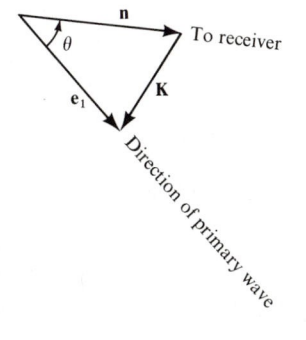

FIGURE **13.7**

Relation between $\mathbf{n}$, the unit vector in the direction to the receiver, and $\mathbf{e}_1$, the unit vector in the propagation direction of the primary wave.

This is the first relation we have obtained between statistical properties of inhomogeneities in the medium (described by $N(\xi - \xi')$; see Box 13.1) and statistical properties of the scattered waves. To obtain an understanding of the symbols $\langle \ \ \rangle$ in $\langle |\Phi^1|^2 \rangle$, we imagine that within the volume V there can be placed very many different velocity perturbations μ, each having the same autocorrelation N. For each sample perturbation, we can write down a deterministic expression like (13.43). But we do not in fact know $\mu(\xi')\mu(\xi)$ in detail: we know only the average $\langle \mu(\xi')\mu(\xi) \rangle = N(\xi - \xi')\langle \mu^2 \rangle$. Thus $\langle |\Phi^1|^2 \rangle$ means the average value of $|\Phi^1|^2$ given deterministically by (13.43) for each of the velocity perturbations in our thought experiment.

A particular observation of scattered waves in the Earth is a measure of some deterministic integral like (13.43), because the Earth imposes its unique values for the fluctuation appearing in the integrand. It will not then be very useful to make just one comparison between $|\Phi^1|^2$ as observed in the Earth and the theory we are developing for $\langle |\Phi^1|^2 \rangle$. But if very many observations are made of scattered waves, our assumption will be that statistical properties of the *observations* will be explained by the properties of the statistical average $\langle |\Phi^1|^2 \rangle$.

To evaluate the above integral (13.44), we change the variables ξ and ξ' to the relative coordinate $\hat{\xi} = \xi - \xi'$ and use the center-of-mass coordinates $\bar{\xi} = (\xi + \xi')/2$. Since the integrand is not a function of $\bar{\xi}$, and since $\int_V d\bar{\xi}_1 \, d\bar{\xi}_2 \, d\bar{\xi}_3 = V$, we have

$$\langle |\Phi^1|^2 \rangle = \frac{A^2 k^4 \langle \mu^2 \rangle V}{(2\pi)^2 |\mathbf{x}|^2} \int_V N(\hat{\xi}) \exp(ik\mathbf{K} \cdot \hat{\xi}) \, d\hat{\xi}_1 \, d\hat{\xi}_2 \, d\hat{\xi}_3. \qquad (13.45)$$

Changing $(\hat{\xi}_1, \hat{\xi}_2, \hat{\xi}_3)$ to the spherical coordinates (r', θ', ϕ'), with $\mathbf{K}$ as the polar axis, we write

$$r' = |\hat{\xi}|,$$

$$\mathbf{K} \cdot \hat{\xi} = |\mathbf{K}| r' \cos \theta'$$

$$d\hat{\xi}_1 \, d\hat{\xi}_2 \, d\hat{\xi}_3 = r'^2 \, dr' \sin \theta' \, d\theta' \, d\phi'.$$

Then we have

$$\int_V N(\hat{\xi}) \exp(ik\mathbf{K}\hat{\xi}) \, d\hat{\xi}_1 \, d\hat{\xi}_2 \, d\hat{\xi}_3$$

$$= \int_V N(r') \exp(ik|\mathbf{K}| \cdot r' \cos \theta') r'^2 \, dr' \sin \theta' \, d\theta' \, d\phi'$$

$$= 4\pi \int_0^\infty N(r') \frac{\sin(k|\mathbf{K}|r')}{k|\mathbf{K}|} r' \, dr', \qquad (13.46)$$

where the integration limit for r' is extended to infinity, assuming that the correlation distance a is much smaller than the linear dimension of V.

The above integral can be evaluated in a compact form for the cases $N(r) = e^{-r/a}$ and $N(r) = e^{-r^2/a^2}$. Putting the result into (13.45), we obtain

$$\langle |\Phi^1|^2 \rangle = \frac{2A^2 k^4 \langle \mu^2 \rangle a^3 V}{\pi |\mathbf{x}|^2}$$

$$\times \frac{1}{\left(1 + 4k^2 a^2 \sin^2 \dfrac{\theta}{2}\right)^2} \qquad \text{for } N(r) = e^{-r/a} \qquad (13.47)$$

and

$$\langle |\Phi^1|^2 \rangle = \frac{A^2 k^4 \langle \mu^2 \rangle a^3 V}{4\sqrt{\pi} |\mathbf{x}|^2}$$

$$\times \exp\left(-k^2 a^2 \sin^2 \frac{\theta}{2}\right) \qquad \text{for } N(r) = e^{-r^2/a^2}. \qquad (13.48)$$

The formula (13.47) was first obtained by Pekeris (1947), though here we have followed the derivation given by Chernov (1960).

In either case, the power of scattered waves is proportional to k^4 when $ka \ll 1$. This is the well-known Rayleigh scattering. If ka is small, the scattered power does not depend on θ. Thus the velocity perturbation with scale length smaller than a wavelength produces an isotropic scattering. However, when ka is small, the gradients of velocity and elasticity perturbation neglected in the present analysis become important, and their effects are directional, as described in Section 13.3.3.

On the other hand, when ka is large, the scattering due to velocity perturbation becomes directed forward, and the scattered power is concentrated within an angle $(ka)^{-1}$ around the direction of primary-wave propagation ($\theta = 0$). Back-scattering power ($\theta = \pi$) becomes very small, especially for the case $N(r) = e^{-r^2/a^2}$.

13.3.2 Energy loss of primary waves during the passage through an inhomogeneous cube of side L

The energy flow carried by the primary waves (13.45) through a face of a cube of side L is proportional to $L^2 |\Phi_0|^2$. Taking γ as the proportionality constant, the energy flow I is given by

$$I = \gamma L^2 |\Phi_0|^2. \qquad (13.49)$$

The total energy flow carried away by the scattered waves can be evaluated as

$$\Delta I = \gamma \int_S \langle |\Phi^1|^2 \rangle \, dS, \tag{13.50}$$

where S is the spherical surface with radius R enclosing the inhomogeneous cube at a far distance. Using the angle θ defined in Figure 13.7, the surface element dS can be written as

$$dS = 2\pi R^2 \sin \theta \, d\theta$$

$$= 2\pi R^2 \cdot 2 \sin \frac{\theta}{2} \cos \frac{\theta}{2} \, d\theta$$

$$= 2\pi R^2 |\mathbf{K}| \cos \frac{\theta}{2} \, d\theta. \tag{13.51}$$

Putting (13.45), (13.46), and (13.51) into (13.50), we find

$$\Delta I = \gamma 2L^3 A^2 k^2 \langle \mu^2 \rangle \int_0^\infty N(r')[1 - \cos 2kr'] \, dr'. \tag{13.52}$$

For $N(r) = e^{-r/a}$, the above formula gives the fractional loss of energy as

$$\frac{\Delta I}{I} = \frac{8 \langle \mu^2 \rangle k^4 a^3 L}{1 + 4k^2 a^2} \tag{13.53}$$

and, for $N(r) = e^{-r^2/a^2}$,

$$\frac{\Delta I}{I} = \sqrt{\pi} \langle \mu^2 \rangle k^2 a L (1 - e^{-k^2 a^2}). \tag{13.54}$$

In deriving the scattering formulas (13.47) and (13.48), we made an assumption that the primary waves Φ^0 are unchanged during their propagation through the inhomogeneous region. Much of this work was pioneered by Max Born, and the above assumption is known as the *Born approximation*. It violates the energy-conservation law, because the scattered energy carries the fraction of primary wave energy given by equations (13.53) and (13.54). Thus our scattering formula should be applicable to cases in which $\Delta I/I$ is very small. The condition can be simplified as (e.g., for the case $N(r) = e^{-r/a}$)

$$2 \langle \mu^2 \rangle k^2 a L \ll 1, \qquad \text{for } ka \gg 1$$

and   (13.55)

$$8 \langle \mu^2 \rangle k^4 a^3 L \ll 1, \qquad \text{for } ka \ll 1.$$

13.3.3 *Contributions from gradients in velocity and elasticity constants.*

For small ka, the scattering formulas derived above may be inadequate, because we neglected the spatial gradient of velocity and elastic constants in deriving the integral expression (13.38) for scattered waves. If one keeps the spatial gradient of velocity perturbation in (13.33), then (13.38) should be replaced by

$$\Phi^1(x, t) = \frac{A\omega^2}{2\pi c_0^2} \int_V \left[-\frac{\delta c}{c_0} + \frac{ic_0}{\omega} \frac{\partial}{\partial \xi_1} \left(\frac{\delta c}{c_0} \right) \right] \exp\left[-i\omega \left(t - \frac{r}{c_0} - \frac{\xi_1}{c_0} \right) \right] \frac{1}{r} \, dV(\xi).$$
$$(13.56)$$

Taking the average of the product of Φ^1 and its complex conjugate, $\langle |\Phi^1|^2 \rangle$ can be expressed as an integral of the form (13.45), in which

$$\left\langle \mu(\xi') \frac{\partial \mu(\xi)}{\partial \xi_1} \right\rangle = \frac{\partial}{\partial \xi_1} \langle \mu(\xi')\mu(\xi) \rangle = \langle \mu^2 \rangle \frac{\partial N(\hat{\xi})}{\partial \xi_1}$$

and

$$\left\langle \frac{\partial \mu(\xi')}{\partial \xi_1'} \frac{\partial \mu(\xi)}{\partial \xi_1} \right\rangle = \frac{\partial^2}{\partial \xi_1 \partial \xi_1'} \langle \mu(\xi')\mu(\xi) \rangle = -\langle \mu^2 \rangle \frac{\partial^2 N(\hat{\xi})}{\partial \hat{\xi}_1^2}$$

appear. These terms in the integrand can be easily evaluated by integration by parts, assuming that $N(\hat{\xi})$ and its derivatives vanish on the surface of inhomogeneous region V, the dimensions of which are large compared to the correlation distance a. Noting that the ξ_1-component of $\mathbf{K}$, K_1, is equal to $2 \sin^2 (\theta/2)$, we obtain a result that is simply an addition of a factor of $\cos^2 \theta$ to (13.45), i.e.,

$$\langle |\Phi^1|^2 \rangle = \frac{A^2 k^4 \langle \mu^2 \rangle \cos^2 \theta \cdot V}{(2\pi)^2 |\mathbf{x}|^2} \int_V N(\hat{\xi}) \exp[i k \mathbf{K} \cdot \hat{\xi}] \, d\hat{\xi}_1 \, d\hat{\xi}_2 \, d\hat{\xi}_3. \quad (13.57)$$

This result may be expected from the equivalent body forces shown in Figures 13.3 and 13.4, because in both P-to-P and S-to-S scattering, the localized velocity inhomogeneity acts like a single force, and their amplitude radiation patterns are described by $\cos \theta$. We expect similar simple corrections for including the effect of $\nabla(\delta\lambda)$ and $\nabla(\delta\mu)$. In fact, Haddon has carried out such a calculation for P-to-P scattering, and obtained the result

$$\langle |\Phi^1|^2 \rangle = \frac{A^2 k^4 V}{(2\pi)^2 |\mathbf{x}|^2} \left(\frac{\gamma_\lambda}{2} - \frac{\gamma_\rho \cos \theta}{2} + \frac{\gamma_\mu \cos^2 \theta}{2} \right)^2 \int_V N(\hat{\xi})$$
$$\times \exp[i k \mathbf{K} \cdot \hat{\xi}] \, d\hat{\xi}_1 \, d\hat{\xi}_2 \, d\hat{\xi}_3. \quad (13.58)$$

where the perturbations $\delta\rho$, $\delta\mu$, and $\delta\lambda$ are assumed to share the same spatial distribution function $H(\xi)$, with the normalized spatial autocorrelation function $N(\hat{\xi})$:

$$\frac{\delta\rho}{\rho_0} = \gamma_\rho H(\xi),$$

$$\frac{\delta\lambda}{\lambda_0 + 2\mu_0} = \gamma_\lambda H(\xi),$$

$$\frac{2\,\delta\mu}{\lambda_0 + 2\mu_0} = \gamma_\mu H(\xi),$$

$$N(\hat{\xi}) = \langle H(\xi)H(\xi + \hat{\xi})\rangle.$$

(13.59)

The θ independence of $\delta\lambda$ and $\cos^2\theta$ dependence of $\delta\mu$ shown in (13.58) are consistent, respectively, with the isotropic source and dipole source shown in Figure 13.3.

For a large ka, we may neglect the spatial gradients of velocity and elastic constants. In that case, the scattering formulas (13.47) and (13.48) show that $\langle|\Phi^1|^2\rangle$ becomes very small except for small θ, and scattered energy is concentrated in the forward direction. Since, for small θ, $\cos\theta \sim 1$ and

$$-\frac{\delta\rho}{\rho_0} + \frac{\delta\lambda}{\lambda_0 + 2\mu_0} + \frac{2\,\delta\mu}{\lambda_0 + 2\mu_0} = 2\frac{\delta\alpha}{\alpha_0},$$

we find that (13.58) agrees with the equation (13.45) obtained by neglecting the contributions from gradients.

Equation (13.58) was used by Haddon and Cleary (1974) in the interpretation of precursors of $PKIKP$ as scattered waves from lateral inhomogeneity near the core-mantle boundary. Observed precursor amplitudes were explained by an inhomogeneous region of thickness $L = 200$ km, with the inhomogeneity scale length $a = 30$ km, and RMS fluctuations of 1% in the density, bulk modulus, and rigidity. (As we mentioned in the Introduction to Volume II, many of these precursors had earlier been explained in terms of a transition region in the fluid core, which would have corresponded to a new branch, the GH branch, of the PKP travel-time curve.)

A similar characterization of the Earth's interior by a random medium has also been applied to structure near the surface. In this case, however, the observation points are close to the scattering sources, and the scattered and primary waves arrive together. Thus, in order to find the nature of inhomogeneity, we must study the fluctuation of observed amplitude and phase due to the contamination of primary waves by scattered waves.

13.3.4 *Fluctuation of amplitude and phase observed by seismic arrays*

Let us consider an inhomogeneous region occupying $x_1 > 0$ and primary plane waves propagating in the $+x_1$-direction. Assuming that ka is sufficiently large to allow neglecting the contributions from gradients of velocity and elastic constants, as well as neglecting the P-to-S and S-to-P conversions, we go back to (13.37) and (13.38) and write the sum of primary and scattered waves as

$$\Phi = \Phi^0 + \Phi^1 = A \exp\left[-i\omega\left(t - \frac{x_1}{c_0}\right)\right]$$

$$+ \frac{A\omega^2}{2\pi c_0^2} \int_V \left(-\frac{\delta c}{c_0}\right) \frac{\exp\left[-i\omega\left(t - \frac{r}{c_0} - \frac{\xi_1}{c_0}\right)\right]}{r} dV(\xi). \quad (13.60)$$

The fluctuation in amplitude and phase may be obtained by putting

$$\Phi = (A + \Delta A) \exp\left[-i\omega\left(t - \frac{x_1}{c_0}\right) + i\Delta\phi\right]. \quad (13.61)$$

Taking the real and imaginary parts of (13.60) and (13.61), we find, for $\Delta A/A \ll 1$ and $\Delta\phi \ll 1$, that

$$\Delta\phi = \frac{k^2}{2\pi} \int_V \frac{1}{r} \sin k[r - (x_1 - \xi_1)]\mu(\xi)\, dV(\xi),$$

$$\frac{\Delta A}{A} = \Delta \ln A = \frac{k^2}{2\pi} \int_V \frac{1}{r} \cos k[r - (x_1 - \xi_1)]\mu(\xi)\, dV(\xi), \quad (13.62)$$

where $k = \omega/c_0$ and $\mu(\xi) = -\delta c/c_0$.

In the case of large-scale inhomogeneities ($ka \gg 1$), we can neglect backscattering, and need to consider contributions only from volume elements $dV(\xi)$ lying in the region in which $\rho < x_1 - \xi_1$ (see Fig. 13.8). Then

$$r = \sqrt{\rho^2 + (x_1 - \xi_1)^2}$$

$$\sim (x_1 - \xi_1) + \frac{1}{2}\frac{\rho^2}{x_1 - \xi_1}, \quad (13.63)$$

which approximates the spherical wave front by a paraboloid. Using (13.63) to substitute for $r - (x_1 - \xi_1)$ in (13.62) and replacing $1/r$ by $1/(x_1 - \xi_1)$, we

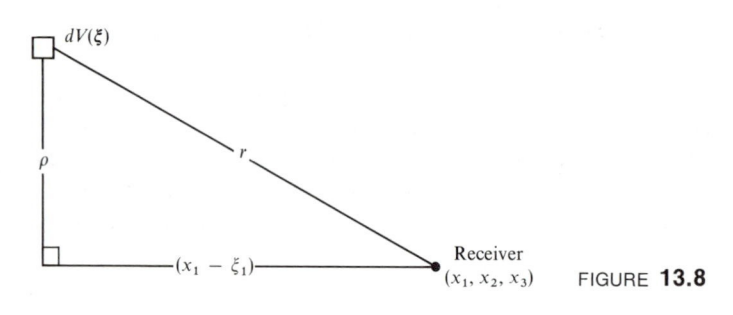

FIGURE **13.8**

obtain

$$\Delta\phi = \frac{k^2}{2\pi} \int_V \frac{\sin[k\rho^2/2(x_1 - \xi_1)]}{x_1 - \xi_1} \mu(\xi)\, dV(\xi),$$

$$\Delta \ln A = \frac{k^2}{2\pi} \int_V \frac{\cos[k\rho^2/2(x_1 - \xi_1)]}{x_1 - \xi_1} \mu(\xi)\, dV(\xi),$$

(13.64)

where the integration region V is a layer between $\xi_1 = 0$ and $\xi_1 = x_1$. The shapes of part of these integrands are shown in Figure 13.9. as functions of ρ.

Using (13.64), we can estimate the average values of $|\Delta\phi|^2$ and $|\Delta \ln A|^2$ basically in the same way as we obtained $\langle|\Phi^1|^2\rangle$ in (13.47) and (13.48). Under the assumption that the travel distance L through the inhomogeneous region is larger than the correlation distance a, Chernov (1960) obtained the following formulas for the case $N(r) = e^{-r^2/a^2}$:

$$\langle|\Delta\phi|^2\rangle = \frac{\sqrt{\pi}}{2} \langle\mu^2\rangle k^2 aL\left(1 + \frac{1}{D}\tan^{-1} D\right),$$

$$\langle|\Delta \ln A|^2\rangle = \frac{\sqrt{\pi}}{2} \langle\mu^2\rangle k^2 aL\left(1 - \frac{1}{D}\tan^{-1} D\right),$$

(13.65)

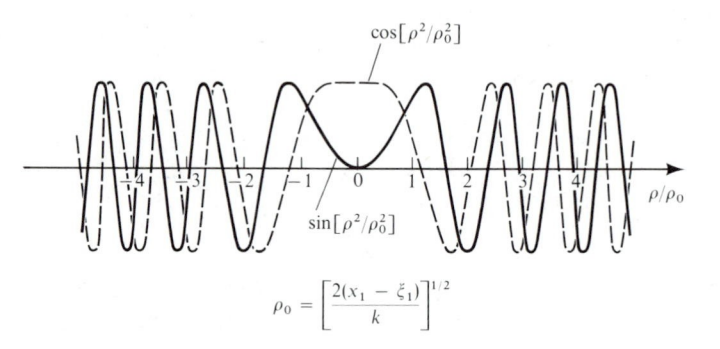

FIGURE **13.9**

where D is called the *wave parameter*, defined by

$$D = \frac{4L}{ka^2}. \tag{13.66}$$

This is the ratio of the size of the first Fresnel zone to the scale length of inhomogeneity. For small D, the amplitude fluctuation is smaller than the phase fluctuation. The case of an infinite slab of uniform velocity perturbation, discussed earlier, corresponds to $D = 0$, for which the phase fluctuation is the one expected from geometrical ray optics and the amplitude fluctuation

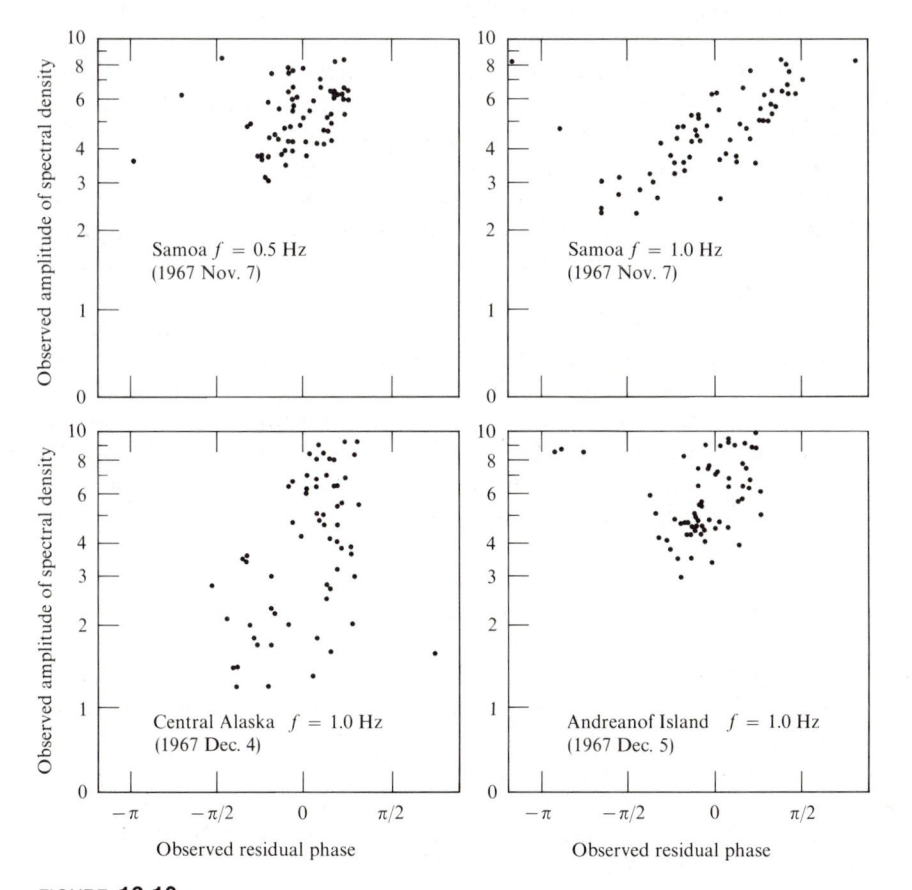

FIGURE **13.10**

Relation between amplitude and phase fluctuation observed at LASA for P-waves from four events. (The phase fluctuation is measured with respect to the arrival time of a plane wavefront determined to fit the first arrivals by a least-squares method.) They show positive correlation, in agreement with Chernov's theory. [After Aki, 1973; copyrighted by the American Geophysical Union.]

vanishes as can be seen in (13.65). The phase fluctuation is largest for $D = 0$, decreases with increasing D, and becomes the same as the amplitude fluctuation as D approaches infinity.

Chernov (1960) gives the correlation coefficient between $\Delta \ln A$ and $\Delta \phi$ for the case $N(r) = e^{-r^2/a^2}$:

$$\frac{\langle \Delta \ln A \cdot \Delta \phi \rangle}{[\langle (\Delta \ln A)^2 \rangle \langle (\Delta \phi)^2 \rangle]^{1/2}} = \frac{1}{2} \frac{\log(1 + D^2)}{[D^2 - (\tan^{-1} D)^2]^{1/2}}. \tag{13.67}$$

No correlation exists for $D = \infty$, but the correlation is always positive for a finite D and approaches $\frac{1}{2}\sqrt{3/2} \sim 0.6$ for small D. The spatial autocorrelation functions for both amplitude and phase fluctuation were also calculated by Chernov. These theoretical results were compared with observations on teleseismic P-waves obtained at the Montana LASA by Aki (1973) and Capon (1974). Figure 13.10 shows $\Delta \ln A$ vs $\Delta \phi$ for P-waves from typical teleseismic events. The plane wave that best fits the first-arrival time at each station is considered as the primary wave, and $\Delta \phi$ was measured relative to the arrival time of the best-fitting plane wave. A good agreement was obtained between observed and calculated variances and correlation functions for the frequency range between 0.5 and 0.7 Hz. It was found that the inhomogeneity under the Montana LASA has a correlation distance of about 10 km, with a fractional RMS velocity fluctuation of about 3–4% extending to a depth of about 60–100 km. For this frequency range, ka is around 6, satisfying the assumption of large ka. The condition for small perturbation, $\Delta I/I < 1$, given in (13.54), is marginally satisfied for 0.5 Hz. However, the condition is clearly violated for frequencies higher than 0.7 Hz. The observed statistical properties of 1-Hz waves show systematic departure from those predicted under the assumption of small perturbation.

13.3.5 Classification of scattering problems

In order to get some perspective on the scattering problems encountered in seismology, we shall classify them according to the two most important non-dimensional numbers controlling the scattering phenomena, namely, ka and kL. ka is 2π times the ratio of inhomogeneity scale length a to wavelength λ, and measures the smoothness or roughness of inhomogeneity within a wavelength. kL is 2π times the number of wavelengths traveled by the primary waves through an inhomogeneous region.

For strong-motion seismograms obtained at a short distance from an earthquake source, the distance traveled by primary waves can be a small fraction of the wavelength. On the other hand, for short-period seismograph records of typical teleseismic P-waves, the travel distance may be 1000 times

the wavelength. Thus kL in seismology varies widely, covering the range from 1 to 10^4.

The largest inhomogeneity scale length is of the order of the Earth's circumference, as manifested in the distribution of oceans and continents. The smallest may be the dimension of the grain size of crystals or the length of micro-cracks in rock. Thus ka in seismology can take almost any value. However, scattering effects can be neglected for very large or very small ka. To see this more clearly, we shall use the wave parameter D defined in (13.66) and the fractional energy loss $\Delta I/I$ by scattering given in (13.55).

As we discussed earlier, when D is small, the amplitude fluctuation is small and the phase fluctuation is close to what is expected from geometric ray theory. In other words, the inhomogeneity is so smooth that one can consider it as a piecewise homogeneous region. This approach has been used extensively in various areas of seismology. For example, in interpreting the phase and group velocities of surface waves, one divides the Earth's surface into several regions, within each of which the crust-mantle structure is assumed to be laterally homogeneous. Then the observed travel time is considered as a sum of travel times occurring in each region. Similar approaches have been taken in interpreting teleseismic body waves and refraction and reflection records. This "ray-theoretical" approach has been most productive in the past. The region of our $ka - kL$ diagram in which this approach is justified is $D < 1$, the upper left side of the line $D = 1$, shown in Figure 13.11. Since the upper left corner of the diagram, in which $a > L$, belongs by definition to the homogeneous region, the region where ray theory is effective is a wedge-shaped region between $a = L$ and $D = 1$.

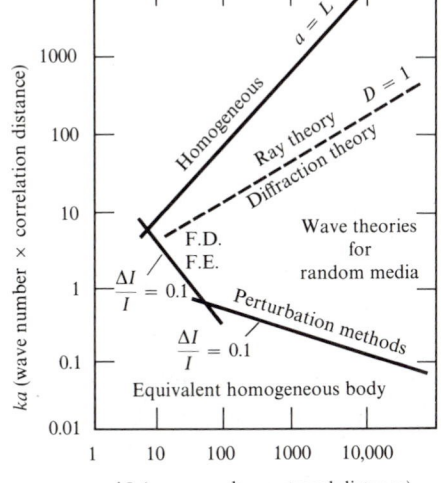

FIGURE **13.11**

Classification of scattering problems and applicable methods in the ka-kL diagram, where k is the wavenumber, a is the correlation distance of inhomogeneity, and L is the travel distance.

On the other hand, the scattering may become negligible when the inhomogeneity scale is much smaller than the wavelength. In this case, the medium behaves like a homogeneous body with some averaged properties, just as an aggregate of different crystals behaves macroscopically like a homogeneous body.

The strength of scattering may be estimated by the fractional loss $\Delta I/I$. If $\Delta I/I$ is small (say, less than 10%), we may neglect the scattering and consider the medium as homogeneous. Figure 13.11 shows two lines from (13.55):

$$\frac{\Delta I}{I} = 8\langle \mu^2 \rangle k^4 a^3 L = 0.1 \qquad \text{for } ka < 1,$$

$$\frac{\Delta I}{I} = 2\langle \mu^2 \rangle k^2 aL = 0.1 \qquad \text{for } ka > 1,$$

BOX **13.2**

Grossly homogeneous media

When the scale length of inhomogeneity is sufficiently smaller than the wavelength, one can neglect the effect of scattering and then replace the inhomogeneous medium by an equivalent homogeneous medium. The elastic constants of the equivalent homogeneous medium are called "effective moduli." In seismology, special attention has been given to the effective moduli of two-phase media, such as a solid containing pores and cracks that are either empty or filled with gas or liquid. The two-phase medium is important as a mechanical model of rocks at shallow depth. It may be applicable to the low-velocity layer of the asthenosphere, where partial melts are believed to exist.

The calculation of effective moduli for a cracked solid was pioneered by Walsh (1965), and his formulas have been used in interpreting various experimental results. More recently, Budiansky and O'Connell (1976) advocated a self-consistent approach that approximately takes account of interaction between cracks. To illustrate the outline of their method, let us find the effective bulk modulus k^* for a solid containing randomly oriented empty cracks. Consider, first, the solid without cracks in a state of uniform hydrostatic stress P maintained by prescribed boundary tractions. The potential energy of the system is $-P^2V/2k$, where k is the bulk modulus of the solid and V is the total volume. Keeping the external loading unchanged, the cracks are introduced into the body, producing a potential energy change ΔW. Then the effective bulk modulus k^* may be defined by

$$-P^2V/(2k^*) = -P^2V/(2k) + \Delta W. \tag{1}$$

The solution is available for energy change ΔW in the case of a single crack in an unbounded homogeneous medium. The self-consistent approach uses this solution, but the homogeneous medium is characterized by the effective moduli instead of the moduli of the solid. Thus ΔW is a function of k^*, and k^* is obtained by solving (1). Similarly, the effective Young's modulus can be determined considering the loading by uniaxial tension.

where we assumed that $\langle \mu^2 \rangle = 0.001$, which is an RMS velocity perturbation of $\sqrt{10\%}$. Below these two lines, the medium may be regarded as homogeneous. In the vicinity of these lines, where the scattering is weak, the method of small perturbations may be applicable. Above these lines and below $D = 1$ lies the region of strong scattering. The problem is reduced as we move to the left corner of this region of strong scattering. The finite-difference, finite-element, and other numerical methods may be effectively used for solving problems in this corner.

The problem becomes increasingly complex as we move to the right along the zone $1 < ka < 10$, where waves travel a long distance in a medium with inhomogeneity scale comparable to wavelength. An example of this extreme complexity is the coda waves of local earthquakes: slowly decaying motions with frequencies in the range 1–50 Hz, lasting 100–1000 oscillations. They have a remarkable simple property in that the temporal decay of oscillations at a particular frequency, recorded on a seismometer, is nearly independent of epicenter location, station location, and the nature of the direct wave path connecting the epicenter and station (Rautian and Khalturin, 1978). What is an extremely complex problem from a deterministic approach turns out to be quite simply treated by statistical methods, based either on back-scattering from numerous scatterers or on diffusion of scattered energy. The diffusion model has been used in the interpretation of another complex record, the lunar seismo-gram (Dainty et al., 1974), which belongs to the same region of the ka-kL diagram as coda waves in the Earth.

With this overall view of seismic scattering problems in mind, we shall now proceed to perturbation methods for more realistic initial models, such as the layered half-space and sphere. Then we shall discuss numerical methods, such as the finite-difference method.

13.4 Lateral Inhomogeneity in a Layered Medium

Very often we use theoretical results based upon laterally homogeneous layered media in interpreting observations on both body and surface waves in the Earth. It is therefore important to know what happens when the layered medium becomes laterally inhomogeneous. The inhomogeneity may be caused by the change in the thickness of a layer or by the change in seismic velocities and density in each layer.

13.4.1 Perturbations in material properties

In this section, we shall follow Kennett (1972a) and consider the two-dimensional case in which material properties and wave fields are functions only of x and z, where z is depth. The time dependence is assumed to be $\exp(-i\omega t)$. Then the

equations of motion for u, v, and w are

$$-\rho\omega^2 u = \frac{\partial \tau_{xx}}{\partial x} + \frac{\partial \tau_{xz}}{\partial z},$$

$$-\rho\omega^2 v = \frac{\partial \tau_{xy}}{\partial x} + \frac{\partial \tau_{yz}}{\partial z}, \tag{13.68}$$

$$-\rho\omega^2 w = \frac{\partial \tau_{xz}}{\partial x} + \frac{\partial \tau_{zz}}{\partial z}.$$

Assuming that the medium is isotropic, the stress-strain relation gives the following equations:

$$\tau_{xx} = (\lambda + 2\mu)\frac{\partial u}{\partial x} + \lambda\frac{\partial w}{\partial z},$$

$$\tau_{xy} = \mu\frac{\partial v}{\partial x},$$

$$\tau_{xz} = \mu\left(\frac{\partial u}{\partial z} + \frac{\partial w}{\partial x}\right), \tag{13.69}$$

$$\tau_{yz} = \mu\frac{\partial v}{\partial z},$$

$$\tau_{zz} = (\lambda + 2\mu)\frac{\partial w}{\partial z} + \lambda\frac{\partial u}{\partial x}.$$

Eliminating τ_{xx} and τ_{xy} from the above eight equations, and arranging them in such a way that the derivatives with respect to z appear on the left-hand side, we can write the equations for displacement and traction components in the following form:

$$\frac{\partial}{\partial z}\begin{pmatrix} u \\ w \\ \tau_{xz} \\ \tau_{zz} \end{pmatrix} = \begin{pmatrix} 0 & -\partial_x & \dfrac{1}{\mu} & 0 \\ -\dfrac{\lambda}{\lambda+2\mu}\partial_x & 0 & 0 & \dfrac{1}{\lambda+2\mu} \\ -(\partial_x\zeta)\partial_x - \zeta\partial_{xx} - \rho\omega^2 & 0 & 0 & -\partial_x\left(\dfrac{\lambda}{\lambda+2\mu}\right) - \dfrac{\lambda}{\lambda+2\mu}\partial_x \\ 0 & -\rho\omega^2 & -\partial_x & 0 \end{pmatrix}\begin{pmatrix} u \\ w \\ \tau_{xz} \\ \tau_{zz} \end{pmatrix},$$

$$\tag{13.70}$$

where $\zeta = 4\mu(\lambda + \mu)/(\lambda + 2\mu)$, $\partial_x = \partial/\partial x$, and $\partial_{xx} = \partial^2/\partial x^2$; and

$$\frac{\partial}{\partial z}\begin{pmatrix} v \\ \tau_{yz} \end{pmatrix} = \begin{pmatrix} 0 & \dfrac{1}{\mu} \\ -(\partial_x\mu)\partial_x - \mu\partial_{xx} - \rho\omega^2 & 0 \end{pmatrix}\begin{pmatrix} v \\ \tau_{yz} \end{pmatrix}. \tag{13.71}$$

As shown above, the equations for the in-plane (P-SV) problem are separated from those for the anti-plane (SH) problem when both the inhomogeneity and the wave field are independent of y.

Let us separate the spatial distribution of material properties into a primary part that depends only on depth z and a perturbation that depends on both z and x. We shall use the subscript 0 for the primary part and write the perturbation part as, for example,

$$\Delta\rho(x, z) = \rho(x, z) - \rho_0(z),$$

$$\Delta\left(\frac{1}{\mu(x, z)}\right) = \frac{1}{\mu(x, z)} - \frac{1}{\mu_0(z)}. \tag{13.72}$$

Then we can rewrite the equations (13.70) and (13.71) in the following form:

$$\frac{\partial}{\partial z}\mathbf{f}(x, z) = \mathbf{A}_0(z)\mathbf{f}(x, z) + \mathbf{A}_1(x, z)\mathbf{f}(x, z), \tag{13.73}$$

where $\mathbf{f}(x, z)$ is the motion-stress vector with components $(u, w, \tau_{xz}, \tau_{zz})^T$ in the P-SV problem and $(v, \tau_{yz})^T$ in the SH problem. $\mathbf{A}_0(z)$ is the matrix encountered in the case of a laterally homogeneous medium (7.28). For the P-SV problem,

$$\mathbf{A}_0(z) = \begin{pmatrix} 0 & -\partial_x & \dfrac{1}{\mu_0} & 0 \\ -\dfrac{\lambda_0}{\lambda_0 + 2\mu_0}\partial_x & 0 & 0 & \dfrac{1}{\lambda_0 + 2\mu_0} \\ -\zeta_0\partial_{xx} - \rho_0\omega^2 & 0 & 0 & -\dfrac{\lambda_0}{\lambda_0 + 2\mu_0}\partial_x \\ 0 & -\rho_0\omega^2 & -\partial_x & 0 \end{pmatrix}, \tag{13.74}$$

and for the SH problem,

$$\mathbf{A}_0(z) = \begin{pmatrix} 0 & \dfrac{1}{\mu_0} \\ -\mu_0\partial_{xx} - \rho_0\omega^2 & 0 \end{pmatrix}. \tag{13.75}$$

The matrix $\mathbf{A}_1(x, z)$ contains the effect of lateral inhomogeneity. For the *P-SV* problem, subtracting (13.74) from (13.70), we have

$$
\mathbf{A}_1(x, z) = \begin{pmatrix}
0 & 0 & \Delta\left(\dfrac{1}{\mu}\right) & 0 \\[2ex]
-\Delta\left(\dfrac{\lambda}{\lambda + 2\mu}\right)\partial_x & 0 & 0 & \Delta\left(\dfrac{1}{\lambda + 2\mu}\right) \\[2ex]
\begin{aligned}-(\partial_x \Delta\zeta)\partial_x \\ -\Delta\zeta\partial_{xx} - \Delta\rho\omega^2\end{aligned} & 0 & 0 & -\partial_x\Delta\left(\dfrac{\lambda}{\lambda + 2\mu}\right) \\[3ex]
 & & & -\Delta\left(\dfrac{\lambda}{\lambda + 2\mu}\right)\partial_x \\[2ex]
0 & -\Delta\rho\omega^2 & 0 & 0
\end{pmatrix},
$$

$$(13.76)$$

and for the *SH* problem, subtracting (13.75) from (13.71) gives

$$
\mathbf{A}_1(x, z) = \begin{pmatrix}
0 & \Delta\left(\dfrac{1}{\mu}\right) \\[2ex]
-(\partial_x\Delta\mu)\partial_x - \Delta\mu\partial_{xx} - \Delta\rho\omega^2 & 0
\end{pmatrix}.
\qquad (13.77)
$$

As seen in earlier chapters, the x-dependence of wave fields in laterally homogeneous media can be separated by the factor e^{ikx}, and an arbitrary field can be constructed by an integral with respect to k, the horizontal wavenumber. Even in the case of lateral inhomogeneity, the concept of wavenumber is still useful. Let us write the Fourier transform of $\mathbf{A}_1(x, z)$ and $\mathbf{f}(x, z)$ with respect to x as

$$
\mathbf{A}_1(k, z) = \int \mathbf{A}_1(x, z)e^{-ikx}\, dx,
$$

$$
\mathbf{f}(k, z) = \int \mathbf{f}(x, z)e^{-ikx}\, dx,
$$

and then the Fourier transform of (13.73) can be written as

$$
\frac{\partial}{\partial z}\mathbf{f}(k, z) = \mathbf{A}_0(z)\mathbf{f}(k, z) + \frac{1}{2\pi}\int_{-\infty}^{\infty} \mathbf{A}_1(k - k', z)\mathbf{f}(k', z)\, dk', \qquad (13.78)
$$

where ∂_x and ∂_{xx} in (13.74) and (13.75) are replaced by ik and $-k^2$, respectively.

The solution of (13.78) was given in (7.48) in terms of the propagator matrix $\mathbf{P}(z, z_0)$, which satisfies

$$\frac{d}{dz} \mathbf{P}(z, z_0) = \mathbf{A}_0(z)\mathbf{P}(z, z_0) \tag{13.79}$$

and

$$\mathbf{P}(z_0, z_0) = \mathbf{I}, \tag{13.80}$$

where $\mathbf{I}$ is the identity matrix. Using $\mathbf{P}(z, z_0)$, the solution of (13.78) can be written as

$$\mathbf{f}(k, z) = \mathbf{P}(z, z_0)\mathbf{f}(k, z_0) + \frac{1}{2\pi} \int_{z_0}^{z} \mathbf{P}(z, z') \, dz' \int_{-\infty}^{\infty} \mathbf{A}_1(k - k', z')\mathbf{f}(k', z') \, dk'. \tag{13.81}$$

Clearly, this gives the correct initial condition at $z = z_0$.

Equation (13.81) shows that the motion-stress vector $\mathbf{f}(k, z)$ consists of two terms. One is obtained by the operation of the propagator matrix for the laterally homogeneous medium. The other represents the contribution of distributed sources with intensity per unit vertical distance given by

$$\frac{1}{2\pi} \int_{-\infty}^{\infty} \mathbf{A}_1(k - k', z)\mathbf{f}(k', z) \, dk'.$$

This convolution represents the scattering effect involving the conversion of a plane wave with the wavenumber k' into one with wavenumber k. If $\mathbf{A}_1(k, z)$ is concentrated at $k = 0$ like $\delta(k)$, then there will be little conversion between different wavenumbers. This is the case when $\mathbf{A}_1(x, z)$ is a smooth function in x. On the other hand, if $\mathbf{A}_1(x, z)$ varies rapidly with x, then $\mathbf{A}_1(z, k)$ will show a broad spectrum in k. Expressing a measure of the breadth as Δk, we see that the conversion will occur between wavenumbers within the range $k - \Delta k$ and $k + \Delta k$. For a large Δk, both $k - \Delta k$ and k (or $k + \Delta k$ and k) can have opposite signs, thus causing back-scattering. Thus we see again that the small-scale inhomogeneity generates scattered waves in all directions, but the scattering is concentrated in the forward direction in the case of large-scale inhomogeneities.

The integral equation (13.81) can be solved by an iterative method if $\mathbf{A}_1$ is in some sense sufficiently smaller than $\mathbf{A}_0$ to ensure convergence of the iteration process. We first set $\mathbf{A}_1$ equal to zero, and the zeroth-order solution is obtained by solving the problem for the laterally homogeneous unperturbed medium.

We then convolve this solution with $\mathbf{A}_1(k, z)$ and put it into the last term of (13.81). Carrying out the depth integration after multiplying by the propagator matrix, we can obtain the first-order solution. The process can be repeated until the solution converges. The first-order solutions have been obtained explicitly by Kennett (1972a) for a *P-SV* problem in which the unperturbed medium is a homogeneous half-space, and for an *SH* problem in which the unperturbed medium is a single layer over a half-space.

13.4.2   *Effect of irregular interfaces*

When discontinuities in material properties occur only across horizontal plane interfaces, the motion-stress vector $\mathbf{f}(u, v, w, \tau_{xz}, \tau_{yz}, \tau_{zz})$ is continuous across the interfaces (Section 5.4). Let us now consider what happens when the interface is not planar, but has an irregular shape. As in the preceding section, we shall assume that neither the wave field nor the medium depends on y. The depth of the irregular interface is taken as $z = h(x)$, which fluctuates around $z = 0$ and separates a homogeneous medium 1 with parameters ρ_1, λ_1 and μ_1 from another homogeneous medium 2 with parameters ρ_2, λ_2, and μ_2, as shown in Figure 13.12.

The condition of continuity in displacement must be imposed now at $z = h(x)$, because that is the boundary between two regions to each of which an appropriate displacement representation is given. Similarly, the condition of continuity in traction must be imposed at $z = h(x)$ on the traction acting along the interface, because that separates two regions to each of which an appropriate stress representation is given. Taking $\mathbf{n} = (n_x, n_y)$ as the unit normal to the interface, we require the continuity of traction (2.10):

$$\tau_{nx} = \tau_{xx}n_x + \tau_{zx}n_z,$$

$$\tau_{nz} = \tau_{xz}n_x + \tau_{zz}n_z, \tag{13.82}$$

$$\tau_{ny} = \tau_{xy}n_x + \tau_{zy}n_z,$$

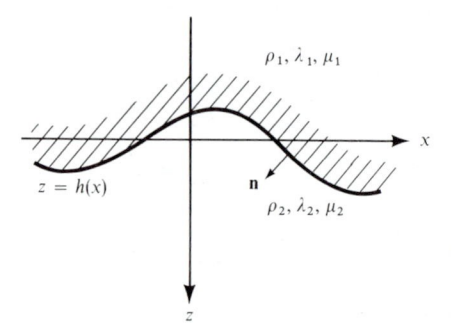

FIGURE **13.12**

An irregular interface between two homogeneous media.

with $(\tau_{nx}, \tau_{ny}, \tau_{nz})$ continuous across $z = h(x)$. Replacing τ_{xx} and τ_{yx} in (13.82) by the derivatives of displacements using the stress-strain relation (13.69), and noting that

$$n_x = \frac{-h'}{(1 + h'^2)^{1/2}} \quad \text{and} \quad n_z = \frac{1}{(1 + h'^2)^{1/2}},$$

we can write

$$\begin{pmatrix} u \\ w \\ (1 + h'^2)^{1/2}\tau_{nx} \\ (1 + h'^2)^{1/2}\tau_{nz} \end{pmatrix} = \begin{pmatrix} 1 & 0 & 0 & 0 \\ 0 & 1 & 0 & 0 \\ -\zeta h'\partial_x & 0 & 1 & \dfrac{-h'\lambda}{\lambda + 2\mu} \\ 0 & 0 & -h' & 1 \end{pmatrix} \begin{pmatrix} u \\ w \\ \tau_{zx} \\ \tau_{zz} \end{pmatrix}, \tag{13.83}$$

$$\begin{pmatrix} v \\ (1 + h'^2)^{1/2}\tau_{ny} \end{pmatrix} = \begin{pmatrix} 1 & 0 \\ -\mu h'\partial_x & 1 \end{pmatrix} \begin{pmatrix} v \\ \tau_{zy} \end{pmatrix}, \tag{13.84}$$

where $\zeta = 4\mu(\lambda + \mu)/(\lambda + 2\mu)$ and $h' = dh/dx$. The left-hand sides of the above two equations are the quantities that are continuous across the interface. Again, the P-SV and SH problems are decoupled in our two-dimensional problem. Summarizing equations (13.83) and (13.84) in a vector form by

$$\mathbf{g} = \mathbf{Qf}, \tag{13.85}$$

we find that $\mathbf{Q}$ can be written as

$$\mathbf{Q} = \mathbf{I} + h'\mathbf{Q}_1, \tag{13.86}$$

where for the P-SV problem

$$\mathbf{Q}_1 = \begin{pmatrix} 0 & 0 & 0 & 0 \\ 0 & 0 & 0 & 0 \\ -\zeta\partial_x & 0 & 0 & -\lambda/(\lambda + 2\mu) \\ 0 & 0 & -1 & 0 \end{pmatrix} \tag{13.87}$$

and the SH problem

$$\mathbf{Q}_1 = \begin{pmatrix} 0 & 0 \\ -\mu\partial_x & 0 \end{pmatrix}. \tag{13.88}$$

Using superscripts (1) or (2) to designate the particular medium, the interface condition can be expressed as

$$(\mathbf{I} + h'\mathbf{Q}_1^{(1)})\mathbf{f}^{(1)}(x, h(x)) = (\mathbf{I} + h'\mathbf{Q}_1^{(2)})\mathbf{f}^{(2)}(x, h(x)). \tag{13.89}$$

Let us now assume that the fluctuation of the interface is small, so that we can approximate $\mathbf{f}(x, h(x))$ by a Taylor expansion as

$$\mathbf{f}(x, h(x)) - f(\mathbf{x}, 0) \sim \left(\frac{\partial \mathbf{f}}{\partial z}\right)_{z=0} h(x) = \mathbf{A}_0(0)\mathbf{f}(x, 0)h(x).$$

Then, from (13.89), omitting terms of order higher than h and h', we get

$$(\mathbf{I} + h'\mathbf{Q}_1^{(1)} + h\mathbf{A}_0^{(1)})\mathbf{f}^{(1)}(x, 0) = (\mathbf{I} + h'\mathbf{Q}_1^{(2)} + h\mathbf{A}_0^{(2)})\mathbf{f}^{(2)}(x, 0). \quad (13.90)$$

In the case of horizontal plane interfaces, $\mathbf{f}$ was continuous at the interface $z = 0$, or $\mathbf{f}^{(1)}(x, 0) = \mathbf{f}^{(2)}(x, 0)$. Because of its irregular interface shape, we now find that the motion-stress vector $\mathbf{f}$ is apparently discontinuous at $z = 0$. The discontinuity is given by

$$\mathbf{f}^{(2)}(x, 0) - \mathbf{f}^{(1)}(x, 0) = (h'\mathbf{Q}_1^{(2)} + h\mathbf{A}_0^{(2)})\mathbf{f}^{(2)}(x, 0) - (h'\mathbf{Q}_1^{(1)} + h\mathbf{A}_0^{(1)})\mathbf{f}^{(1)}(x, 0). \quad (13.91)$$

As shown in Chapter 7, the discontinuity in the motion-stress vector acts like a seismic source. When the discontinuity in $\mathbf{f}$ is sufficiently smaller than $\mathbf{f}$ itself, we can replace $\mathbf{f}^{(1)}$ and and $\mathbf{f}^{(2)}$ on the right-hand side of (13.91) by $\mathbf{f}^{(0)} \sim \mathbf{f}^{(1)} \sim \mathbf{f}^{(2)}$. Then we can rewrite (13.91) as

$$\mathbf{f}^{(2)}(x, 0) - \mathbf{f}^{(1)}(x, 0) = [h'(x)(\mathbf{Q}_1^{(2)} - \mathbf{Q}_1^{(1)}) + h(x)(\mathbf{A}_0^{(2)} - \mathbf{A}_0^{(1)})]\mathbf{f}^{(0)}(x, 0). \quad (13.92)$$

Taking the Fourier transform of the above equation, and recognizing the product in x-space as the convolution in k-space, we get

$$\mathbf{f}^{(2)}(k, 0) - \mathbf{f}^{(1)}(k, 0) = \frac{1}{2\pi} \int_{-\infty}^{\infty} h(k - k')\mathbf{L}_{21}(k, k')\mathbf{f}^{(0)}(k') \, dk', \quad (13.93)$$

where

$$\mathbf{L}_{21} = \begin{pmatrix} 0 & 0 & \dfrac{1}{\mu_2} - \dfrac{1}{\mu_1} & 0 \\ -ik'\left(\dfrac{\lambda_2}{\lambda_2 + 2\mu_2} - \dfrac{\lambda_1}{\lambda_1 + 2\mu_1}\right) & 0 & 0 & \dfrac{1}{\lambda_2 + 2\mu_2} - \dfrac{1}{\lambda_1 + 2\mu_1} \\ -(\rho_2 - \rho_1)\omega^2 + (\zeta_2 - \zeta_1)kk' & 0 & 0 & -\left(\dfrac{\lambda_2}{\lambda_2 + 2\mu_2} - \dfrac{\lambda_1}{\lambda_1 + 2\mu_1}\right)ik \\ 0 & -(\rho_2 - \rho_1)\omega^2 & 0 & 0 \end{pmatrix}$$

$$(13.94)$$

for the *P-SV* problem and

$$
\mathbf{L}_{21} = \begin{pmatrix} 0 & \dfrac{1}{\mu_2} - \dfrac{1}{\mu_1} \\ -(\rho_2 - \rho_1)\omega^2 + (\mu_2 - \mu_1)kk' & 0 \end{pmatrix} \qquad (13.95)
$$

for the *SH* problem.

Similar to the result for material inhomogeneity shown in (13.81), the scattering effect of an irregular interface is represented by the convolution of the primary wave field with the interface shape in k-space. If $h(x)$ is smooth and $h(k)$ is concentrated at $k = 0$, the scattered waves will have the same wavenumber as the primary waves. If $h(x)$ is sharply changing and $h(k)$ has a broad spectrum, an interaction between different wavenumbers will occur, including back-scattering.

The scattered surface waves at far field, due to body or surface waves incident upon a localized irregular interface shape, can be calculated by the method described in (7.132)–(7.136), where the solution was given for seismic sources with a prescribed discontinuity in the motion-stress vector. Kennett (1972b) applied this method to the problems of inclusions near the surface of a half-space, and at an interface of a single layer and a half-space.

13.4.3   Rayleigh-ansatz method for the irregular-interface problem

The perturbation method described in the preceding section is restricted to a weakly irregular interface, because the Taylor expansion is carried only to the first power of interface height $h(x)$ and because a scattered field in the right-hand side of (13.92) is omitted. Let us find a way to eliminate these restrictions, first going back to the exact interface condition

$$
(\mathbf{I} + h'\mathbf{Q}_1^{(1)})\mathbf{f}^{(1)}(x, h(x)) = (\mathbf{I} + h'\mathbf{Q}_1^{(2)})\mathbf{f}^{(2)}(x, h(x)), \qquad (13.89 \text{ again})
$$

where $\mathbf{Q}_1$ was given in (13.87) and (13.88).

Assuming that both mediums 1 and 2 are homogeneous and isotropic, we can express the displacement field as a superposition of plane P-, SV-, and SH-waves, as shown in Chapter 5 and 6. We use

$$
\mathbf{f}(x, z) = \frac{1}{2\pi} \int_{-\infty}^{\infty} \mathbf{f}(k, z)e^{ikx} \, dk \qquad (13.96)
$$

and

$$
\mathbf{f}(k, z) = \mathbf{F}(k, z)\mathbf{w}(k). \qquad (13.97)
$$

$\mathbf{F}$ is the layer matrix, with each column giving the displacements and stresses in either an upgoing or a downgoing wave, and $\mathbf{w}$ weights the columns of $\mathbf{F}$. The layer matrix for SH is given in (5.63), and for P-SV in (5.65).

Inserting (13.96) into (13.89), we have the following equation:

$$(\mathbf{I} + h'\mathbf{Q}_1^{(1)}) \int_{-\infty}^{\infty} \mathbf{f}^{(1)}(k, h(x))e^{ikx}\,dk$$
$$= (\mathbf{I} + h'\mathbf{Q}_1^{(2)}) \int_{-\infty}^{\infty} \mathbf{f}^{(2)}(k, h(x))e^{ikx}\,dk. \tag{13.98}$$

If we insert (13.97) into the above equation, the result is a vector integral equation that relates the weighting vector $\mathbf{w}^{(1)}(k)$ in the medium above the interface to that below, $\mathbf{w}^{(2)}(k)$. In the case of a flat interface, $h(x)$ and $h'(x)$ are zero and (13.98) reduces to

$$\mathbf{f}^{(1)}(k, 0) = \mathbf{f}^{(2)}(k, 0). \tag{13.99}$$

Thus there is no coupling between a wavenumber k on one side of the flat interface and a different wavenumber $k'(\neq k)$ on the other. For irregular interfaces, however, the coupling among different wavenumbers occurs not only between two sides of the interface, but also between upgoing and downgoing waves on each side, since each of $\mathbf{w}^{(1)}$ and $\mathbf{w}^{(2)}$ will in general contain both types of waves.

In addition to the interface condition (13.89), the source and radiation conditions must be prescribed for a given problem. Consider, for example, a plane SH-wave with wavenumber k_0 that is emergent from medium 2, as shown in Figure 13.13. The radiation condition in medium 1 and the source condition in medium 2 are, respectively, given as

$$\mathbf{w}^{(1)} = \begin{pmatrix} 0 \\ \grave{S}_1(k) \end{pmatrix} \quad \text{and} \quad \mathbf{w}^{(2)} = \begin{pmatrix} \grave{S}_2(k) \\ \delta(k - k_0) \end{pmatrix}. \tag{13.100}$$

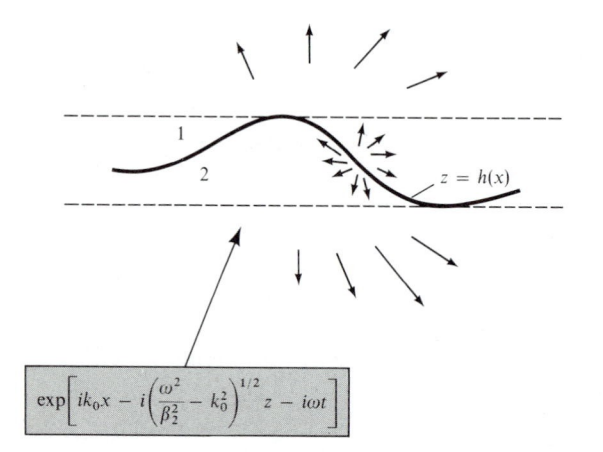

FIGURE **13.13**

The Rayleigh ansatz (13.100) cannot describe the downgoing waves in medium 1 and upgoing waves in medium 2 in the vicinity of the irregular interface.

This will allow downgoing scattered SH-waves in medium 2 and upgoing scattered SH-waves in medium 1. But the above ansatz cannot permit downgoing scattered SH-waves in medium 1 or upgoing scattered SH-waves in medium 2 (the incident wave is given by $\delta(k - k_0)$), because the radiation condition and source condition are violated in these cases. On the other hand, we need those prohibited waves to express the wave field near the interface, as depicted in Figure 13.13 by arrows nearly parallel to the interface. This intrinsic problem was pointed out by Lippman (1953) and subsequently came to be called the Rayleigh ansatz error. In practice, however, the ansatz appears to be quite satisfactory for a variety of problems involving moderatedly irregular interfaces.

A numerical solution of the integral equation (13.98) may be obtained by replacing the integral with a finite sum over discrete wavenumbers. Instead of (13.96) we can use the following discrete wavenumber representation:

$$\mathbf{f}(x, z) = \frac{1}{L} \sum_n \mathbf{F}(k_n, z)\mathbf{w}(k_n) \exp(ik_n x), \tag{13.101}$$

where $k_n = 2\pi n/L$ and the summation is made over all integers n. The above representation implies that the wave field is periodic in x with the repetition interval L. This requires that the interface shape $h(x)$ be periodic in x with the same period. Thus the effects from the neighboring intervals are sources of error. These effects may be negligible for a station located at a distance sufficiently less than L from an interface irregularity, so that the disturbances from the neighboring interval arrive later with smaller amplitude. For this reason, the present method is most effective for studying the near field of a scattering source.

The effect from neighboring intervals can be minimized by introducing a small imaginary part into the frequency. Writing $\omega = \text{Re } \omega + i \text{ Im } \omega$, we can express the Fourier transform of a one-sided function $f(t)$ as

$$f(\omega) = \int_0^\infty f(t) \exp(i\omega t) \, dt$$

$$= \int_0^\infty [f(t) \exp(- \text{Im } \omega \cdot t)] \exp(i \text{ Re } \omega \cdot t) \, dt. \tag{13.102}$$

The Fourier transform extended to a complex frequency is nothing but a usual Fourier transform for a function that has been tapered with an exponentially decaying window. Thus, by making the frequency complex, we can attenuate the undesirable late arrivals.

Putting (13.101) into (13.98) reduces the interface condition to an infinite set of simultaneous linear equations for $\mathbf{w}^{(1)}(k_n)$ and $\mathbf{w}^{(2)}(k_n)$. To solve these equations, we must truncate the summation at a certain number N, assuming that the contributions from wavenumbers higher than the truncation limit are negligible. We then Fourier-transform the interface equation (13.98) at as many discrete wavenumbers as are contained in the truncated sum, giving for SH

problems a set of $2N$ equations at each interface ($4N$ for $P - SV$). Since the number of unknowns in $\mathbf{w}^{(i)}(k_n)$ ($n = 1, 2, \ldots, N$) is $2N$ per layer ($4N$ for $P - SV$), they are completely determined by the interface condition plus the radiation and source condition given by (13.100).

As an example, let us consider the scattering of plane SH-waves at an irregular interface that simulates a dented Moho, as shown in Figure 13.14. Here we eliminate the complex layering effects by considering two half-spaces in contact at the interface. The waves are incident from depth in the lower half-space at the incidence angle $\theta_0 = 55°$ from vertical, and the interface shape is a full-cycle cosine with wavelength 50 km with maximum depth of dent 5 km. The repetition interval L is chosen as 256 km. Amplitude and phase delay are computed for incident waves with three wavelengths 5, 10, and 20 km in the lower half-space. Figure 13.14 shows the amplitude normalized to the unperturbed case and the time delay relative to the unperturbed case. The above calculations were made by Larner (1970). He compared them with the ray-theoretical solution obtained by ray-tracing using Snell's law. The ray-theoretical amplitudes are computed using transmission coefficients for plane waves at plane interfaces with the assumption that energy flow is constant along tubes bounded by given rays. The jump discontinuities in amplitude at $x = -5$ km and $x = 45$ km correspond to discontinuities in the second derivative of the interface shape. The large amplitude near $x = 23$ km is caused by the focusing effect. The result for $\lambda_2 = 5$ km is very similar to that obtained by ray theory. As the incident wavelength increases, the range of perturbed amplitude and phase delay broadens. This is consistent with the prediction of perturbation theory described in the preceding section. As shown in (13.93), the irregular interface shape acts as a seismic source characterized by a convolution of the primary wave with the interface shape in k-space. For the cosine dent with width W and the plane primary waves with $k = k_0$, we expect that the scattered waves will be distributed over the range $k_0 - 2\pi/W < k < k_0 + 2\pi/W$. The range of angular spreading may be measured by $k/\omega/\beta_2 = k\lambda_2/2\pi = \sin \theta$, where θ is the incidence angle of a ray with wavenumber k in medium 2. The corresponding range is $\sin \theta_0 - \lambda_2/W < \sin \theta < \sin \theta_0 + \lambda_2/W$. Thus the angular spread of scattered waves will be greater for longer waves. For $W = 50$ km and $\lambda_2 = 5, 10,$ and 20 km, we expect that the spread in $\sin \theta$ will be $\pm 0.1, \pm 0.2,$ and ± 0.4, respectively. This is exactly what we see in Figure 13.15, in which the absolute values of $\acute{S}_1(k_n)$ of scattered waves in the upper half-space, normalized to that of primary waves, are plotted against $k_n/\omega/\beta_2 = \sin \theta$. The spread of the wavenumber spectrum to the value 6 db down from the peak is about $\pm 0.1, \pm 0.2,$ and ± 0.4 for $\lambda_2 = 5, 10,$ and 20 km, respectively.

The scattering we see in this example is predominantly forward, because, as shown in Figure 13.15, the spread in wavenumber is not enough to change the sign of k. That the scattering is predominantly forward is clearly demonstrated in the amplitude and time-delay distribution in space shown in Figure 13.14.

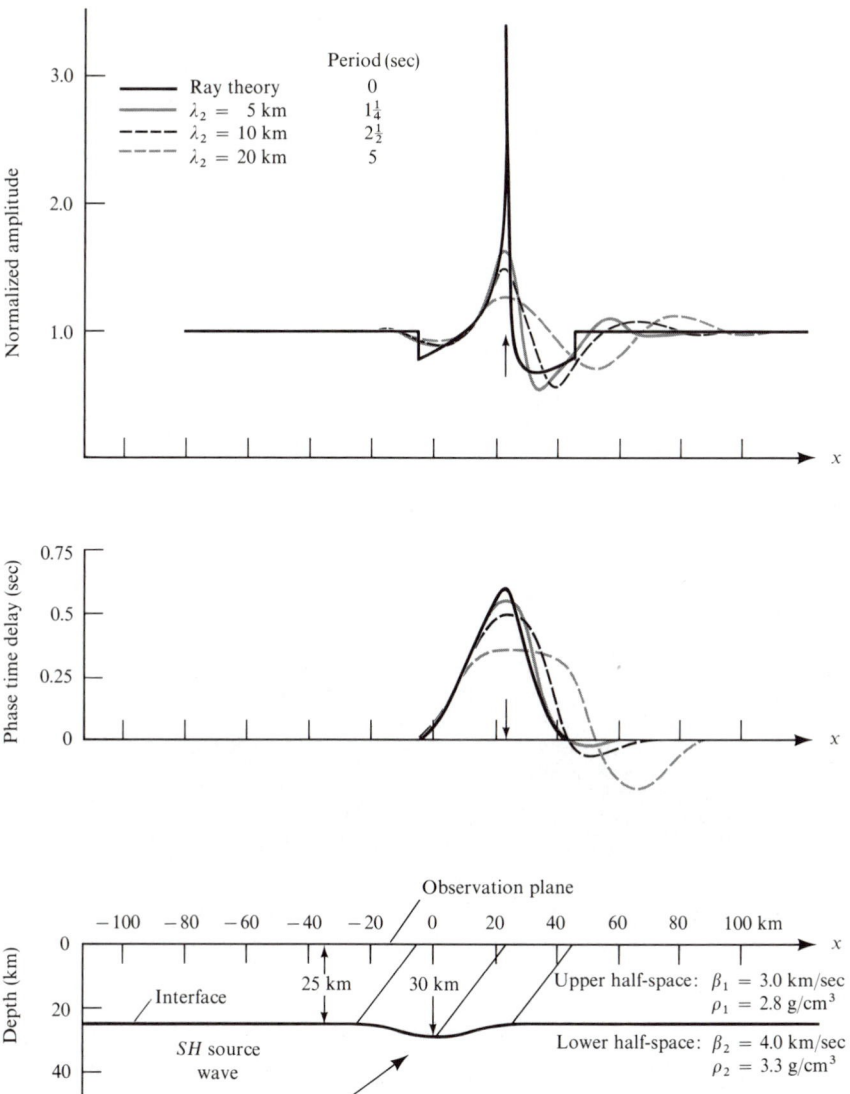

FIGURE **13.14**

Effects of wavelength on *SH*-wave scattering is shown for a dented interface joining two half-spaces. Displacement amplitude and time delay are plotted along a plane in the upper half-space. The arrows at $x = 23$ km indicate the intersection of the geometric ray path that passes through the trough of the dent with the observation plane. Results are for three different wavelengths and one angle of incidence ($55°$). [From Larner, 1970.]

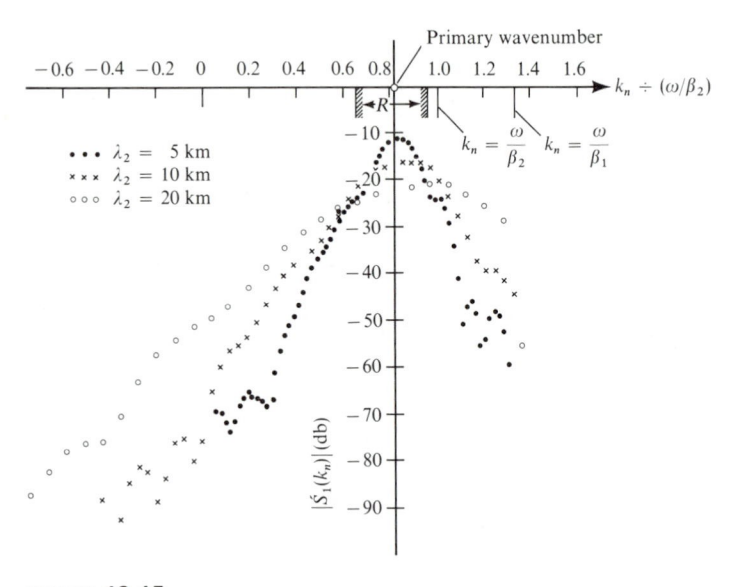

FIGURE **13.15**

Wavenumber spectra for the upgoing waves in the upper half-space normalized to the primary wave for the three cases shown in Figure 13.14. The region denoted by R is the wavenumber interval over which we expect scattered waves according to the ray theory. The regions $|k_n| > \omega/\beta_2$ and $|k_n| > \omega/\beta_1$ correspond to inhomogeneous waves in lower and upper half-spaces, respectively. They are needed to match the boundary condition. [From Larner, 1970.]

The Rayleigh-ansatz method can be extended to a variety of applications. Aki and Larner (1970) first applied the method to the response of soft basins of various cross sections to incident plane SH-waves. Larner (1970) used the method to determine the depth and shape of the Moho under the Montana LASA using the fluctuation of amplitudes and arrival times of teleseismic P-waves across the array. Bouchon (1973) extended the method to study the seismic motion at a surface with irregular topography. The boundary condition at the irregular free surface can be obtained by setting $\tau_{nx} = \tau_{ny} = \tau_{nz} = 0$ in (13.83) and (13.84).

Bouchon and Aki (1977a) included an arbitrary seismic source in the problem. As discussed in Chapter 6, a two-dimensional source with $\exp(-i\omega t)$ time dependence in an unbounded homogeneous medium can be represented by an integral with respect to the horizontal wavenumber k. We replace the integral by a finite sum over discrete wavenumbers and add it in with the terms in (13.101). These will have the necessary singularities for the given source, and unknown $\mathbf{w}(k)$ coefficients are determined to satisfy the interface and boundary conditions. The numerical difficulties caused by singularities such as Rayleigh-wave roots can be avoided by introducing a small

imaginary part in frequency. This effect can be removed from the final time-domain solution by multiplying by $\exp(\text{Im}\,\omega \cdot t)$ (see (13.102)). Figure 13.16 shows an example of the result of such a calculation. The medium consists of a finite soft basin overlying a layered half-space. The seismic source is a double couple corresponding to a dip-slip earthquake located below the basin. The amplitudes of vertical and horizontal displacements at the surface are calculated for two different frequencies. To separate the source and medium effects, the results for the cases with and without basins are shown together.

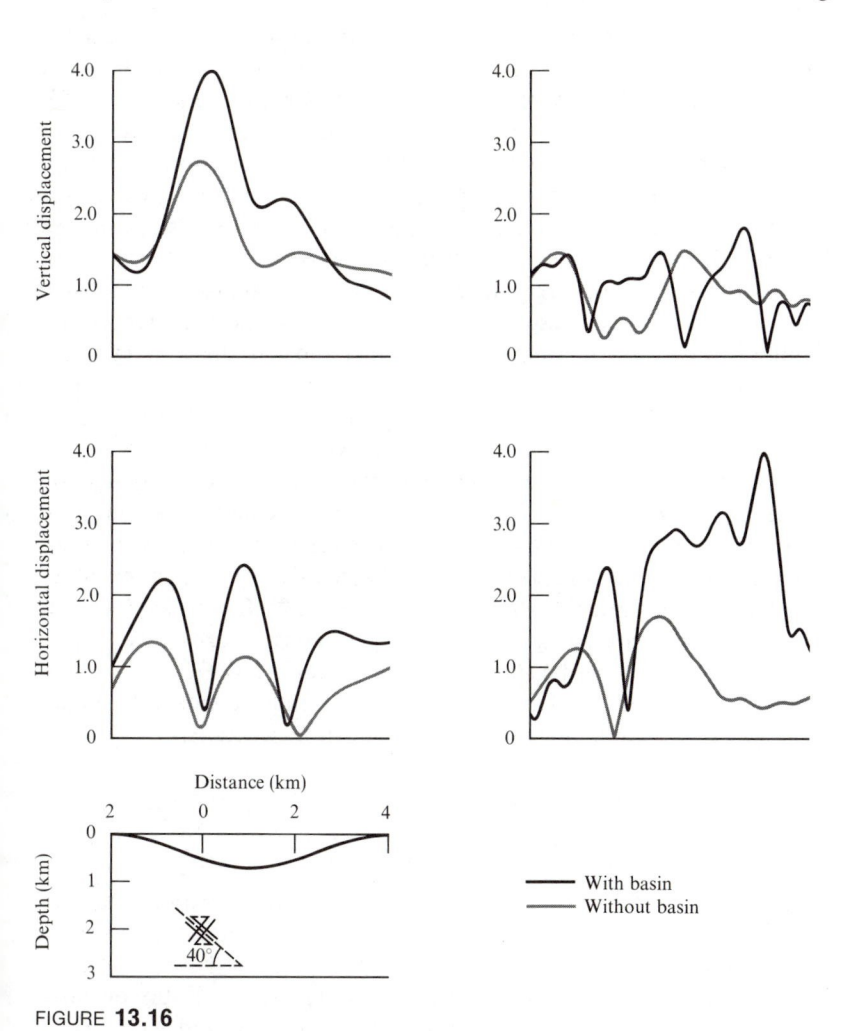

FIGURE **13.16**

Seismic motion above a soft sediment basin due to a dip-slip earthquake source located below the basin. The sediment effect amplifies the surface motion at 0.5 Hz (left side of figure), but can attenuate it at 2.0 Hz (right side). [From Bouchon and Aki, 1977a.]

Although the patterns are very complex, the horizontal motions appear to be systematically more amplified by the basin structure than the vertical motion for all frequencies. A comparison of various methods applied to the soft-basin problem will be given in Section 13.6.4.

13.5 Spectral Splitting of Free Oscillations due to Lateral Inhomogeneity of the Earth's Structure

In the previous section, we studied the scattering of seismic waves by localized lateral inhomogeneity in a layered half-space. The results obtained there are useful for interpreting seismograms observed near the inhomogeneity or in the far field outside the localized inhomogeneous region. We know, however, that the inhomogeneities exist all around the Earth. The oceans, continents, mid-ocean ridges, trenches, transform faults, mountain ranges and other large-scale inhomogeneities are distributed over the whole Earth. Consider surface waves generated from an earthquake. Primary waves will spread out, sweep the Earth's surface, and converge at the antipode. During this sweep, all the inhomogeneities will act as secondary sources and generate scattered waves. These scattered waves will interfere with each other and with the primary waves making a return trip from the antipode to the epicenter, as well as with scattered waves generated by the returning primary waves. If the Earth were a loss-less medium, the above process would go on forever, and the result would be a very complex, completely coupled pattern of surface waves propagating around the Earth. What would happen to the free-oscillation periods? To answer this question, we shall introduce small perturbations to the parameters of the laterally homogeneous Earth models, of which the free-oscillation problem was studied in Chapter 8. Here we shall start with the basic equations derived in Chapter 8 for an arbitrarily heterogeneous, hydrostatically pre-stressed medium and follow the steps taken by Madariaga (1972).

From (8.54)–(8.59) and (8.68), the equation of motion for an arbitrary heterogeneous medium with initial stress field $-P\delta_{ij}$ can be written as

$$-\rho\omega^2 u_i - 2i\rho\omega(\mathbf{\Omega} \times \mathbf{u})_i = (\lambda e_{kk}\delta_{ij} + 2\mu e_{ij})_{,j} + \rho(\Phi_{,j}u_j)_{,i} - P_{,i}e_{kk} + \rho\phi_{,i} \tag{13.103}$$

where $\mathbf{u}\exp(-i\omega t)$ is the displacement of a small elastic vibration about its equilibrium state, and $\phi\exp(-i\omega t)$ is the variation of gravitational potential about its equilibrium value Φ. The second term on the left-hand side is the Coriolis force due to the Earth's rotation $\mathbf{\Omega}$. λ, μ, and ρ are the elastic constants and density, e_{ij} is the strain component associated with $\mathbf{u}$, and $e_{kk} = u_{k,k}$ is dilatation. The gravitational potential ϕ satisfies

$$\nabla^2\phi = 4\pi\gamma(\rho u_i)_{,i}. \tag{13.104}$$

Using the spherical coordinates (r, θ, ϕ), we shall separate the elastic constants and density into a spherically symmetric part and a small arbitrarily inhomogeneous part

$$\rho(r, \theta, \phi) = \rho_0(r) + \delta\rho(r, \theta, \phi),$$

$$\lambda(r, \theta, \phi) = \lambda_0(r) + \delta\lambda(r, \theta, \phi), \qquad (13.105)$$

$$\mu(r, \theta, \phi) = \mu_0(r) + \delta\mu(r, \theta, \phi).$$

Putting (13.105) into (13.103) and (13.104) and keeping only first-order terms of perturbation, we have the following equations for $\mathbf{u}$ and ϕ:

$$H_i(\mathbf{u}, \phi) = H_{0i}(\mathbf{u}, \phi) + \delta H_i(\mathbf{u}, \phi) = -\rho_0\omega^2 u_i,$$

$$\mathcal{L}(\mathbf{u}, \phi) = \mathcal{L}_0(\mathbf{u}, \phi) + \delta\mathcal{L}(\mathbf{u}, \phi) = 0, \qquad (13.106)$$

where

$$H_{0i}(\mathbf{u}, \phi) = (\lambda_0 e_{kk}\delta_{ij} + 2\mu_0 e_{ij})_{,j} - \rho_0\Phi_{0,r}e_{kk}\delta_{ir}$$
$$+ \rho_0(\Phi_{0,r}u_r)_{,i} + \rho_0\phi_{,i}, \qquad (13.107)$$

$$\mathcal{L}_0(\mathbf{u}, \phi) = \frac{1}{4\pi\gamma}\nabla^2\phi - (\rho_0 u_i)_{,i}, \qquad (13.108)$$

$$\delta H_i(\mathbf{u}, \phi) = \delta\rho\omega^2 u_i + 2i\rho_0\omega(\mathbf{\Omega} \times \mathbf{u})_i + (\delta\lambda e_{kk}\delta_{ij} + 2\delta\mu e_{ij})_{,j}$$
$$+ \delta\rho[(\Phi_{0,r}u_r)_{,i} - \Phi_{0,r}\delta_{ri}e_{kk}]$$
$$+ \rho_0[(\delta\Phi_{,j}u_j)_{,i} - \delta\Phi_{,j}e_{kk}] + \delta\rho\phi_{,i}, \qquad (13.109)$$

$$\delta\mathcal{L}(\mathbf{u}, \phi) = -(\delta\rho u_i)_{,i}, \qquad (13.110)$$

and $\delta\Phi$ is the perturbation of the gravity field due to the density perturbation. $\delta\Phi$ satisfies Poisson's equation:

$$\nabla^2\,\delta\Phi = -4\pi\gamma\,\delta\rho.$$

Equations (13.106), together with the appropriate homogeneous boundary conditions, define an eigenvalue problem that gives the free-oscillation periods for a laterally inhomogeneous Earth. Just as we did for the ansatz used in the preceding section to express the perturbed field as a linear combination of solutions of the unperturbed problem, we shall assume that the eigenfunction of the laterally inhomogeneous Earth can be expressed as a linear combination of the eigenfunctions (sharing the same order numbers (n, l)) of the spherically symmetric Earth. Let us consider a spheroidal mode $_nS_l$ or a torsional mode

$_nT_l$ in a spherically symmetric Earth. As shown in Chapter 8, the eigenfunctions corresponding to $_nS_l$ are

$$_n\mathbf{u}^{0Sm} = U(r)\mathbf{R}_l^m(\theta, \phi) + V(r)\mathbf{S}_l^m(\theta, \phi),$$

$$_n\phi_l^{0m} = K(r)Y_l^m(\theta, \phi),$$

(13.111)

and that corresponding to $_nT_l$ is

$$_n\mathbf{u}^{0Tm} = W(r)\mathbf{T}_l^m(\theta, \phi),$$

(13.112)

for which the vector surface harmonics are given in (8.13). For fixed l and n, all $2l + 1$ spheroidal modes $_lS_n^m(-l \le m \le l)$ have the same angular eigenfrequency $_n\omega_l^S$, and all $2l + 1$ torsional modes $_nT_l^m$ have the same angular eigenfrequency $_n\omega_l^T$. The lateral inhomogeneities will, in general, remove this degeneracy and split the unperturbed frequency into $2l + 1$ frequencies, which we write as

$$_n\omega_l^{Si} = {}_n\omega_l^S + \delta_n\omega_l^{Si} \qquad (-l \le i \le l),$$

$$_n\omega_l^{Ti} = {}_n\omega_l^T + \delta_n\omega_l^{Ti} \qquad (-l \le i \le l).$$

(13.113)

To approximate the corresponding eigenfunctions, we use the following ansatz:

$$_n\mathbf{u}_l^{Si} = \sum_m c_m^i {}_n\mathbf{u}_l^{0Sm} \qquad (-l \le i \le l),$$

$$_n\phi_l^i = \sum_m c_m^i {}_n\phi_l^{0m} \qquad (-l \le i \le l),$$

(13.114)

BOX **13.3**

Quasi-degeneracy

The use of the linear combination of unperturbed eigenfunctions with fixed (n, l) as a trial function for the eigenfunction for a perturbed medium ((13.114) and (13.115)) may be allowed if the degenerate eigenfrequencies of multiplets of the unperturbed problem are sufficiently different from each other, so that no coupling occurs. On the other hand, if the degenerate eigenvalues of multiplets of the unperturbed problem are so close to each other that the first-order corrections within any single multiplet due to perturbation are larger than the separation of different multiplets, the above form of trial function is no longer allowed. This is the case of *quasi-degeneracy*, and was studied by Dahlen (1969) for a rotating, elliptical Earth and by Luh (1973) for a laterally inhomogeneous Earth.

and

$$_n\mathbf{u}_l^{Ti} = \sum_m D_{m\,n}^i \mathbf{u}_l^{0Tm} \qquad (-l \le i \le l). \tag{13.115}$$

Putting the above ansatz into (13.106) and omitting the labels S, l, and n, we find for spheroidal modes that

$$\sum_m c_m^i [H_{0j}(\mathbf{u}^{0m}, \phi^{0m}) + \delta H_j(\mathbf{u}^{0m}, \phi^{0m})] = -\rho_0(\omega^i)^2 \sum_m c_m^i \mathbf{u}_j^{0m},$$

$$\sum_m c_m^i [\mathscr{L}_0(\mathbf{u}^{0m}, \phi^{0m}) + \delta\mathscr{L}(\mathbf{u}^{0m}, \phi^{0m})] = 0. \tag{13.116}$$

The eigenfunctions $(\mathbf{u}^{0m}, \phi^{0m})$ for the unperturbed medium satisfy the equations

$$H_{0i}(\mathbf{u}^0, \phi^0) = -\rho_0\omega^2 u_i^0,$$

$$\mathscr{L}_0(\mathbf{u}^0, \phi^0) = 0. \tag{13.117}$$

Subtracting (13.117) from (13.116), we obtain

$$\sum_m c_m^i \, \delta H_j(\mathbf{u}^{0m}, \phi^{0m}) = -\rho_0 \, \delta(\omega^i)^2 \sum_m c_m^i \mathbf{u}_j^{0m},$$

$$\sum_m c_m^i \, \delta\mathscr{L}(\mathbf{u}^{0m}, \phi^{0m}) = 0. \tag{13.118}$$

Taking the inner product of (13.118) with each eigenfunction for the unperturbed medium, we find

$$\sum_{m=-l}^l c_m^i \, \delta H_{m'm} = \delta(\omega^i)^2 c_{m'}^i \qquad (-l \le i \le l) \tag{13.119}$$

where

$$\delta H_{m'm} = \frac{-\int_v [u_j^{0m'*} \, \delta H_j(\mathbf{u}^{0m}, \phi^{0m}) + \phi^{0m'*} \, \delta\mathscr{L}(\mathbf{u}^{0m}, \phi^{0m})] \, dV}{\int_v \rho_0 |\mathbf{u}^{0m'}|^2 \, dV} \tag{13.120}$$

(* indicating the complex conjugate).

The system of linear equations (13.119) defines an eigenvalue problem. For nonvanishing c_m^i, $\delta(\omega^i)^2$ must be the eigenvalue of matrix $\delta H_{m'm}$. Once the eigenvalue is determined, the corresponding eigenfunction can be obtained by putting the solution c_m^i of (13.119) into (13.114).

For toroidal oscillations, for example, the frequency perturbation is the eigenvalue of matrix $\delta \mathbf{H}$ with elements

$$\delta H_{m'm} = -\frac{\int_v u_i^{0m'*}(2\,\delta \mu e_{ij}^{0m})_{,j}\, dV + (\omega)^2 \int \delta \rho u_i^{0m'*} u_i^{0m}\, dV}{\int_v \rho_0 |\mathbf{u}^{0m'}|^2\, dV}, \quad (13.121)$$

where $\delta \mathbf{H}$ is shown explicitly in terms of $\delta \mu$ and $\delta \rho$, as given in (13.109); the effect of the Earth's rotation, discussed in Chapter 8, is neglected. One can show, by using Gauss's theorem and the traction-free condition on the Earth's surface, that the matrix $\delta H_{m'm}$ is Hermitian:

$$\delta H_{m'm} = \delta H_{mm'}^*. \quad (13.122)$$

Hence the eigenvalues $\delta(\omega^i)^2 \sim 2\omega\,\delta\omega^i$ of (13.119) will be real.

In order to evaluate $\delta H_{m'm}$, one must know eigenfunctions u_i^{0m} for the unperturbed medium. Since their dependence on (θ, ϕ) are given by $Y_l^m(\theta, \phi)$ as in (13.111) and (13.112), it is natural to expand the medium perturbations $\delta\lambda$, $\delta\mu$, and $\delta\rho$ also by spherical harmonics:

$$\delta\lambda(r, \theta, \phi) = \sum_{s=1}^{\infty} \sum_{t=-s}^{s} \delta\lambda_s^t(r) Y_s^t(\theta, \phi),$$

$$\delta\mu(r, \theta, \phi) = \sum_{s=1}^{\infty} \sum_{t=-s}^{s} \delta\mu_s^t(r) Y_s^t(\theta, \phi), \quad (13.123)$$

$$\delta\rho(r, \theta, \phi) = \sum_{s=1}^{\infty} \sum_{t=-s}^{s} \delta\rho_s^t(r) Y_s^t(\theta, \phi)$$

Writing the volume element as $dV = r^2\, dr \cdot d\Omega$, the integral in (13.121) can be separated into radial and angular integrals. The radial part must be evaluated numerically. The angular part is reduced to the integral of products of three spherical harmonics:

$$I = \int_{\Omega_0} d\Omega\, Y_l^{m'*}(\theta, \phi) Y_s^t(\theta, \phi) Y_l^m(\theta, \phi), \quad (13.124)$$

where Ω_0 is the surface of a sphere of unit radius. The above integral may be expressed in terms of Clebsch-Gordon coefficients, which are familiar from the quantum theory of angular momentum (Edmonds, 1960),

$$I = \sqrt{\frac{2s+1}{4\pi}}\,(stlm|sllm')(s0l0|sll0). \quad (13.125)$$

The bracketed terms are Clebsch-Gordon coefficients. They can be expanded in an integer series that is easily evaluated by a computer.

The angular integral I is different from zero only if

$$\begin{array}{lll} \text{(i)} & s \text{ is even,} & \\ \text{(ii)} & s < 2l, \text{ and} & (13.126) \\ \text{(iii)} & m' = t + m. & \end{array}$$

These three relations between the degree and order of the spherical harmonics are called selection rules. They have important and interesting implications on the effect of lateral heterogeneity on the free-oscillation periods.

Rule (i) shows that the lateral inhomogeneity expressed by spherical harmonics of odd order number does not show an observable effect on free-oscillation periods. Obviously, then, we cannot uncover the odd harmonics of lateral inhomogeneity from observed free-oscillation periods. This is true, however, only when the assumptions used in the perturbation method are valid. They may break down if two different modes of the unperturbed medium have very close frequencies. In such a case, the method of quasi-degeneracy is applicable (Box 13.3).

Rule (ii) shows that free oscillations of order l are affected only by the lateral inhomogeneity harmonics of orders lower than $2l$. In other words, those inhomogeneities whose scale length is shorter than half the wavelength of the free oscillation do not affect the eigenfrequency. Although from Figure 13.11 we expected weak scattering for small ka, the value of ka corresponding to the inhomogeneity scale length equal to half the wavelength is π, a rather high value. It is interesting to note that, as can be seen in Figure 13.11, the scattering effect for plane waves in an unbounded random medium may be neglected only for kL smaller than about 10 if ka is equal to π.

Rule (iii) shows that the coupling between the unperturbed eigenfunctions $_n\mathbf{u}_l^m$ and $_n\mathbf{u}_l^{m'}$ occurs only through inhomogeneity harmonics with the azimuthal order number t equal to $m' - m$. We have seen a similar relation earlier in this chapter. For example, we found in (13.81) that the wavenumber spectra of primary waves, scattered waves, and inhomogeneity in laterally inhomogeneous layered media are related by a convolution in k-space. If the inhomogeneity and primary waves had single wavenumbers k_0 and k_1, respectively, the convolution would give $\delta(k - k_0) * \delta(k - k_1) = \delta(k - k_0 - k_1)$. Therefore, the scattered wave will have the single wavenumber $k_0 + k_1$. This is essentially rule (iii).

From rule (iii), it follows that the diagonal element δH_{mm} contains only $Y_s^0(\theta, \phi)$ and the sum of the diagonal elements will have a factor of the form

$$\int_{\Omega_0} d\Omega\, Y_s^0(\theta, \phi) \sum_{m=-l}^{l} Y_l^{m'*}(\theta, \phi) Y_l^m(\theta, \phi).$$

Then the addition theorem for spherical harmonics gives

$$\sum_{m=-l}^{l} Y_l^{m*}(\theta, \phi) Y_l^m(\theta, \phi) = \frac{2l+1}{4\pi} P_l(0) = \frac{2l+1}{4\pi},$$

and since

$$\int_{\Omega_0} d\Omega Y_s^0(\theta, \phi) = 0,$$

we have

$$\sum_{m=-l}^{l} \delta H_{mm} = 0. \tag{13.127}$$

Because of the invariance of the trace of a matrix, we find that the sum of eigenvalues $\delta_n\omega_l^i$ is also zero:

$$\sum_{i=-l}^{l} \delta_n\omega_l^i = 0. \tag{13.128}$$

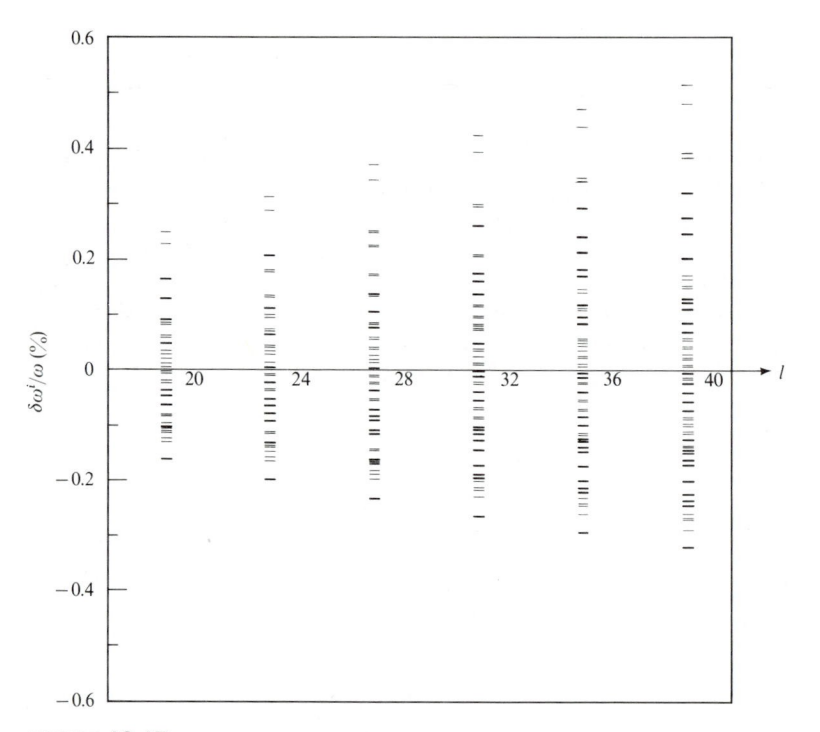

FIGURE **13.17**

Spectral splitting of toroidal modes for several different order numbers for the ocean-continent heterogeneity with the model CANSD under continents (Brune and Dorman, 1963) and 8099 under oceans (Dorman et al., 1960). [From Madariaga and Aki, 1972; copyrighted by the American Geophysical Union.]

This is called the *diagonal sum rule*, and was first applied to the perturbations of free oscillations by Gilbert (1971a). Because of this rule, we can be sure that the average of frequencies split by lateral heterogeneity is equal to the frequency for the unperturbed, spherically symmetric Earth model. If we work with observations of free-oscillation frequencies that are an average of the frequencies split by lateral heterogeneity, then these observations are relevant to a spherically symmetric Earth model obtained from the actual Earth by lateral averaging.

The actual calculation of $\delta\omega^i$ was made by Madariaga (1972) for torsional oscillations and by Luh (1973) for spheroidal oscillations. Figure 13.17 shows an example.

13.6 Finite-difference Method

The finite-difference method offers a most direct and straightforward path from the problem formulated in terms of basic equations, initial and boundary conditions to the digital computer, for solution with a minimum of analytical effort. The method is general and flexible, and may be applied to an arbitrarily inhomogeneous body of any shape. The method is essentially similar to laboratory simulation using a scale model, but has greater advantage over the latter in accuracy of the result, in ease of preparing a model and a seismic source without any special skill other than computer programming, and in reducing the effect of artificial boundaries imposed by the finiteness inherent to models.

The size and complexity of a problem that can be solved by the finite-difference method is limited by the capability of the available computer. It is, therefore, not a trivial matter to find an algorithm that minimizes the required memory size and computer time. An efficient algorithm must be based upon maximum exploitation of symmetry of a given problem, simplification of basic equations and boundary conditions allowable for the intended accuracy, and optimal choices of grid configurations, finite-difference formulas, and conditions at the artificial boundaries.

Let us start with the simplest case of a one-dimensional wave equation to illustrate the elements of the finite-difference method as applied in seismology.

13.6.1 *One-dimensional wave equation*

Consider a medium in which the elastic constants and density vary only in one direction, x. The equation of motion for displacement associated with plane waves propagating in the same direction is

$$\rho(x)\frac{\partial^2 u}{\partial t^2} = \frac{\partial}{\partial x}\left[E(x)\frac{\partial u}{\partial x}\right], \tag{13.129}$$

where $E(x) = \lambda(x) + 2\mu(x)$ for P-waves and $E(x) = \mu(x)$ for S-waves.

To avoid taking space derivatives of $E(x)$, we shall use the particle velocity $\dot{u} = \partial u/\partial t$ and stress $\tau = E(x)\,\partial u/\partial x$ as variables. Then equation (13.129) is replaced by the following simultaneous equations:

$$\frac{\partial \dot{u}}{\partial t} = \frac{1}{\rho(x)}\frac{\partial \tau}{\partial x},$$

$$\frac{\partial \tau}{\partial t} = E(x)\frac{\partial \dot{u}}{\partial x}. \tag{13.130}$$

There are many ways to approximate the above equations by finite-difference formulas. To show the importance of the choice of a proper scheme, we shall first give an example of a bad scheme that is always unstable. Let us sample the xt-plane at points $(l\,\Delta t, m\,\Delta x)$, where l and m are integers. As shown in Figure 13.18, we shall use the values of τ and $\dot{u}$ at three space points $(l, m - 1)$, (l, m), and $(l, m + 1)$ at the same time to find their values at the next time point $(l + 1, m)$. Equation (13.130) can be approximated by central differences for x-derivatives and forward differences for t-derivatives:

$$\frac{\dot{u}_m^{l+1} - \dot{u}_m^l}{\Delta t} = \frac{1}{\rho_m}\frac{\tau_{m+1}^l - \tau_{m-1}^l}{2\,\Delta x},$$

$$\frac{\tau_m^{l+1} - \tau_m^l}{\Delta t} = E_m\frac{\dot{u}_{m+1}^l - \dot{u}_{m-1}^l}{2\,\Delta x}. \tag{13.131}$$

To find the stability of the above scheme, we consider initial errors in τ and $\dot{u}$ at $t = 0$ and see if the errors will grow with time. Assuming a disturbance of the form $\exp(ikx - i\omega t)$, we write the error at (l, m) as

$$E(\tau_m^l) = A\,\exp(-i\omega l\,\Delta t + ikm\,\Delta x),$$

$$E(\dot{u}_m^l) = B\,\exp(-i\omega l\,\Delta t + ikm\,\Delta x). \tag{13.132}$$

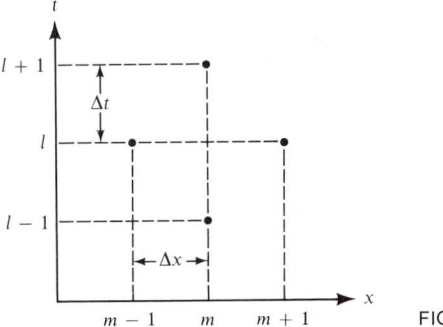

FIGURE **13.18**

BOX **13.4**

Finite-difference approximations to derivatives

For a function $\phi(x)$ that can be expanded in a Taylor series, we have

$$\phi(x \pm \Delta x) = \phi(x) \pm \frac{\partial \phi}{\partial x} \Delta x + \frac{1}{2} \frac{\partial^2 \phi}{\partial x^2} (\Delta x)^2$$

$$\pm \frac{1}{6} \frac{\partial^3 \phi}{\partial x^3} (\Delta x)^3 + \text{higher-order terms.}$$

We define the forward-difference formula by

$$\frac{\partial \phi}{\partial x} = \frac{1}{\Delta x} [\phi(x + \Delta x) - \phi(x)]$$

and the backward-difference formula by

$$\frac{\partial \phi}{\partial x} = \frac{1}{\Delta x} [\phi(x) - \phi(x - \Delta x)].$$

The truncation error for both formulas has a leading term proportional to Δx. On the other hand, the central-difference formula defined by

$$\frac{\partial \phi}{\partial x} = \frac{1}{2 \Delta x} [\phi(x + \Delta x) - \phi(x - \Delta x)]$$

has a leading error of order $(\Delta x)^2$.

By adding the Taylor series for $\phi(x + \Delta x)$ and $\phi(x - \Delta x)$, we can easily find a formula for $\partial^2 \phi / \partial x^2$:

$$\frac{\partial^2 \phi}{\partial x^2} = \frac{1}{(\Delta x)^2} [\phi(x + \Delta x) - 2\phi(x) + \phi(x - \Delta x)].$$

The leading term of error for this formula is also of order $(\Delta x)^2$. See, e.g., Abramowitz and Stegun (1965) for formulas of higher order.

For fixed real k, we shall find the dispersion relation $\omega(k)$, which will describe the time dependence of the error.

Since the errors satisfy the same equation as the solution, we put (13.132) into (13.131) to find

$$B[\exp(-i\omega \Delta t) - 1] = \frac{\Delta t}{2\rho_m \Delta x} A[\exp(ik \Delta x) - \exp(-ik \Delta x)],$$

$$A[\exp(-i\omega \Delta t) - 1] = \frac{E_m \Delta t}{2 \Delta x} B[\exp(ik \Delta x) - \exp(-ik \Delta x)].$$

Eliminating A/B, we get

$$[\exp(-i\omega\,\Delta t) - 1]^2 = -\frac{E_m}{\rho_m}\left(\frac{\Delta t}{\Delta x}\right)^2 (\sin k\,\Delta x)^2$$

or

$$\exp(-i\omega\,\Delta t) = 1 \pm i\sqrt{\frac{E_m}{\rho_m}}\left(\frac{\Delta t}{\Delta x}\right)\sin k\,\Delta x.$$

Thus $|\exp(-i\omega\,\Delta t)| > 1$ for any choice of $\Delta t/\Delta x$, so ω must be complex and the error given by (13.132) will grow exponentially with time step l.

Let us now use the central-difference formula also for the t-derivative. The new formula shown below will give the solution for $(l + 1, m)$ from the values at $(l, m - 1)$, $(l, m + 1)$, and $(l - 1, m)$, as shown in Figure 13.19:

$$\frac{\dot{u}_m^{l+1} - \dot{u}_m^{l-1}}{2\,\Delta t} = \frac{1}{\rho_m}\frac{\tau_{m+1}^l - \tau_{m-1}^l}{2\,\Delta x},$$

$$\frac{\tau_m^{l+1} - \tau_m^{l-1}}{2\,\Delta t} = E_m\frac{\dot{u}_{m+1}^l - \dot{u}_{m-1}^l}{2\,\Delta x}. \tag{13.133}$$

Using the same ansatz (13.132), we get from (13.133)

$$B[\exp(i\omega\,\Delta t) - \exp(-i\omega\,\Delta t)] = \frac{-\Delta t}{\rho_m\,\Delta x}A[\exp(ik\,\Delta x) - \exp(-ik\,\Delta x)],$$

$$A[\exp(i\omega\,\Delta t) - \exp(-i\omega\,\Delta t)] = \frac{-E_m\,\Delta t}{\Delta x}B[\exp(ik\,\Delta x) - \exp(-ik\,\Delta x)].$$

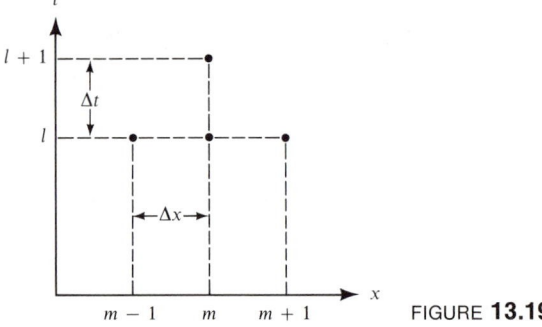

FIGURE **13.19**

Eliminating A/B, we have

$$\sin \omega \, \Delta t = \pm \sqrt{\frac{E_m}{\rho_m} \frac{\Delta t}{\Delta x}} \sin k \, \Delta x \qquad (13.134)$$

In this case, if $\sqrt{E_m/\rho_m} \, (\Delta t/\Delta x) \leq 1$, then $|\sin \omega \, \Delta t|$ is always less than or equal to 1. Consequently, ω is real, and the error will not grow with time. Thus the condition for stability is to make the time-mesh interval Δt smaller than or equal to $\Delta x/c_m$, where $c_m = \sqrt{E_m/\rho_m}$ is the local wave velocity. In the above discussion, it is implicitly assumed that the ansatz (13.132) applies to a region in which E_m and ρ_m are nearly constant.

A still better scheme can be constructed by the use of staggered grids in which the grid for $\dot{u}$ is shifted from the grid for τ by half a grid length both in t and x, as shown in Figure 13.20. We then use the following finite-difference approximation to equation (13.130):

$$\frac{\dot{u}_m^{l+\frac{1}{2}} - \dot{u}_m^{l-\frac{1}{2}}}{\Delta t} = \frac{1}{\rho_m} \frac{\tau_{m+\frac{1}{2}}^l - \tau_{m-\frac{1}{2}}^l}{\Delta x},$$

$$\frac{\tau_{m+\frac{1}{2}}^{l+1} - \tau_{m+\frac{1}{2}}^l}{\Delta t} = E_{m+\frac{1}{2}} \frac{\dot{u}_{m+1}^{l+\frac{1}{2}} - \dot{u}_m^{l+\frac{1}{2}}}{\Delta x}. \qquad (13.135)$$

The leading term of error in the above approximation is four times smaller than that for the preceding case given by (13.133), because the error is proportional to the square of sampling interval, and the interval was shortened by one-half. In this case,

$$\sin \frac{\omega \, \Delta t}{2} = \sqrt{\frac{E_{m+\frac{1}{2}}}{\rho_m} \frac{\Delta t}{\Delta x}} \sin \frac{k \, \Delta x}{2}. \qquad (13.136)$$

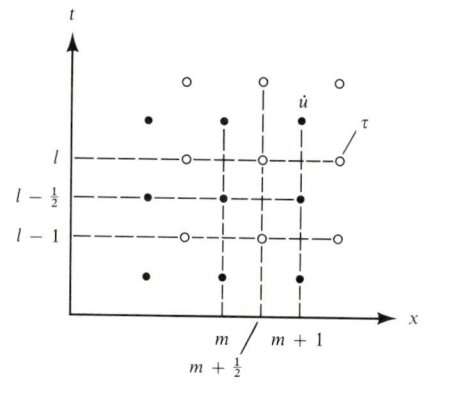

FIGURE **13.20**

Thus the stability requirement for $\Delta t/\Delta x$ is the same as that in the preceding case.

Finally, let us find the finite-difference equation for $\dot{u}$ alone by eliminating τ from (13.135). Consider four neighboring rhombuses, as shown in Figure 13.21. The following equations apply to these rhombuses:

$$\dot{u}_m^{l+\frac{1}{2}} - \dot{u}_m^{l-\frac{1}{2}} = \frac{1}{\rho_m}\frac{\Delta t}{\Delta x}(\tau_{m+\frac{1}{2}}^l - \tau_{m-\frac{1}{2}}^l),$$

$$\tau_{m-\frac{1}{2}}^l - \tau_{m-\frac{1}{2}}^{l-1} = E_{m-\frac{1}{2}}\frac{\Delta t}{\Delta x}(\dot{u}_m^{l-\frac{1}{2}} - \dot{u}_{m-1}^{l-\frac{1}{2}}),$$

$$\tau_{m+\frac{1}{2}}^l - \tau_{m+\frac{1}{2}}^{l-1} = E_{m+\frac{1}{2}}\frac{\Delta t}{\Delta x}(\dot{u}_{m+1}^{l-\frac{1}{2}} - \dot{u}_m^{l-\frac{1}{2}}),$$ (13.137)

$$\dot{u}_m^{l-\frac{1}{2}} - \dot{u}_m^{l-\frac{3}{2}} = \frac{1}{\rho_m}\frac{\Delta t}{\Delta x}(\tau_{m+\frac{1}{2}}^{l-1} - \tau_{m-\frac{1}{2}}^{l-1}).$$

Eliminating all τ from the above equations and rewriting the subscript $l - \frac{1}{2}$ as l, we have

$$\rho_m\frac{(\dot{u}_m^{l+1} - 2\dot{u}_m^l + \dot{u}_m^{l-1})}{(\Delta t)^2} = \frac{E_{m-\frac{1}{2}}}{(\Delta x)^2}(\dot{u}_m^l - \dot{u}_{m-1}^l) + \frac{E_{m+\frac{1}{2}}}{(\Delta x)^2}(\dot{u}_{m+1}^l - \dot{u}_m^l).$$ (13.138)

The right-hand side of the above equation can be obtained by applying the formula

$$\left(\frac{\partial \dot{u}}{\partial x}\right)_m = \frac{\dot{u}_{m+\frac{1}{2}} - \dot{u}_{m-\frac{1}{2}}}{\Delta x}$$ (13.139)

consecutively in $\partial/\partial x[E(x)\,\partial\dot{u}/\partial x]$. Since the left-hand side is the central-difference approximation to $\rho(\partial^2\dot{u}/\partial t^2)$, equation (13.138) can be directly obtained from the equation of motion for u using (13.139). The truncation error and stability condition for equation (13.137) should be the same as in the staggered-grid method.

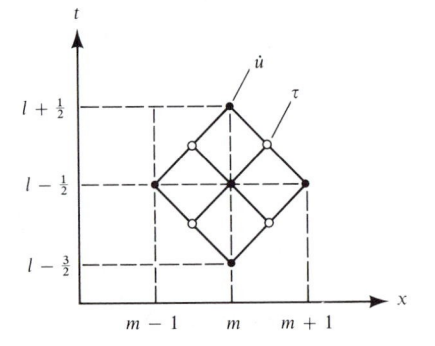

FIGURE **13.21**

The stability conditions obtained for finite-difference equations (13.133), (13.135), and (13.138) are physically understandable. Since any disturbance in the medium is propagated with the velocity c, what happens at the time-space point $(l + 1, m)$ cannot be affected by the values at (l, m'), where $m' > m + 1$ or $m' < m - 1$ if Δt is less than $\Delta x/c$. In that case, we have a physical ground for the fact that the value at $(l + 1, m)$ may be determined uniquely by the values at $(l, m - 1)$, $(l, m + 1)$, and $(l - 1, m)$, as given, e.g., in (13.133). On the other hand, if Δt is greater than $\Delta x/c$, the effects from values at (l, m') other than $(l, m - 1)$ and $(l, m + 1)$ are expected, and the unique determination of the value at $(l + 1, m)$ by equation (13.133) is impossible.

From the above physical arguments, it may be expected that if we use grid points $(l + 1, m - 1)$ and $(l + 1, m + 1)$ in the formula for determining $(l + 1, m)$, there will be no stability problem, because these two points contain all the information about the effects on $(l + 1, m)$. The formula based on this idea is used by Crank and Nicolson (1947) for a diffusion equation. Since each finite-difference equation contains three unknowns, a large system of linear equations with a tri-diagonal matrix must be solved. The application of Crank-Nicolson's method to seismic problems is described by Claerbout (1976).

The stability condition (13.136) also gives the relation between frequency ω and wavenumber k to be satisfied by the solution. The phase and group velocities determined by (13.136) approach the correct local values if Δt and Δx are small, because for small Δt and Δx we can approximate $\sin \theta$ by θ and obtain, from (13.136),

$$\frac{\omega}{k} = \sqrt{\frac{E_{m+\frac{1}{2}}}{\rho_m}} = c_0.$$

It is important to know how small Δx should be in order for the above relation to be approximately correct. The limit depends on the wavelength $\lambda = 2\pi/k$. Rewriting (13.136), we find

$$\frac{\omega}{k} = \frac{\Delta x}{\pi \Delta t} \frac{\lambda}{\Delta x} \sin^{-1}\left(c_0 \frac{\Delta t}{\Delta x} \sin \frac{\pi \Delta x}{\lambda}\right).$$

The corresponding group velocity is given by

$$\frac{\partial \omega}{\partial k} = \frac{c_0 \cos \frac{\pi \Delta x}{\lambda}}{\left[1 - \left(c_0 \frac{\Delta t}{\Delta x} \sin \frac{\pi \Delta x}{\lambda}\right)^2\right]^{1/2}}.$$

The group velocity becomes zero when the wavelength becomes as short as twice the grid interval. Both phase and group velocities are shown in Figure 13.22 as a function of $\Delta x/\lambda$ for several values of $\Delta t/\Delta x$. It is clear that the grid

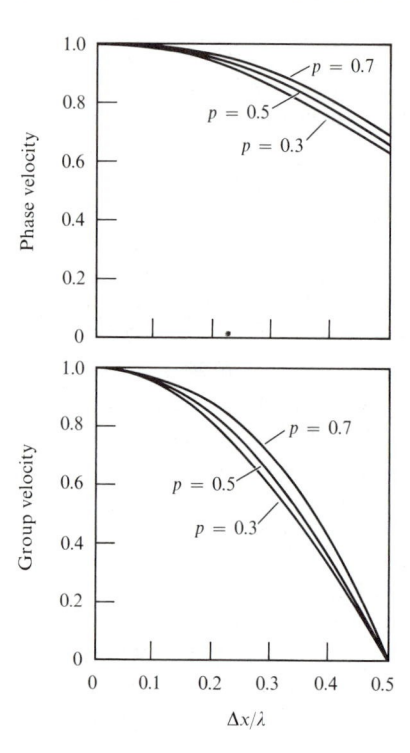

FIGURE **13.22**

Phase and group velocities normalized to c_0 for different stability ratios: $p = c_0 \, \Delta t / \Delta x$. [From Alford et al. 1974.]

interval must be less than one-tenth of the wavelength to avoid a dispersion of more than a few percent. In other words, we need about 10 grid points per wavelength. This is why the finite-difference method is applicable only to a small region in the $ka - kL$ diagram (Fig. 13.11). It is most useful for finding the wave field in the vicinity of seismic source and/or inhomogeneity.

13.6.2 Two- and three-dimensional problems

The extension of finite-difference approximations to equations of motion in two- and three-dimensional problems is straightforward, including estimation of stability and dispersion. Here we shall summarize the formulas applied successfully to some specific problems in seismology.

(i) The simplest two-dimensional problem is the anti-plane problem dealing with SH and Love waves. The basic equation for the displacement v in the y-direction is

$$\rho \frac{\partial^2 v}{\partial t^2} = \frac{\partial}{\partial x}\left(\mu \frac{\partial v}{\partial x}\right) + \frac{\partial}{\partial z}\left(\mu \frac{\partial v}{\partial z}\right), \tag{13.140}$$

where $\mu(x, z)$ is the rigidity of the medium and $\rho(x, z)$ is the density. If the medium is homogeneous, we can use the finite-difference approximation for the second derivatives given in Box 13.4 to approximate (13.140) by

$$v_{m,n}^{l+1} = 2v_{m,n}^{l} - v_{m,n}^{l-1} + \frac{\mu}{\rho}(\Delta t)^2$$

$$\times \left(\frac{v_{m+1,n}^{l} - 2v_{m,n}^{l} + v_{m-1,n}^{l}}{(\Delta x)^2} + \frac{v_{m,n+1}^{l} - 2v_{m,n}^{l} + v_{m,n-1}^{l}}{(\Delta z)^2} \right), \quad (13.141)$$

where Δz is the grid interval in the z-direction, and n is the integer specifying the z-coordinate of the grid. The above formula was used by Boore (1970, 1972) in studying scattering of SH and Love waves in a layered medium with irregular interface and topography.

The extension to the inhomogeneous medium can be done easily by applying the formula (13.136) for x- and z-derivatives. The result is

$$v_{m,n}^{l+1} = 2v_{m,n}^{l} - v_{m,n}^{l-1}$$

$$+ \frac{(\Delta t)^2}{\rho_{m,n}} \left[\frac{\mu_{m+\frac{1}{2},n} v_{m+1,n} - (\mu_{m+\frac{1}{2},n} + \mu_{m-\frac{1}{2},n}) v_{m,n} + \mu_{m-\frac{1}{2},n} v_{m-1,n}}{(\Delta x)^2} \right]$$

$$+ \frac{(\Delta t)^2}{\rho_{m,n}} \left[\frac{\mu_{m,n+\frac{1}{2}} v_{m,n+1} - (\mu_{m,n+\frac{1}{2}} + \mu_{m,n-\frac{1}{2}}) v_{m,n} + \mu_{m,n-\frac{1}{2}} v_{m,n-1}}{(\Delta z)^2} \right],$$

$$(13.142)$$

which approximates (13.140) correctly to second order in increments. The density at (m, n) and rigidity at four points surrounding (m, n) enter into the above equation, which predicts the future value of v at (m, n).

The stability condition for the two-dimensional equation is given by $\Delta t \leq \Delta x/(\sqrt{2}\beta)$, where β is the shear velocity $\sqrt{\mu/\rho}$ and Δz is taken equal to Δx. In general, the stability condition for a wave equation in a uniform medium of n spatial dimensions is given by $\Delta t \leq \Delta x/(\sqrt{n}\beta)$.

(ii) Let us next consider the two-dimensional in-plane problems (P-SV and Rayleigh waves) in which the displacement components u, w are functions of x and z and the y-component v vanishes. For an in-plane problem in a homogeneous isotropic body, the equation of motion (4.1) reduces to

$$\frac{\partial^2 u}{\partial t^2} = \alpha^2 \frac{\partial^2 u}{\partial x^2} + (\alpha^2 - \beta^2) \frac{\partial^2 w}{\partial x\, \partial z} + \beta^2 \frac{\partial^2 u}{\partial z^2},$$

$$\frac{\partial^2 w}{\partial t^2} = \alpha^2 \frac{\partial^2 w}{\partial z^2} + (\alpha^2 - \beta^2) \frac{\partial^2 u}{\partial x\, \partial z} + \beta^2 \frac{\partial^2 w}{\partial z^2},$$

$$(13.143)$$

where α and β are velocities of compressional and shear waves, respectively.

We write the above equation in a vector form as

$$\frac{\partial^2 \mathbf{u}}{\partial t^2} = \mathbf{A}\frac{\partial^2 \mathbf{u}}{\partial x^2} + \mathbf{B}\frac{\partial^2 \mathbf{u}}{\partial x\,\partial z} + \mathbf{C}\frac{\partial^2 \mathbf{u}}{\partial z^2}, \tag{13.144}$$

where

$$\mathbf{u} = \begin{pmatrix} u \\ w \end{pmatrix}, \quad \mathbf{A} = \begin{pmatrix} \alpha^2 & 0 \\ 0 & \beta^2 \end{pmatrix}, \quad \mathbf{B} = \begin{pmatrix} 0 & \alpha^2 - \beta^2 \\ \alpha^2 - \beta^2 & 0 \end{pmatrix},$$

and

$$\mathbf{C} = \begin{pmatrix} \beta^2 & 0 \\ 0 & \alpha^2 \end{pmatrix}.$$

Then, using central finite-difference approximations correct to second order in the increments, we obtain the formula

$$
\begin{aligned}
\mathbf{u}(x, z, t + \Delta t) =\ & 2\mathbf{u}(x, z, t) - \mathbf{u}(x, z, t - \Delta t) \\
& + \mathbf{A}\left(\frac{\Delta t}{\Delta x}\right)^2 [\mathbf{u}(x + \Delta x, z, t) - 2\mathbf{u}(x, z, t) + \mathbf{u}(x - \Delta x, z, t)] \\
& + \mathbf{B}\frac{(\Delta t)^2}{4\Delta x\,\Delta z}[\mathbf{u}(x + \Delta x, z + \Delta z, t) - \mathbf{u}(x + \Delta x, z - \Delta z, t) \\
& \quad - \mathbf{u}(x - \Delta x, z + \Delta z, t) + \mathbf{u}(x - \Delta x, z - \Delta z, t)] \\
& + \mathbf{C}\left(\frac{\Delta t}{\Delta z}\right)^2 [\mathbf{u}(x, z + \Delta z, t) - 2\mathbf{u}(x, z, t) + \mathbf{u}(x, z - \Delta z, t)].
\end{aligned}
\tag{13.145}
$$

This equation was used by Alterman and Rotenberg (1969) in studying seismic waves in a quarter-space, a classic problem in theoretical seismology. The application was extended to the problem of welded two-quarter-spaces by Ottaviani (1971). If we take $\Delta x = \Delta z$, the stability condition for the formula (13.145) is given by $\Delta t \leq \Delta x/(\alpha^2 + \beta^2)^{1/2}$.

(iii) Some three-dimensional seismic problems having especially simple symmetry have been solved by the finite-difference method. One such example is the problem of seismic-wave generation from an expanding circular shear crack. This problem, studied by Madariaga (1976), is described in Section 15.1.5. Here we shall give another example from Alterman and Karal (1968), who first demonstrated the use of the finite-difference method to the seismological community by successfully applying it to the problem of an explosion source in a layered medium.

Assuming that the direct waves from the explosion source have a spherical symmetry, the resultant wave field in the layered medium will have an azimuthal symmetry around the vertical axis through the center of the explosion. Taking the axis of symmetry as the z-axis, the equation of motion for radial displacement $A(r, z)$ and vertical displacement $B(r, z)$ in a homogeneous isotropic body can be reduced to

$$\frac{\partial^2 A}{\partial t^2} = \alpha^2 \left[\frac{\partial^2 A}{\partial r^2} + \frac{1}{r}\frac{\partial A}{\partial r} - \frac{1}{r^2}A + \frac{\partial^2 B}{\partial r\,\partial z} \right] + \beta^2 \left[\frac{\partial^2 A}{\partial z^2} - \frac{\partial^2 B}{\partial r\,\partial z} \right],$$

$$\frac{\partial^2 B}{\partial t^2} = \alpha^2 \left[\frac{\partial^2 B}{\partial z^2} + \frac{1}{r}\frac{\partial A}{\partial r} - \frac{\partial^2 A}{\partial z\,\partial r} \right] + \beta^2 \left[\frac{\partial^2 B}{\partial r^2} - \frac{\partial^2 A}{\partial z\,\partial r} - \frac{1}{r}\frac{\partial A}{\partial z} + \frac{1}{r}\frac{\partial B}{\partial r} \right].$$

$$(13.146)$$

The azimuthal component of displacement vanishes. Expressing the grid intervals in r, z, and t, respectively, as Δr, Δz, and Δt, and the grid point locations as $(m\,\Delta r, n\,\Delta z, l\,\Delta t)$, the finite-difference approximation to (13.146) may be written as follows:

$$A_{m,n}^{l+1} = 2A_{m,n}^{l} - A_{m,n}^{l-1} + \alpha^2 \left(\frac{\Delta t}{\Delta r}\right)^2 [A_{m+1,n}^{l} - 2A_{m,n}^{l} + A_{m-1,n}^{l}]$$

$$+ \frac{\alpha^2}{2}\left(\frac{\Delta t}{\Delta r}\right)^2 \frac{1}{m}[A_{m+1,n}^{l} - A_{m-1,n}^{l}] - \alpha^2 \left(\frac{\Delta t}{\Delta r}\right)^2 \frac{1}{m^2} A_{m,n}^{l}$$

$$+ \frac{1}{4}\frac{(\Delta t)^2}{\Delta r\,\Delta z}(\alpha^2 - \beta^2)[B_{m+1,n+1}^{l} - B_{m+1,n-1}^{l} - B_{m-1,n+1}^{l} + B_{m-1,n-1}^{l}]$$

$$+ \beta^2 \left(\frac{\Delta t}{\Delta z}\right)^2 [A_{m,n+1}^{l} - 2A_{m,n}^{l} + A_{m,n-1}^{l}], \qquad (13.147)$$

$$B_{m,n}^{l+1} = 2B_{m,n}^{l} - B_{m,n}^{l-1} + \alpha^2 \left(\frac{\Delta t}{\Delta z}\right)^2 [B_{m,n+1}^{l} - 2B_{m,n}^{l} + B_{m,n-1}^{l}]$$

$$+ \frac{1}{2}\frac{(\Delta t)^2}{\Delta r\,\Delta z}(\alpha^2 - \beta^2)\frac{1}{m}[A_{m,n+1}^{l} - A_{m,n-1}^{l}]$$

$$+ \frac{1}{4}\frac{(\Delta t)^2}{\Delta r\,\Delta z}(\alpha^2 - \beta^2)[A_{m+1,n+1}^{l} - A_{m+1,n-1}^{l} - A_{m-1,n+1}^{l} + A_{m-1,n-1}^{l}]$$

$$+ \frac{\beta^2}{2}\left(\frac{\Delta t}{\Delta r}\right)^2 \frac{1}{m}[B_{m+1,n}^{l} - B_{m-1,n}^{l}]$$

$$+ \beta^2 \left(\frac{\Delta t}{\Delta r}\right)^2 [B_{m+1,n}^{l} - 2B_{m,n}^{l} + B_{m-1,n}^{l}]. \qquad (13.148)$$

All derivatives appearing in the above equations are again correct to second order in the increments of respective variables.

The above formulas cannot apply to $m = 0$, because they contain the factor $1/m$; $m = 0$ corresponds to $r = 0$, i.e., the axis of azimuthal symmetry. Because of symmetry and the continuity of displacement at $r = 0$, the radial component must vanish at $r = 0$. Thus, at $m = 0$, instead of (13.147) we have

$$A_{0,n}^{l+1} = 0. \tag{13.149}$$

The terms containing $1/m$ in (13.148) correspond to $1/r(\partial B/\partial r)$ and $1/r(\partial A/\partial z)$ in the equation of motion. Except at the source point, physically, they should not become infinity at $r = 0$. For example, we expect that

$$\lim_{r \to 0} \frac{1}{r} \frac{\partial B}{\partial r} = \text{const.} = C_1$$

or, for small r,

$$\frac{\partial B}{\partial r} = C_1 r \quad \text{and} \quad \frac{\partial^2 B}{\partial r^2} = C_1.$$

Thus we can replace $1/r(\partial B/\partial r)$ by $\partial^2 B/\partial r^2$, and $1/r(\partial A/\partial z)$ by $\partial^2 A/\partial r\, \partial z$ for small r. Putting the finite-difference formulas for $\partial^2 B/\partial r^2$ and $\partial^2 A/\partial r\, \partial z$ in (13.148), we find for $m = 0$,

$$B_{0,n}^{l+1} = 2B_{0,n}^{l} - B_{0,n}^{l-1} + \alpha^2 \left(\frac{\Delta t}{\Delta z}\right)^2 [B_{0,n+1}^{l} - 2B_{0,n}^{l} + B_{0,n-1}^{l}]$$

$$+ \frac{(\Delta t)^2}{\Delta r\, \Delta z} (\alpha^2 - \beta^2)[A_{1,n+1}^{l} - A_{1,n-1}^{l}]$$

$$+ 4\beta^2 \left(\frac{\Delta t}{\Delta r}\right)^2 (B_{1,n}^{l} - B_{0,n}^{l}), \tag{13.150}$$

where we have used symmetry relations $A_{-1,n}^{l} = -A_{1,n}^{l}$ and $B_{-1,n}^{l} = B_{1,n}^{l}$.

The stability condition for (13.147) and (13.148) was obtained by Alterman and Karal (1968) as

$$\Delta t \leq \Delta r/(\alpha^2 + \beta^2)^{1/2}$$

in the case $\Delta r = \Delta z$.

(iv) When the waves are propagating primarily in a fixed direction, and the wide-angle scattering, back-scattering, and reflection are negligible, one can

modify the wave equation to allow a considerable reduction in the number of grid points per wavelength. Let us here consider only the most simple case of a two-dimensional wave equation with velocity $c(x, z)$ varying in space:

$$-\omega^2 P = c^2(x, z)\left(\frac{\partial^2 P}{\partial x^2} + \frac{\partial^2 P}{\partial z^2}\right), \tag{13.151}$$

where the time dependence of $\exp(-i\omega t)$ is assumed.

Taking the primary direction of wave propagation along the z-axis, and denoting the spatial average of $\omega/c(x, z)$ as $\bar{k}$, we write

$$P(x, z) = Q(x, z)e^{i\bar{k}z}, \tag{13.152}$$

where $\bar{k} = \langle\omega/c(x, z)\rangle$. If the medium is homogeneous and the wave field consists of plane waves propagating in the z-direction, $Q(x, z)$ is a constant. For a weakly heterogeneous medium and waves propagating primarily in the z-direction, we then expect that $Q(x, z)$ will be a slowly varying function of x, z and may be easily handled by the finite-difference method.

Putting (13.152) into (13.151), we have

$$\frac{\partial^2 Q}{\partial x^2} + \frac{\partial^2 Q}{\partial z^2} + 2i\bar{k}\frac{\partial Q}{\partial z} + \left[\left(\frac{\omega}{c(x, z)}\right)^2 - \bar{k}^2\right]Q = 0 \tag{13.153}$$

We now make an important assumption that $\partial^2 Q/\partial z^2$ may be neglected as compared to other terms in (13.153). Then (13.153) becomes

$$\frac{\partial^2 Q}{\partial x^2} + 2i\bar{k}\frac{\partial Q}{\partial z} + \left[\left(\frac{\omega}{c(x, z)}\right)^2 - \bar{k}^2\right]Q = 0. \tag{13.154}$$

The dropping of $\partial^2 Q/\partial z^2$ is sometimes called the parabolic approximation, and is justified for waves propagating in a direction not far from the z-axis. To find the effect of this approximation, we shall consider the case of a plane wave propagating in a homogeneous medium, for which $\omega/c(x, z) = \bar{k}$. Then (13.154) becomes

$$\frac{\partial^2 Q}{\partial x^2} + 2i\bar{k}\frac{\partial Q}{\partial z} = 0. \tag{13.155}$$

Since $P(x, z) = \exp(ik_x x + ik_z z)$, we have

$$Q(x, z) = \exp(ik_x x + ik_z z - i\bar{k}z).$$

Putting this into (13.155), we get

$$-k_x^2 - 2\bar{k}(k_z - \bar{k}) = 0$$

or

$$k_z = \bar{k} - \frac{k_x^2}{2\bar{k}}. \tag{13.156}$$

The above equation can be considered as an approximation to the solution of the correct wave equation,

$$k_z = (\bar{k}^2 - k_x^2)^{1/2}$$

$$= \bar{k}\left(1 - \frac{1}{2}\frac{k_x^2}{\bar{k}^2} + \cdots\right). \tag{13.157}$$

Equation (13.156) is a good approximation for small $k_x/\bar{k}$; in other words, for waves propagating in directions making small angles with the z-axis. In a wavenumber integral expression for a cylindrical wave, if the k_z in the exponent is approximated by (13.156), the resultant integral gives a parabolic wavefront.

The finite-difference formula for the homogeneous case can be obtained using the Crank-Nicolson method as

$$2i\bar{k}\left(\frac{Q_m^{n+1} - Q_m^n}{\Delta z}\right) = -\frac{Q_m^n - 2Q_m^n + Q_{m-1}^n}{2(\Delta x)^2}$$
$$-\frac{Q_{m+1}^{n+1} - 2Q_m^{n+1} + Q_{m-1}^{n+1}}{2(\Delta x)^2}, \tag{13.158}$$

where Δx and Δz are grid intervals in the x- and z-directions, $x = m\,\Delta x$, and $z = n\,\Delta z$. The formula for the inhomogeneous case (13.154) can be written similarly. As usual, we put the solution of the form

$$Q_m^n = \exp[ik_x m\,\Delta x + i(k_z - \bar{k})n\,\Delta z]$$

into (13.158) to obtain the relation,

$$\tan\left[\frac{(k_z - \bar{k})\,\Delta z}{2}\right] = \frac{-1}{\bar{k}}\frac{\Delta z}{(\Delta x)^2}\sin^2\left(\frac{k_x\,\Delta x}{2}\right). \tag{13.159}$$

We can rewrite the above equation in an explicit form for k_z as

$$\frac{k_z}{k} = 1 - \frac{1}{\pi}\left(\frac{\lambda}{\Delta z}\right) \tan^{-1}\left\{\frac{1}{2\pi}\frac{(\Delta z/\lambda)}{(\Delta x/\lambda)^2} \sin^2\left[\pi\left(\frac{\Delta x}{\lambda}\right)\sin\theta\right]\right\}, \quad (13.160)$$

where $\lambda = 2\pi/k$ is the wavelength. The above equation reduces to (13.156) as Δz and Δx approach zero. Table 13.1 shows the values of k_z/k using the correct solution (13.157), the parabolic approximation (13.156), and the finite-difference approximation to parabolic approximation (13.160) for two different grid-wavelength ratios. θ is the angle between the direction of wave propagation and the z-axis, given by $k_x = k \sin\theta$. Even with only three grid points in a wavelength for both x- and z-directions, the finite-difference approximation shows little dispersion for $\theta < 30°$. If $\lambda/\Delta x$ is taken to be 10, while keeping $\lambda/\Delta z$ at 3, we can extend the approximation to $40°$ with only a few percent dispersion. The method is expected to be useful for the interpretation of reflection data, in which waves propagating in near-vertical directions are important. The approximate differential equation (13.154) has long been well known in the U.S.S.R. (Leontovich and Fok, 1946). Claerbout (1970, 1976) used the equation as the starting point of his attempt (see Section 12.2) to apply the wave equation to the processing of seismic reflection data. Landers and Claerbout (1972) also extended the method to various elastic-wave problems involving P, S, and Rayleigh waves. Numerical advantages of the parabolic equation

TABLE **13.1**
Values of k_z/k for various approximations.

			Finite-difference approx.	
			$\dfrac{\Delta x}{\lambda} = \dfrac{1}{10}$	$\dfrac{\Delta x}{\lambda} = \dfrac{1}{3}$
θ	Correct value	Parabolic approx.	$\dfrac{\Delta z}{\lambda} = \dfrac{1}{3}$	$\dfrac{\Delta z}{\lambda} = \dfrac{1}{3}$
0	1.000	1.000	1.000	1.000
10	0.985	0.985	0.985	0.985
20	0.940	0.942	0.942	0.944
30	0.866	0.875	0.877	0.887
40	0.766	0.794	0.799	0.825
50	0.643	0.707	0.720	0.769
60	0.500	0.625	0.651	0.725
70	0.342	0.558	0.597	0.695
80	0.173	0.515	0.564	0.677
90	0.000	0.500	0.552	0.672

allow its application to the study of waves that have propagated over great distances in a laterally varying medium (i.e., for large kL in Fig. 13.11).

13.6.3 Source, interface, and boundary conditions

From the uniqueness theorem (Chapter 2), we know that the elastic dynamic field in a body is uniquely determined if (i) the initial displacement and particle velocity are known at $t = t_0$ throughout the body, (ii) the body force is given for $t > t_0$ throughout the body, and (iii) displacement *or* traction is given for $t > t_0$ on the surface enclosing the body.

Incorporating these conditions into the finite-difference scheme is usually straightforward. A common practice in introducing the seismic source is to make a part of the body laterally homogeneous so that we can use known analytic expressions for the source in that part.

For example, if we wish to study the scattering of plane body waves incident upon an inhomogeneous region, such as that shown in Figure 13.23, the analytic expressions for displacement and particle velocity for a plane wave can be used to describe the initial condition in the homogeneous part of the body. The initial values of displacement and velocity in the inhomogeneous region are zero if t_0 is taken before the arrival of the wavefront at the inhomogeneous region. The homogeneous region must be large enough so that the incident wave train can pass completely through the artificial boundary before the waves scattered back from the inhomogeneous region reach it. After the passage of incident waves, the transparent boundary condition (discussed below) may be assigned at the artificial boundary.

Similarly, in the case of surface waves incident from left or right, the initial conditions in the laterally homogeneous region and the boundary conditions that prevail at the artificial boundary before the arrival of back-scattering waves can be easily found by using the eigenfunctions of given modes Fourier-transformed into the time domain with desired frequency contents. Strictly

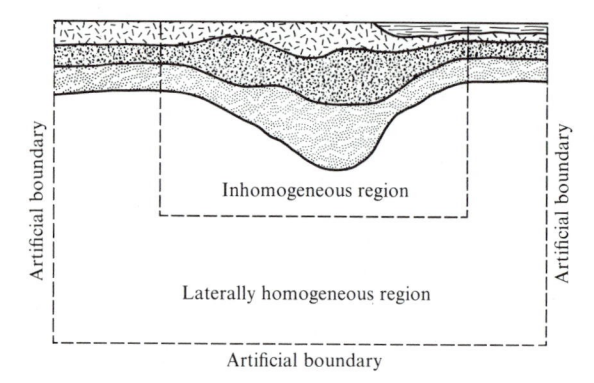

FIGURE **13.23**

The environment of an inhomogeneous region that is to be analyzed by the finite-difference method.

speaking, because of the noncausal character of a normal mode, the surface waveform cannot have a finite length. However, one can make it negligibly small outside a spatial-temporal window determined by the extent of the homogeneous region and then assign the source condition with desired accuracy.

The case of a seismic source existing in the body can be handled in essentially the same way. We shall define an inner artificial boundary S_i that encloses a homogeneous body containing a source. Then we can write the solution inside S_i as the sum of the analytic solution $\mathbf{u}_0$ for the given source in a homogeneous unbounded body and an additional term $\mathbf{u}_1$. Putting the solution outside the inner artificial boundary as $\mathbf{u}_2$, the continuity of displacement and traction at the boundary will require that

$$\mathbf{u}_2 - \mathbf{u}_1 = \mathbf{u}_0$$

and

$$\mathbf{T}_2 - \mathbf{T}_1 = \mathbf{T}_0$$

on the inner artificial boundary. Here $\mathbf{T}_0$, $\mathbf{T}_1$, and $\mathbf{T}_2$ are the tractions corresponding to $\mathbf{u}_0, \mathbf{u}_1,$ and $\mathbf{u}_2$, respectively. The values of $\mathbf{u}_0$ and $\mathbf{T}_0$ at the boundary are known as a function of time, and they will act as the seismic source for $\mathbf{u}_1$ and $\mathbf{u}_2$, because they represent discontinuities in displacement and traction (Section 3.1). By choosing the artificial boundary sufficiently far from the source singularity, we can avoid large values in the solution that may cause numerical difficulty.

When the stress must be specified at the boundary or interface, the space derivative in the normal direction may enter the condition. The derivative can be approximated by the forward difference (Box 13.4) as $(u_1 - u_0)/\Delta x$, where 0 refers to the boundary plane. This formula has error of the order of Δx. Besides, if we use this formula to determine u_0, then the equation of motion is not explicitly satisfied by u_0. An alternative method is to introduce a fictitious plane beyond the boundary plane and designate the solution there as u_{-1}. Then the value of u_{-1} can be determined from the boundary condition in which the derivative is approximated by the second-order formula $(u_1 - u_{-1})/2\,\Delta x$, and u_0 can be determined by the equation of motion applied to the segment at the boundary, including a point on the fictitious plane. At an interface, the fictitious plane can be introduced by extending one or the other medium. According to Alterman and Karal (1968), the extension of the medium with higher elastic moduli is preferred for stability of computation.

The formulation of conditions at a sloped boundary was discussed in detail by Boore (1972). In the case of an interface at which both stress and displacement are continuous, the best method apparently is to eliminate the interface and use the equation of motion, such as (13.138), for a heterogeneous medium in which the elastic constants and densities at grid points are estimated from their appropriate average around each point.

The outer artificial boundary, such as the one shown in Figure 13.23, will generate reflections and contaminate the solution unless it is made transparent or absorbing. Smith (1974) introduced an interesting method of eliminating reflection from a plane artificial boundary. Let us consider here the two-dimensional problems of *SH*- and *P-SV* waves.

For *SH*-waves, we solve the problem twice for each boundary plane. In one problem, the stress is assumed to vanish, and in the other, the displacement is fixed to zero. The sum of the two solutions will contain no reflections from the boundary. The reason is simple; the reflection coefficient for the displacement of any plane *SH*-waves is 1 for the stress-free boundary and -1 for the rigid boundary.

For *P*- and *SV*-waves, we solve the problem again twice for each boundary plane. Taking the particle displacement as well as the wave path for *P*- and *SV*-waves in the *xz*-plane, and putting the artificial boundary normal to the *x*-axis, we assign the following boundary condition at the artificial boundary:

$$u = 0,$$
$$\tau_{xy} = 0 \tag{13.161}$$

in one of the problems, and

$$w = 0,$$
$$\tau_{xx} = 0 \tag{13.162}$$

in the other. Using the appropriate wave potentials (5.15)–(5.19) for incident plane *P*- and *SV*-waves, we can calculate reflected *P*- and *SV*-waves for boundary conditions (13.161) and (13.162). For the case of incident *P*-waves, there is no *SV* reflection under either set of boundary conditions. The reflection coefficient for *P* is 1 for case (13.161) and -1 for case (13.162). Thus the sum of two solutions will contain no reflection from the artificial boundary. For the case of incident *SV*-waves, there is no *P* reflection, and the reflection coefficients for *SV* have opposite sign in the two problems. The sum of two solutions is again free of contamination by refraction. Similar results are given in Problem 5.3 for a boundary normal to the *z*-axis.

When more than one boundary plane is required to be nonreflecting, more solutions must be added to eliminate multiple reflections. For example, consider *SH*-wave reflections at two boundary planes *A* and *B*. To make each a nonreflector, we must solve two problems in which the boundary is rigid or free. If we choose *A* free and *B* rigid in one problem and *B* free and *A* rigid in the other, the wave reflected successively at *A* and *B* will have the same sign in both problems. We get the same result if we choose *A* free and *B* free in one problem and *A* rigid and *B* rigid in the other. Obviously, we need to sum solutions of four problems: *A* free, *B* rigid; *A* free, *B* free; *A* rigid, *B* rigid; and *A*

rigid, B free. In general, we must solve 2^n problems for n boundary planes. Thus, for large n, Smith's exact method will require a large computer time.

13.6.4 Examples

The finite-difference method has been successfully tested using problems with known analytic solutions, such as a point explosive source in a layered half-space (Alterman and Karal, 1968); Love waves propagating through Higuchi's medium (a special case of adjoined two-layered half-spaces; see Problem 13.2) (Boore, 1970); a line source in a medium with a quarter-space wedge (Alford et al., 1974); and others. For problems with no analytic solutions, the method may be tested by comparison with the results obtained by other approximate numerical methods such as those described earlier in this chapter.

BODY WAVES

As an example of body wave scattering, consider the seismic motion caused by SH-waves vertically incident upon a basin structure filled with low-velocity sediment, as shown in Figure 13.24 (Boore et al., 1971). The basin has the shape of a full-cycle cosine with wavelength 5 km (because of symmetry only the right half of the body is shown) and is 600 m deep at the deepest point. The shear velocity in the half-space is 3.5 km/sec, five times higher than the sediment velocity.

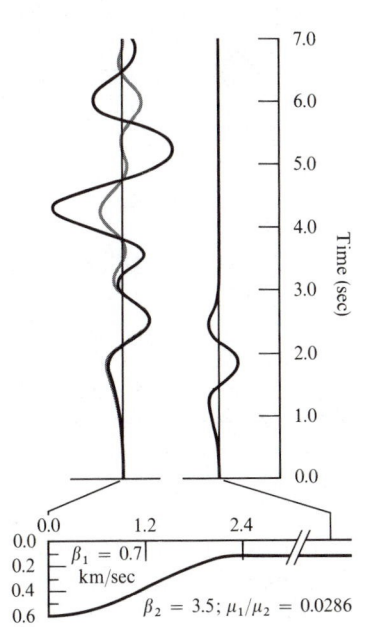

FIGURE **13.24**

Seismic motion at the surface of a basin structure due to vertical incident SH-waves. Comparison of the finite-difference result (black line) with the flat layer approximation (gray line). [From Boore, 1972.]

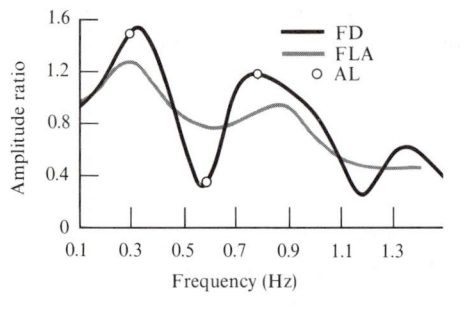

FIGURE **13.25**

The effect of sediment on surface motion as a function of frequency. The finite-difference method (solid line) gives excellent agreement with the Rayleigh method (marked AL) described in Section 13.4.3. The flat layer approximation (FLA) shows a considerable departure from them. [From Boore, 1972.]

In the finite-difference calculation, a plane SH-wave traveling vertically upward is assigned to the initial condition as well as to the subsequent condition at artificial boundaries. The displacement at the center of the basin is calculated in two different ways. The black curve shows the motion in the presence of the basin, calculated by the finite-difference method. The gray curve shows the motion when the basin structure is replaced by a flat layer with the same velocity and thickness as the basin beneath the observation point. Absolute values of the Fourier transforms of these displacements, after multiplication by an exponential window $e^{-\varepsilon t}$ (to eliminate the undesirable reflections from the artificial boundaries), are shown in Figure 13.25. They are normalized to the Fourier amplitudes of the reference signal (shown in Fig. 13.24) calculated for a surface point outside the basin in the absence of the basin. There are considerable differences between the result of finite-difference calculation and the flat-layer approximation (FLA). On the other hand, the finite-difference result agrees well with those (shown by circles) obtained by the method (AL) due to Aki and Larner (1970), which is based on the Rayleigh ansatz and a discrete-wavenumber representation (equation (13.101)), both of which were described in detail in Section 13.4.3. In applying the AL method, a complex frequency with the imaginary part ε is used to obtain the result corresponding to the exponential window (13.102) in the time domain. The excellent agreement between the two results gives confidence in both methods.

A similar basin problem was studied by Hong and Helmberger (1978) using a new method called "Glorified Optics," which is based on a generalized ray theory adapted to wave propagation in nonplanar structures. Their synthetic seismograms calculated for many multi-reflected ray paths using approximations valid for high frequencies are in surprisingly good agreement with the results obtained by the finite-difference and finite-element methods, as shown in Figure 13.26. In the figure, we added synthetic seismograms computed by P. Y. Bard and M. Bouchon using the discrete-wavenumber representation method (DW), which gives Fourier synthesis of the results obtained by the AL method. Results obtained by the DW method, the finite difference (FD), and finite-element (FE) methods are in good agreement with each other. The Glorified Optics (GO) method also gives satisfactory agreement with other

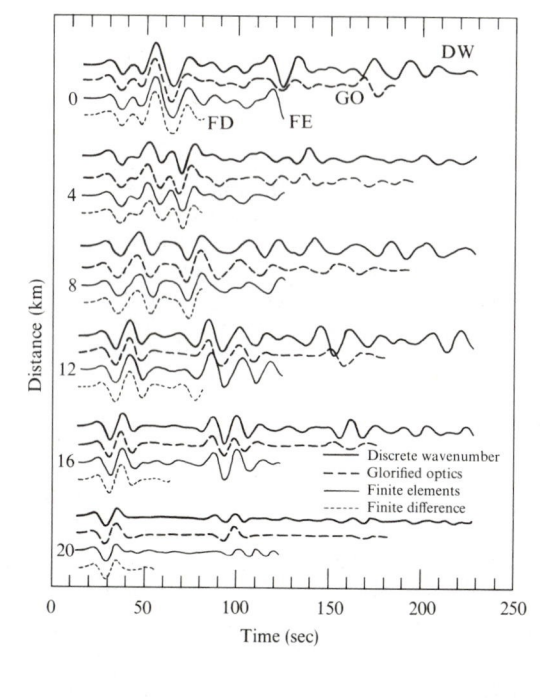

FIGURE **13.26**

Synthetic seismograms computed by various methods for motions on the surface of a basin structure due to *SH*-waves impinging vertically from the half-space. The shape of the basin and the velocities are identical to those shown in Figure 13.24. The only difference is in scale; the time and distance scales are both 10 times greater than those in Figure 13.24. The *SH*-displacement is computed at equal intervals from the center of the basin (0 km) to a horizontal distance of 20 km. This figure is adapted from Hong and Helmberger (1978) to include the results obtained by the DW method.

methods for early arrivals, although some departures are noticed for later arrivals. The GO method should work well for higher frequencies, for which other methods become time consuming.

SURFACE WAVES

As an example of the finite-difference calculation of surface-wave scattering, let us consider Love waves in a sloping layered medium, as shown in Figure 13.27, which may roughly represent the transition zone from oceanic to continental structure. Boore (1970) calculated the motion at several points on the surface above the transition zone when Love waves are incident from left or right. The space-time function for incident Love waves is computed by the Fourier synthesis of eigenfunctions that will give a desired waveform at the surface. Several different schemes were used for finite-difference approximation to the boundary condition at the sloped interface.

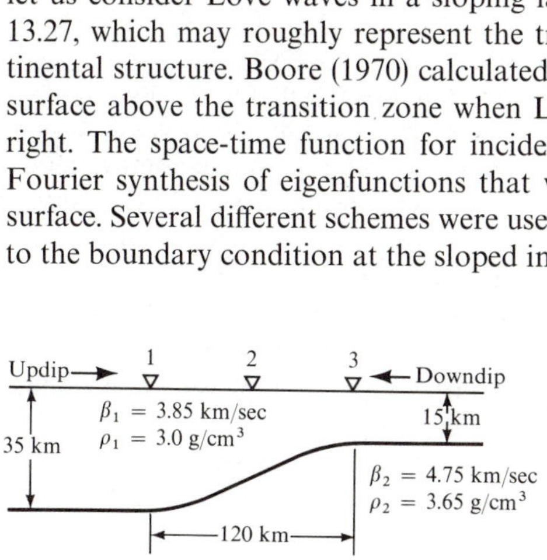

FIGURE **13.27**

Model of continent-ocean transition. [From Boore, 1972.]

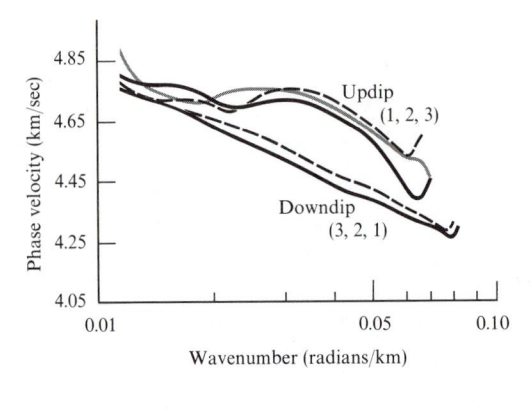

FIGURE **13.28**

Phase velocities measured using the finite-difference solutions for Love waves propagating in both directions across the continent-ocean transition model of Figure 13.27. The three different updip curves and two different downdip curves were generated by a variety of attempts to model the boundary conditions across the interface. Close similarity of the results, for each direction of propagation, is a strong indication that phase-velocity anisotropy is real, and not a numerical artifact. [From Boore, 1972.]

Simulating the actual procedure of phase-velocity measurement (see Chapter 11) over the transition zone, the records calculated for stations at the surface are Fourier-transformed, and the phase velocity is determined as the angular frequency times the travel distance between two stations divided by the increment of Fourier phase during the travel. The resultant phase-velocity curves are shown when incident waves arrive from thick crust (updip case) and from thin crust (downdip case). We find a significant difference between the two cases, as shown in Figure 13.28. The measured phase velocity is higher in the updip case than in the downdip. Let us find if other approximate methods also give a similar anisotropy in apparent phase velocity.

An approach similar to the Rayleigh ansatz used in the previous example of body waves was proposed for surface waves by Alsop (1966). Consider two quarter-spaces welded at the vertical plane. Each quarter-space is a horizontally layered medium. The idea is to use a linear combination of normal modes—which would have existed if the quarter-space were a half-space—as an ansatz (trial solution) in the quarter-space and to determine the linear coefficients that satisfy the continuity condition at the vertical interface.

This idea works perfectly if the continuity condition is met by a finite number of nonleaky normal modes, as in the case of Higuchi's medium cited earlier. This medium consists of two quarter-spaces each having a single homogeneous layer of the same thickness overlying a homogeneous half space. The rigidities and densities of the two layers and two half-spaces are restricted such that the shape of the eigenfunction for the fundamental Love-wave mode is identical for the two quarter-spaces. In this case, the continuity of traction and displacement across the vertical interface can be achieved simply by including reflected Love waves in the quarter-space from which the primary waves are incident, and transmitted Love waves in the other quarter-space.

For more general cases, however, it is difficult to satisfy the continuity condition with a linear combination of normal modes because, as shown in

Section 7.1, there are only a finite number of normal modes at a given frequency in a layered half-space.

A complete solution can be obtained with this approach if we introduce a rigid bottom at a finite depth in the half-space. Then the medium becomes finite in the vertical direction, all the modes become nonleaky, and one can describe an arbitrary motion as a linear combination of normal modes. Lysmer and Drake (1972) considered such a medium consisting of a transition zone sandwiched between two different, layered quarter-spaces with a rigid lower boundary. They applied the finite-element method (Zienkiewicz and Cheung, 1968) to the propagation of Rayleigh-type normal modes through such a medium, and obtained the total motion in the time domain as well as the decomposition into normal modes. The phase velocity for a separate mode measured over the transition zone is the same whether the mode is incident from left or right of the transition zone. The phase velocity measured for the total motion, however, showed the same anisotropy found by Boore.

SUGGESTIONS FOR FURTHER READING

Boore, D. M. Finite difference methods for seismic wave propagation in heterogeneous materials. *In* B. A. Bolt (editor), *Seismology: Seismic Waves and Earth Oscillations* (Methods in *Computational Physics*, Vol. 11). New York: Academic Press, 1972.

Chernov, L. A. *Wave Propagation in a Random Medium*. New York: McGraw-Hill, 1960.

Claerbout, J. F. *Fundamentals of Geophysical Data Processing With Applications to Petroleum Prospecting*. New York: McGraw-Hill, 1976.

Davies, D., and R. M. Sheppard. Lateral heterogeneity in the earth's mantle. *Nature*, **239**, 318–323, 1972.

Flatté, S. M., R. Dashen, W. H. Munk, K. M. Watson, and F. Zachariasen. *Sound Transmission Through a Fluctuating Ocean*. Cambridge University Press, 1979.

Haddon, R. A. W., and J. R. Cleary. Evidence for scattering of seismic PKP waves near the mantle-core boundary. *Physics of the Earth and Planetary Interiors*, **8**, 211–234, 1974.

Hudson, J. A. Scattered waves in the coda of P. *Journal of Geophysics*, **43**, 359–374, 1977.

Julian, B. R., and M. K. Sengupta. Seismic travel time evidence for lateral inhomogeneity in the deep mantle. *Nature*, **242**, 443–447, 1974.

Landers, T., and J. F. Claerbout. Numerical calculations of elastic waves in laterally inhomogeneous media. *Journal of Geophysical Research*, **77**, 1476, 1482, 1972.

McDaniel, S. T. Parabolic approximations for underwater sound propagation. *Journal of the Acoustical of America*, **58**, 1178–1185, 1975.

Smith, W. D. The application of finite element analysis to body wave propagation problems. *Geophysical Journal of the Royal Astronomical Society*, **42**, 747–768, 1975.

Tatarski, V. I. *Wave Propagation in a Turbulent Medium*. New York: McGraw-Hill, 1961.

Tolstoy, I., and C. S. Clay. *Ocean Acoustics*. New York: McGraw-Hill, 1966.

Zienkiewicz, O. C. *The Finite Element Method in Engineering Science*. New York: McGraw-Hill, 1971.

PROBLEMS

13.1   Pick an inhomogeneous zone of your interest in the Earth, such as the core-mantle boundary, transition zones in the mantle, low-velocity layer, downgoing lithosphere, mid-ocean ridge, ocean-continent boundary, root of mountains, magma chambers in volcanic areas. Assume that the inhomogeneous zones other than the one you have picked have no effect on your seismograms. Design experiments by choosing seismic sources, receivers, and frequency ranges in any way you like. There are five lengths involved in this problem.

(1) Size of the inhomogeneous zone.

(2) Wave length or dimension of inhomogeneity.

(3) Wave length of seismic wave.

(4) Distance to the station from the inhomogeneous zone.

(5) Distance to the source from the inhomogeneous zone.

Discuss what mathematical models and what approximate methods are appropriate for analyzing the observation from a given experiment. The analysis method appropriate for a particular experiment depends on the relative size of the above five lengths (cf. Fig. 13.11). What would be the most efficient experiment (choice of source, station, and wavelength) for a given "inhomogeneous zone" in view of the feasibility of the experiment, the computer time required for analysis, expected resolution, and accuracy of results?

13.2   Consider a single layer overlying a half-space divided by a vertical plane. On one side of the vertical plane, the shear velocity and rigidity are β_1, μ_1 for the layer and β_2, μ_2 for the half-space. On the other side, they are β_3, μ_3 for the layer and β_4, μ_4 in the half-space. The layer thickness is the same on both sides. Higuchi (1932) showed that if the following conditions are satisfied,

$$\frac{1}{\beta_1^2} - \frac{1}{\beta_2^2} = \frac{1}{\beta_3^2} - \frac{1}{\beta_4^2}$$

and

$$\frac{\mu_1}{\mu_2} = \frac{\mu_3}{\mu_4},$$

plane Love waves normally incident on the vertical plane generate only reflected and transmitted Love waves without mode conversion or body-wave scattering. Determine the reflection and transmission coefficients for displacements at the surface.

13.3 Suppose that the ray path between two points A and B is known in a medium with velocity $c = c(\mathbf{x})$. Suppose now that a slight velocity perturbation $\delta c(\mathbf{x})$ is introduced into the medium. Then the original travel time between A and B,

$$T = \int_A^B \frac{ds}{c(\mathbf{x})} \qquad \text{(taken along the ray path)},$$

is affected in two ways. One, the integrand becomes $1/(c + \delta c)$; two, the ray path itself is slightly changed. Give a statement of Fermat's principle, and show that (to first order) the effect of the velocity perturbation on T is given by

$$\delta T = - \int_A^B \frac{\delta c \, ds}{c^2},$$

where the path of integration is the *original* ray path. (This result can be used to obtain a travel-time correction for the Earth's ellipticity of figure. Concise results are given by Dziewonski and Gilbert, 1976.)

13.4 For a medium with scale length of inhomogeneity given by a, describe intuitively why the scattering tends to zero in the limit both as $a \to 0$ and $a \to \infty$.

CHAPTER **14**

The Seismic Source: Kinematics

In Section 2.5 and Chapter 3, we studied seismic sources so that we could begin to determine what aspects of the theory of wave propagation are needed in seismology. Having now completed this study of wave propagation (Chapters 4–9, 13), we return to a more thorough examination of seismic sources. Of all the various types of source that can generate seismic waves (explosions, rapid phase transformations, etc.), the principal source we shall study is that involving a surface (the fault plane) across which shearing motions develop. In Section 3.1, we showed that if the displacement discontinuity across a fault surface is known as a time-dependent function of position on the fault, then motions throughout the medium are completely determined. This is a kinematic result (see Box 5.3) of great importance, since in practice it enables us to interpret the observable motions that radiate from the source region in terms of particle motions on a fault plane. This is the main subject of the present chapter, in which we shall characterize what may be learned from far-field and near-field observations about the kinematics of motion at an earthquake source.

To understand the physical processes actually occurring in the source region, one must study stress-dependent material properties. That is, one examines the way in which material failure nucleates and spreads (e.g., over a fault plane), rapidly relieving stresses that had slowly risen (due to long-term tectonic processes) to exceed the strength of material in the source region. This is a

dynamic problem, and a very difficult one, which we take up in the next chapter. But as we develop now the kinematics of fault motion, we are guided to some extent by the constraint that faulting is a process of failure in shear.

Our starting point is the representation theorem (equation (3.2)) of Chapter 3. Neglecting body forces and stress discontinuities, recall that the elastic displacement **u** caused by a displacement discontinuity $[\mathbf{u}(\boldsymbol{\xi}, \tau)]$ across an internal surface Σ has the components

$$u_i(\mathbf{x}, t) = \int_{-\infty}^{\infty} d\tau \iint_{\Sigma} [u_j(\boldsymbol{\xi}, \tau)] c_{jkpq} G_{ip,q}(\mathbf{x}, t; \boldsymbol{\xi}, \tau) v_k \, d\Sigma(\boldsymbol{\xi}), \qquad (14.1)$$

where c_{jkpq} are the elastic constants defined in equation (2.18); $G_{ip}(\mathbf{x}, t; \boldsymbol{\xi}, \tau)$ is the Green function defined in Section 2.4; $\boldsymbol{v}$ is the normal to Σ as shown in Figure 3.1; and $G_{ip,q}(\mathbf{x}, t; \boldsymbol{\xi}, \tau)$ is the derivative of G_{ip} with respect to ξ_q. In a homogeneous, isotropic, unbounded medium, the Green function can be stated explicitly. Thus, using (2.37) and (4.23), with body force taken as a unit impulse, it follows that

$$G_{ip}(\mathbf{x}, t; \boldsymbol{\xi}, \tau) = \frac{1}{4\pi\rho} (3\gamma_i\gamma_p - \delta_{ip}) \frac{1}{r^3} \int_{r/\alpha}^{r/\beta} t' \, \delta(t - \tau - t') \, dt'$$

$$+ \frac{1}{4\pi\rho\alpha^2} \gamma_i\gamma_p \frac{1}{r} \delta\left(t - \tau - \frac{r}{\alpha}\right)$$

$$- \frac{1}{4\pi\rho\beta^2} (\gamma_i\gamma_p - \delta_{ip}) \frac{1}{r} \delta\left(t - \tau - \frac{r}{\beta}\right), \qquad (14.2)$$

where γ is the unit vector from the source point $\boldsymbol{\xi}$ to the receiver point $\mathbf{x}$, and $r = |\mathbf{x} - \boldsymbol{\xi}|$ is the distance between those two points.

14.1 Kinematics of an Earthquake as Seen at Far Field

We shall obtain formulas for the displacement waveforms of P- and S-waves in the far field for faulting in a homogeneous, isotropic, unbounded medium. We shall use ray theory to extend this result for faulting in an inhomogeneous medium. After outlining some general properties of these waveforms, we shall look in particular at their low-frequency component. Specializing next to the case of unidirectional propagation, we shall study waveforms due to a source characterized by five parameters: the fault length; the fault width; the rupture velocity; the final offset; and the "rise time," which characterises the time taken for the offset, at a particular point on the fault, to reach its final value. This simple five-parameter characterization is often adequate to interpret the waves leaving a finite source. Where it is inadequate, and a more detailed description

of the fault motions is required, the next step is to analyze various stages, such as nucleation of motion, the spreading of rupture, and the stopping of motion. We describe several examples of these stages and conclude the analysis of far-field waveforms by an examination of their intermediate-frequency and high-frequency content.

14.1.1 *Far-field displacement waveforms observed in a homogeneous, isotropic, unbounded medium*

A homogeneous, isotropic, unbounded medium is a simple one with which to work; we choose it in order to minimize the complication of path effects.

If the receiver position $\mathbf{x}$ is sufficiently far from all points $\boldsymbol{\xi}$ on the fault surface Σ, then only the far-field terms in the Green function (14.2) are significant. From (14.1), after carrying out the integral with respect to τ, we obtain the far-field displacement

$$
u_i(\mathbf{x}, t) = -\frac{1}{4\pi\rho\alpha^2} \frac{\partial}{\partial x_q} \iint_{\Sigma} c_{jkpq} \frac{\gamma_i \gamma_p}{r} \left[u_j\left(\boldsymbol{\xi}, t - \frac{r}{\alpha}\right) \right] v_k \, d\Sigma
$$
$$
+ \frac{1}{4\pi\rho\beta^2} \frac{\partial}{\partial x_q} \iint_{\Sigma} c_{jkpq} \left(\frac{\gamma_i \gamma_p - \delta_{ip}}{r} \right) \left[u_j\left(\boldsymbol{\xi}, t - \frac{r}{\beta}\right) \right] v_k \, d\Sigma, \quad (14.3)
$$

in which we have used the relation $\partial/\partial \xi_q = -\partial/\partial x_q$, valid for operations on quantities such as γ and r, which are dependent only on the difference between $\mathbf{x}$ and $\boldsymbol{\xi}$. Carrying out the differentiation with respect to x_q and ignoring all terms that attenuate more rapidly than r^{-1}, we obtain

$$
\text{far-field of } u_i(\mathbf{x}, t) = \iint_{\Sigma} \frac{c_{jkpq}}{4\pi\rho\alpha^3 r} \gamma_i \gamma_p \left[\dot{u}_j\left(\boldsymbol{\xi}, t - \frac{r}{\alpha}\right) \right] \frac{\partial r}{\partial x_q} v_k \, d\Sigma
$$
$$
- \iint_{\Sigma} \frac{c_{jkpq}}{4\pi\rho\beta^3 r} (\gamma_i \gamma_p - \delta_{ip}) \left[\dot{u}_j\left(\boldsymbol{\xi}, t - \frac{r}{\beta}\right) \right] \frac{\partial r}{\partial x_q} v_k \, d\Sigma. \quad (14.4)
$$

Obviously, the first term corresponds to P-waves and the second to S-waves. Note that $\partial r/\partial x_q$ is merely γ_q.

If the station is far enough away, as compared to the linear dimension of fault surface Σ, we can safely assume that the distance r and direction cosine γ_i are approximately constant, independent of $\boldsymbol{\xi}$, and that such slowly varying factors can be taken outside the integral. For simplicity, we shall further assume that the fault surface Σ is a plane and that the direction of the displacement discontinuity is the same everywhere on the fault. We write

$$
[u_j(\boldsymbol{\xi}, t)] = n_j \cdot \Delta u(\boldsymbol{\xi}, t), \quad (14.5)
$$

where Δu is a scalar function, which we shall call the "source function," or the "slip function" in the case of a shear fault. Under these assumptions, equation (14.4) reduces to

$$\text{far-field of } u_i(\mathbf{x}, t) = \frac{\gamma_i}{4\pi\rho\alpha^3 r} \cdot c_{jkpq}\gamma_p\gamma_q v_k n_j \cdot \iint_\Sigma \Delta\dot{u}\left(\xi, t - \frac{r}{\alpha}\right) d\Sigma$$

$$+ \frac{\delta_{ip} - \gamma_i\gamma_p}{4\pi\rho\beta^3 r} \cdot c_{jkpq}\gamma_q v_k n_j \cdot \iint_\Sigma \Delta\dot{u}\left(\xi, t - \frac{r}{\beta}\right) d\Sigma. \quad (14.6)$$

The above equation permits a remarkably simple exposition of far-field displacement due to P- and S-waves from an earthquake source. Since $\gamma_i\gamma_i = 1$ and $\gamma_i(\delta_{ip} - \gamma_i\gamma_p) = 0$, we see immediately that the particle motion of P-waves is parallel to γ, and that of S-waves is perpendicular to γ. Wave amplitudes attenuate with distance as r^{-1}, and are inversely proportional to the cube of their propagation velocities. Since other factors are comparable between them, the S-wave amplitude is roughly α^3/β^3 (~ 5) times larger than the P-wave amplitude.

The factor $(c_{jkpq}\gamma_p\gamma_q v_k n_j)$ represents the radiation pattern of P-waves, determined by the orientation of the fault plane (v_k), the direction of displacement discontinuity (n_j), and the direction to the station (γ_p) from the fault. Similarly, taking γ' and γ'' as orthogonal unit vectors in the plane perpendicular to γ, the radiation of S-waves is described by their amplitude $(c_{jkpq}\gamma'_p\gamma_q v_k n_j)$ in the γ'-direction and $(c_{jkpq}\gamma''_p\gamma_q v_k n_j)$ in the γ''-direction. Relative amplitudes between the two directions determine the polarization angle for S-waves, which has been widely used in addition to the first motion for P-waves for determining fault-plane solutions.

Finally, the displacement waveforms of P- and S-waves are described by a simple integral of the form

$$\Omega(\mathbf{x}, t) = \iint_\Sigma \Delta\dot{u}\left(\xi, t - \frac{r}{c}\right) d\Sigma, \quad (14.7)$$

where c is the velocity of wave propagation.

14.1.2 Far-field displacement waveforms for inhomogeneous isotropic media, using the geometrical-spreading approximation

In Chapter 4, equations (4.79)–(4.81), we obtained the far-field displacement at $\mathbf{x}$ under the assumption that the whole fault (with area A) was acting effectively

as a point source. This is the case when wavelengths of interest are much longer than the fault dimensions, but are much shorter than the distance from source region to receiver. It should be emphasized that our concern throughout all of Section 14.1 is with the case in which wavelengths are comparable with (and possibly less than) the fault dimensions, so that there is interference between waves radiated from different parts of the fault surface. An alternative way to obtain the far-field displacement waveform (14.4) is therefore to integrate the effect of each area element $d\Sigma$, using (4.79) for the P-waves and (4.80) for S-waves, but making allowance for different travel times to $\mathbf{x}$ from different parts of the fault surface. For example, for P-waves, using $[\dot{\mathbf{u}}] \, d\Sigma$ now instead of $\ddot{\mathbf{u}}A$, we find from (4.79) that

far-field P-wave of $\mathbf{u}_i(\mathbf{x}, t)$

$$= \frac{1}{4\pi\rho\alpha^3} \iint_\Sigma 2\mu \frac{\gamma_i\gamma_j}{r} \left[\dot{u}_j \left(\xi, t - \frac{|\mathbf{x} - \xi|}{\alpha} \right) \right] \gamma_k \nu_k \, d\Sigma(\xi), \quad (14.8)$$

which is just the same as the P-wave component in (14.4), since $c_{jkpq}[\dot{u}_j]\gamma_q\nu_k = 2\mu\gamma_j[\dot{u}_j]\gamma_k\nu_k$ for slip in isotropic media.

The virtue of this approach is that it can so easily be extended to inhomogeneous media, since in Chapter 4 we identified the far-field approximation with the geometrical-spreading approximation, and in equations (4.88)–(4.90) we gave what now can be regarded as the integrand of an integral over a finite fault surface. For inhomogeneous media, then, the equation that generalizes the P-wave component of (14.6) is

far-field P-wave of $\mathbf{u}(\mathbf{x}, t)$

$$= \frac{\mathbf{l}\mu(\xi_0)\mathscr{F}^P}{4\pi\rho^{1/2}(\xi_0)\rho^{1/2}(\mathbf{x})\alpha^{5/2}(\xi_0)\alpha^{1/2}(\mathbf{x})\mathscr{R}^P(\mathbf{x}, \xi_0)} \iint_\Sigma \Delta\dot{u}(\xi, t - T^P(\mathbf{x}, \xi)) \, d\Sigma, \quad (14.9)$$

in which $\mathbf{l}$ (the direction at $\mathbf{x}$ of the ray from ξ), $\mathscr{F}^P$ (the radiation pattern), and $\mathscr{R}^P$ (the geometrical spreading factor) are evaluated for some reference point ξ_0 on Σ. Formulas similar to (14.9) but for the waveforms of SV and SH can immediately be written down from (4.89) and (4.90).

14.1.3   General properties of displacement waveforms in the far field

For convenience in presentation, we shall continue to use homogeneous media for development of the theory, giving an example in Box 14.1 of how our source theory can be merged with wave-propagation theory for realistic media.

Taking the origin of coordinates at a reference point on the fault, the distance r between the surface element $d\Sigma$ and the receiver point $\mathbf{x}$ can be written as

$$r = r_0 \left[1 + \frac{|\xi|^2}{r_0^2} - \frac{2(\xi \cdot \gamma)}{r_0} \right]^{1/2}$$

$$= r_0 \left\{ 1 + \frac{1}{2} \left[\frac{|\xi|^2}{r_0^2} - \frac{2(\xi \cdot \gamma)}{r_0} \right] - \frac{1}{8} [\quad]^2 \cdots \right\}$$

$$= r_0 - (\xi \cdot \gamma) + \frac{1}{2} \frac{|\xi|^2}{r_0} - \frac{(\xi \cdot \gamma)^2}{2r_0} + \cdots, \tag{14.10}$$

where r_0 is the distance to receiver from the origin, $r = |\mathbf{x}|$, γ is the unit vector pointing to the receiver, and ξ is the location vector of $d\Sigma$ measured from the origin (Fig. 14.1).

For r_0 large as compared with the linear dimension of Σ, we may approximate equation (14.10) by

$$r \sim r_0 - (\xi \cdot \gamma). \tag{14.11}$$

The error δr in path length due to this approximation may be measured by the largest terms neglected in the series expansion in equation (14.10),

$$\delta r = \frac{1}{2r_0} [|\xi|^2 - (\xi \cdot \gamma)^2].$$

If this error is equal to or greater than a quarter wavelength $\lambda/4$, a serious error will be introduced in the result of integration. Therefore, the approximation by equation (14.11) is justified only for

$$\frac{1}{2r_0} [|\xi|^2 - (\xi \cdot \gamma)^2] \ll \frac{\lambda}{4}$$

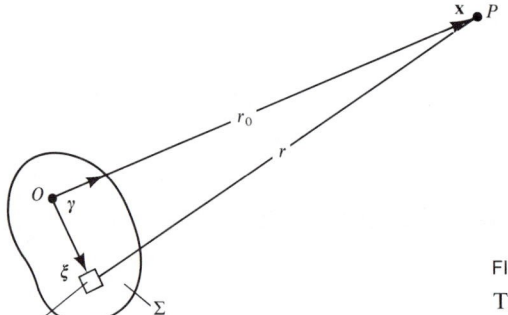

FIGURE **14.1**

The origin of coordinates is taken on a finite fault surface.

or, conservatively,

$$L^2 \ll \frac{\lambda r_0}{2},$$ (14.12)

where L is the maximum of $|\xi|$ on Σ. This is the same as the condition to be satisfied for the region of Fraunhofer diffraction in optics. For comparison, note that the condition we assumed in Chapter 4, in which the whole fault was regarded as a point source, amounted to $L \ll \lambda$, which is a much more restrictive condition on the applicable frequency range than (14.12). Under condition (14.12), we can rewrite the displacement waveform given in equation (14.7) as

$$\Omega(\mathbf{x}, t) = \iint_{\Sigma} \Delta \dot{u} \left[\xi, t - \frac{r_0 - (\xi \cdot \gamma)}{c} \right] d\Sigma.$$ (14.13)

Taking the Fourier transform of the above equation with respect to t, we get

$$\Omega(\mathbf{x}, \omega) = \iint_{\Sigma} \Delta \dot{u}(\xi, \omega) \exp \left\{ i\omega \frac{[r_0 - (\xi \cdot \gamma)]}{c} \right\} d\Sigma$$

$$= \exp \left(i\omega \frac{r_0}{c} \right) \iint_{\Sigma} \Delta \dot{u}(\xi, \omega) \exp \left[-i\omega \frac{(\xi \cdot \gamma)}{c} \right] d\Sigma,$$

where $\Omega(\mathbf{x}, \omega)$ and $\Delta \dot{u}(\xi, \omega)$ are the Fourier transforms of $\Omega(\mathbf{x}, t)$ and $\Delta \dot{u}(\xi, t)$, respectively.

The above equation shows that the Fourier transform $\Omega(\mathbf{x}, \omega)$ of the observed displacement waveform with phase correction for the delay $\omega r_0 / c$ due to propagation can be expressed as a superposition of plane waves: $\exp[-i\omega(\xi \cdot \gamma)/c]$.

$$\Omega(\mathbf{x}, \omega) e^{-i\omega r_0 / c} = \iint_{\Sigma} \Delta \dot{u}(\xi, \omega) \exp[-i\omega(\xi \cdot \gamma)/c] \, d\Sigma.$$ (14.14)

The right-hand side of the above equation has the form of a double Fourier transform in space, $\iint_{\Sigma} \Delta \dot{u}(\xi, \omega) \exp[-i(\xi \cdot \mathbf{k})] \, d\Sigma$. If this transform were known for all $\mathbf{k}$ in the two-dimensional wavenumber space, we could determine $\Delta \dot{u}(\xi, \omega)$ completely from far-field observations. Unfortunately, as shown in (14.14), the transform is known only for the projection of $\omega \gamma / c$ on Σ. Since γ is a unit vector, the range of $\mathbf{k}$ we can recover from far-field observation is restricted to $|\mathbf{k}| \le \omega/c$. In other words, we cannot find details of the seismic source with scale lengths shorter than the shortest wavelength observed. This is because only waves with phase velocity (along the plane Σ) $\omega/|\mathbf{k}|$ greater than the medium velocity c can radiate to the far-field. The waves with phase velocity smaller than c are inhomogeneous waves trapped near Σ. Thus, for a complete recovery of source function $\Delta u(\xi, t)$, we need observations at near field.

14.1.4   *Behavior of the seismic spectrum at low frequencies*

As the frequency ω approaches zero, the Fourier transform $\Omega(\mathbf{x}, \omega)$ of the far-field displacement waveform approaches a constant value:

$$\Omega(\mathbf{x}, \omega \to 0) = \iint_{\Sigma} \Delta\dot{u}(\xi, \omega \to 0) \, d\Sigma.$$

Since

$$\Delta\dot{u}(\xi, \omega) = \int \Delta\dot{u}(\xi, t) \exp(i\omega t) \, dt$$

and

$$\Delta\dot{u}(\xi, \omega \to 0) = \int \Delta\dot{u}(\xi, t) \, dt = \Delta u(\xi, t \to \infty),$$

we have

$$\Omega(\mathbf{x}, \omega \to 0) = \iint_{\Sigma} \Delta u(\xi, t \to \infty) \, d\Sigma.$$

Thus $\Omega(\mathbf{x}, \omega \to 0)$ approaches a constant equal to the integral of the final slip over the fault area. In other words, the spectrum (absolute value of Fourier transform) of the far-field displacement waveform becomes flat at low frequencies, and its height is proportional to the seismic moment defined in equation (3.16). This result is true for any $\Delta u(\xi, t)$ and is therefore independent of details of the process on the fault plane.

If we make another assumption that the fault slip never reverses its direction (i.e., that $\Delta\dot{u}$ does not change sign during an earthquake—a reasonable assumption if a large static friction operates at the time when $\Delta\dot{u} = 0$), then, from equation (14.13), we find that $\Omega(\mathbf{x}, t)$ will have the same sign for all t. In that case, the Fourier transform $\Omega(\mathbf{x}, \omega)$ takes its maximum value at $\omega = 0$. This result, which is again independent of the details of faulting, has been pointed out by Savage (1972), Molnar, Jacob, and McCamy (1973), and Randall (1973) in a controversy triggered by Archambeau (1968) on the behavior of the seismic spectrum at low frequencies.

If the area of fault surface Σ is infinitesimally small and if the slip $\Delta u(\xi, t)$ varies as a step function in time, then we find from equation (14.13) that the far-field waveform is a delta function and that the spectrum is flat for the whole frequency range. Therefore, we may say that for low frequencies where the spectrum is flat, the seismic source is equivalent to a point source with a step-function slip. This simple source has been extensively used in the single-station method of determining the phase velocities for Love and Rayleigh waves from relatively small earthquakes. Weidner (1972) compared the results

obtained by the single-station method with those by the two-station method for the North Atlantic paths and found that the step-function assumption was correct within a phase error of less than about 0.02–0.1 cycle for periods of 20–80 sec and for earthquakes with magnitude 6 or less. The same assumption was used also successfully in the determination of focal depth of earthquakes with magnitude less than about 6 using the amplitude (Tsai and Aki, 1971) and the phase (Weidner and Aki, 1973) of Rayleigh waves. The step-function assumption using free-oscillation data in the period range 150–1200 sec was supported by Mendiguren (1972) for a deep earthquake in Colombia. A summary of measurements on seismic moment and fault area was given by Kanamori and Anderson (1975) for many earthquakes of various magnitude.

14.1.5 *A fault model with unidirectional propagation*

As shown in the preceding section, the observed waveforms at far field alone cannot uniquely determine the source function $\Delta u(\xi, t)$. In such a case, it is essential to have a small number of source parameters that can adequately describe the source function. Then limited observations at far field can be effectively used to determine these parameters.

Let us first describe the fault plane as a rectangle with length L and width W. The rupture initiates at one end of the length and propagates along the length with velocity v. Setting the coordinate system (ξ_1, ξ_2) parallel to the length and width of the fault plane as shown in Figure 14.2, we specify the rupture propagation by

$$\Delta u(\xi, t) = f(t - \xi_1/v) \qquad 0 < \xi_1 < L, 0 < \xi_2 < W,$$
$$= 0 \qquad \xi_1 < 0, \xi_1 > L, 0 < \xi_2 < W, \qquad (14.15)$$
$$= 0 \qquad \xi_2 < 0, \xi_2 > W.$$

Putting this into equation (14.13), we get

$$\Omega(\mathbf{x}, t) = \int_0^W \int_0^L \dot{f}\left(t - \frac{r_0}{c} - \frac{\xi_1}{v} + \frac{\xi_1 \gamma_1 + \xi_2 \gamma_2}{c}\right) d\xi_1 \, d\xi_2. \qquad (14.16)$$

Assuming that W and $\xi_2 \gamma_2$ are small, and taking Ψ as the angle between the direction to the receiver and the direction of rupture propagation, we can rewrite equation (14.16) as

$$\Omega(\mathbf{x}, t) = W \int_0^L \dot{f}\left[t - \frac{r_0}{c} - \xi_1\left(\frac{1}{v} - \frac{\cos \Psi}{c}\right)\right] d\xi_1. \qquad (14.17)$$

This is essentially a moving average of $\dot{f}(t - r_0/c)$ over a time interval of $L[1/v - (\cos \Psi)/c]$. Since $\dot{f}(t - r_0/c)$ is the far-field displacement waveform

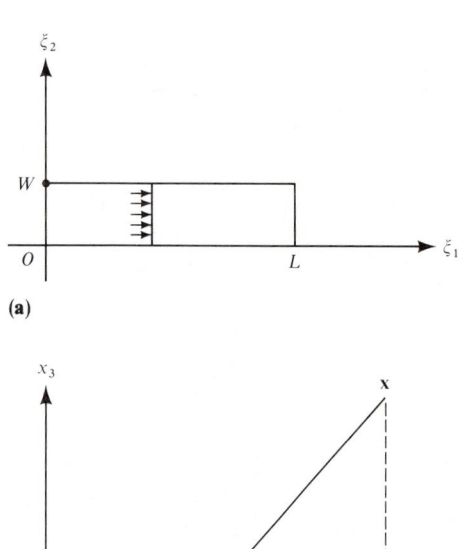

(a)

(b)

FIGURE **14.2**

Unidirectional faulting on a rectangular fault. **(a)** The fault plane. **(b)** Direction $\theta = 0$ is normal to the fault plane, and $\Psi = 0$ is the direction of rupture propagation.

expected for an infinitesimally small fault, we see that the rupture propagation over a finite length of fault has a smoothing effect on the waveform.

To find this effect on the spectrum, we take the Fourier transform of (14.7). Writing the Fourier transform of $f(t)$ as $f(\omega)$, we get

$$\Omega(\mathbf{x}, \omega) = -i\omega W f(\omega) e^{i\omega r_0/c} \int_0^L \exp\left[i\omega\xi_1\left(\frac{1}{v} - \frac{\cos \Psi}{c}\right)\right] d\xi_1$$

$$= \omega f(\omega) WL \frac{\sin X}{X} \exp\left[i\left(\frac{\omega r_0}{c} - \frac{\pi}{2} + X\right)\right], \tag{14.18}$$

where $X = (\omega L/2)[1/v - (\cos \Psi)/c]$. The effect of the finiteness of the fault on the amplitude spectrum is expressed by $X^{-1} \sin X$, which is depicted in Figure 14.3. This effect, first discussed by Ben-Menahem (1961), produces nodes at $X = \pi, 2\pi, \ldots$. The first node corresponds to the period $2\pi/\omega = L[1/v - (\cos \Psi)/c]$. An example of observed spectral nodes is shown in Figure 14.4

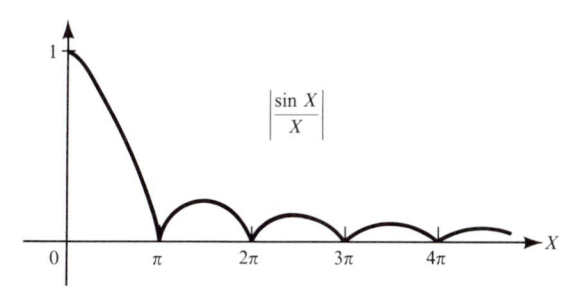

FIGURE **14.3**

The factor $|X^{-1} \sin X|$ in the spectrum of an observed pulse shape can be thought of in the time domain as convolution with a box function of temporal duration $L(1/v - c^{-1} \cos \Psi) = L[1/v - c^{-1}(\sin \theta \cos \phi)]$. This is the *apparent* duration of rupture, as detected by receivers along the direction (θ, ϕ), for rupture speed v and wave-propagation speed c.

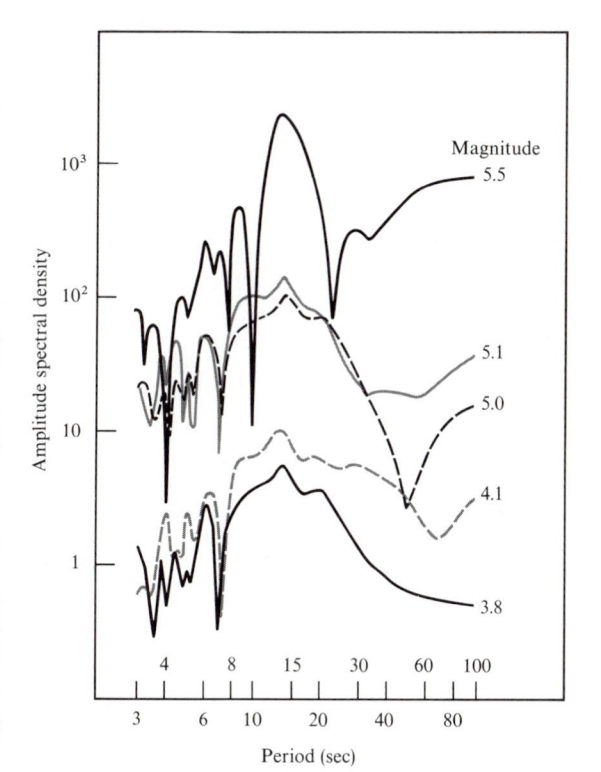

FIGURE **14.4**

Amplitude spectra of Love waves from a series of earthquakes in Parkfield, Calif., recorded at Berkeley, Calif., at a distance of 270 km. The main shock, with local magnitude 5.5 ($M_S = 6.4$), shows nodes at 22.5, 9.8, and 7.6 sec. The last node is probably a path effect, because it shows up for all earthquakes, independent of magnitude. [From Filson and McEvilly, 1967.]

for Love waves from the Parkfield earthquake of 1966 June, observed at Berkeley, California (Filson and McEvilly, 1967). The first node at $T = 22.5$ sec is explained by the rupture velocity of 2.2 km/sec, the fault length of about 30 km, and cos Ψ of about 1.

For higher frequencies, the envelope of $X^{-1} \sin X$ is proportional to ω^{-1}. This smoothing effect is weakest in the direction of rupture propagation ($\Psi = 0$) and strongest in the opposite direction ($\Psi = \pi$). As a result, we observe more high-frequency waves at $\Psi = 0$ than at $\Psi = \pi$. This effect is similar to the Doppler effect, which shifts the frequency ω of a moving oscillator to $\omega[1 - (v/c)(\cos \Psi)]^{-1}$. In our case, the nodes of $X^{-1} \sin X$ are shifted as a function of Ψ as $\omega_{nodes} = (2N\pi v/L)[1 - (v/c) \cos \Psi]^{-1}$, $N = 1, 2, \ldots$. We must, however, recognize an essential difference between the Doppler effect and the effect of $X^{-1} \sin X$. The latter effect is due to the destructive interference between waves coming from different parts of the fault plane, and tends to smooth out high frequencies. The Doppler effect has no such smoothing effect, because there is only one oscillator. The Doppler effect can be obtained by using

$$\Delta \dot{u}(\xi, t) = \exp(i\omega t) \delta(t - \xi_1/v) \delta(\xi_2)$$

instead of equation (14.15).

Two additional source parameters are needed to complete the unidirectional, rectangular fault model. They are the final slip D and the rise time T characterizing the slip function $f(t)$. Ben-Menahem and Toksöz (1963) used an exponential function

$$\begin{aligned} f(t) &= 0 & t < 0, \\ &= D(1 - e^{-t/T}) & t > 0, \end{aligned}$$

and Haskell (1964) used a ramp function

$$\begin{aligned} f(t) &= 0 & t < 0, \\ &= Dt/T & 0 < t < T, \\ &= D & t > 0, \end{aligned} \tag{14.19}$$

as shown in Figure 14.5. Taking the Fourier transform of $\dot{f}(t)$ and putting it into (14.18), we obtain

$$|\Omega(\mathbf{x}, \omega)| = WLD \frac{\sin X}{X} \left| \frac{1}{1 - i\omega T} \right|,$$

$$|\Omega(\mathbf{x}, \omega)| = WLD \frac{\sin X}{X} \left| \frac{1 - e^{i\omega T}}{\omega T} \right| \tag{14.20}$$

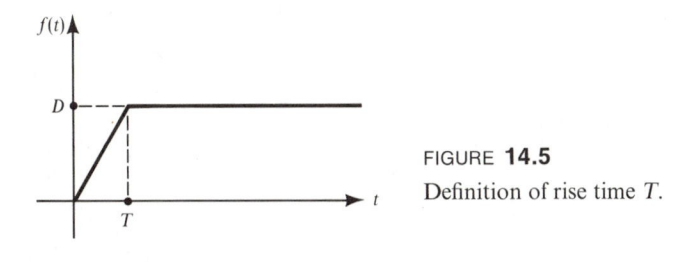

FIGURE **14.5**
Definition of rise time T.

for the exponential function and ramp function, respectively. In either case, the effect of a finite rise time T introduces additional smoothing of the waveform. For high frequencies, it attenuates the spectrum proportional to ω^{-1}. Thus the finite length of the fault over which the rupture is propagated, together with the finite time needed to complete the slip at a point on the fault, make the spectrum attenuate as ω^{-2} at high frequencies.

So far, we have introduced the following five source parameters:

i) Fault length L.
ii) Fault width W.
iii) Rupture velocity v.
iv) Permanent slip D.
v) Rise time T.

The corresponding far-field displacement spectrum is given by (14.20). The spectrum is flat near $\omega = 0$, and the height of the flat part is proportional to DWL, or seismic moment. For frequencies higher than the reciprocals of rupture time $L[1/v - (\cos \Psi)/c]$ and rise time T, the spectrum decays with frequency as ω^{-2}. If the effect of finite width is also taken into account, as done by Hirasawa and Stauder (1965), the spectral decay of ω^{-3} should be expected. Another extension of the model often used in practice is to assume that the rupture propagates from a point in the fault plane to positive and negative directions along its length. This is called bilateral faulting and the corresponding seismic spectrum can be easily obtained by appropriately superposing the results for two unilateral fault motions propagating in opposite directions.

The seismic energy, spectral density, and near-field effects of the above five-parameter model were studied in detail by Haskell (1964, 1966, 1969), and it is often called Haskell's model. This model has been used extensively in the interpretation of observed body waves, surface waves, free oscillations, and near-field seismograms in the late 1960's and the 1970's. For many earthquakes, reliable estimates of the product of L, W, and D have been made. The measurement of L is easier than that of W or D, because its effect can be studied using longer waves, which suffer less from the complex path effect. The reliable

estimation of D and T requires near-field data, which are usually difficult to obtain.

Once L, W, and D are determinined, one can make a rough estimate of stress drop associated with the faulting, by referring to the crack problem of similar geometry. The results so far indicate that for most earthquakes the stress drop is 10 to 100 bars, independent of magnitude. The slip velocity (D/T) across the fault also appears to be a constant: 10 to 100 cm/sec. According to Abe's (1975) summary, the stress drop $\Delta\sigma$ and the slip velocity (D/T) appear to be proportional to each other with the coefficient β (shear velocity) times ρ (density), which is nothing but the impedance for plane shear waves. We shall discuss the implication of these results later in this chapter.

An extensive study of source parameters of major earthquakes in and near Japan was made by Kanamori and his colleagues using the Haskell model. The result, as summarized by Kanamori (1973), showed that the amount of slip and the extent of the fault area obtained by the seismic method are in good agreement with those obtained by a static method, using geodetic measurements for earthquakes caused by "brittle elastic" rebound. On the other hand, for earthquakes attributed to "visco-elastic" rebound, the slip and fault area were found to be significantly greater by the static method than by the seismic method, indicating that the seismic event does not totally represent tectonic processes associated with an earthquake.

14.1.6   Nucleation, spreading, and stopping of rupture

The unidirectional propagation of rupture in Haskell's rectangular fault model appears to be an oversimplification of the process when we look closer at the nucleation of the rupture process. To make the model more realistic, it may be necessary to allow rupture to initiate at a point (rather than simultaneously everywhere along a line segment) and then spread out radially (rather than propagate in a single direction) at a uniform velocity until it covers an arbitrary two-dimensional surface on the fault plane. Far-field waveforms from this type of source model were first studied by Savage (1966) using equation (14.13).

As shown in Figure 14.6, we shall place the fault in the plane $x_3 = 0$ and assume that rupture propagates from the origin in all directions with uniform velocity v and stops at the periphery of fault plane Σ. Initially the rupture front is a circle described by $\rho = vt$, but the final fault will have a periphery given by $\rho = \rho_b(\phi')$, where (ρ, ϕ') are cylindrical coordinates in the fault plane.

Savage (1966) assumed the displacement discontinuity was a step function in time with final value $\Delta U(\rho, \phi')$. In our notation, the model can be expressed as

$$\Delta u(\xi, t) = \Delta U(\rho, \phi')H(t - \rho/v)[(1 - H(\rho - \rho_b)].   \tag{14.21}$$

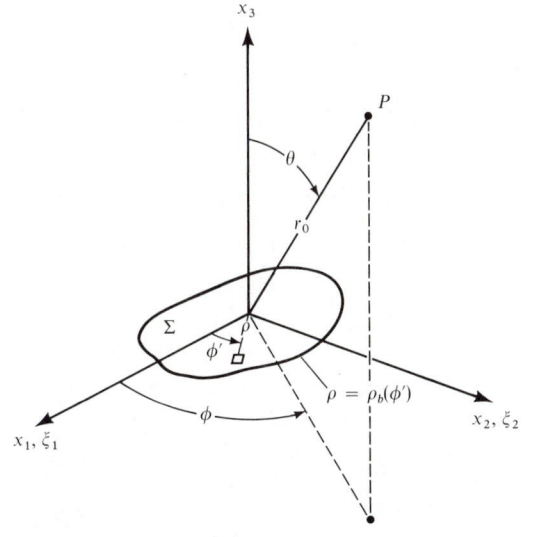

FIGURE **14.6**

The rupture starts from the origin and spreads in the x_1x_2-plane with a constant velocity v. Initially, the rupture front is a circle $\rho = vt$, but the final fault plane has a periphery given by $\rho = \rho_b(\phi')$. P is the observation point, and an element $d\Sigma$ of the fault is shown at (ρ, ϕ').

Putting this into (14.13), we find

$$\Omega(\mathbf{x}, t) = \iint_{\Sigma} \Delta\dot{u}\left(\boldsymbol{\xi}, t - \frac{r_0 - (\boldsymbol{\xi} \cdot \boldsymbol{\gamma})}{c}\right) d\Sigma_{\xi}$$

$$= \iint \delta\left(t - \frac{r_0}{c} + \frac{\rho \sin\theta \cos(\phi - \phi')}{c} - \frac{\rho}{v}\right) \Delta U(\rho, \phi')$$

$$\times [1 - H(\rho - \rho_b)]\rho \, d\rho \, d\phi', \qquad (14.22)$$

where we used the spherical coordinates shown in Figure 14.6 for expressing $(\boldsymbol{\xi} \cdot \boldsymbol{\gamma})$. Since $\int f(x) \delta(ax - b) \, dx = f(b/a)/a$, the integration with respect to ρ gives

$$\int \delta\left(t - \frac{r_0}{c} - \frac{\rho q_c}{v}\right) \Delta U(\rho, \phi')[1 - H(\rho - \rho_b)]\rho \, d\rho$$

$$= \left(t - \frac{r_0}{c}\right) \Delta U\left(\frac{t - \dfrac{r_0}{c}}{q_c/v}, \phi'\right) \frac{v^2}{q_c^2} \qquad \text{for } 0 < \frac{t - \dfrac{r_0}{c}}{q_c/v} < \rho_b$$

$$= 0 \qquad \text{for } \frac{t - \dfrac{r_0}{c}}{q_c/v} > \rho_b,$$

where $q_c = 1 - (v/c) \sin \theta \cos(\phi - \phi')$ is assumed positive everywhere; in other words, a subsonic rupture propagation $(v < c)$ is assumed. If $v > c$, we shall observe waves arriving before r_0/c in the directions (θ, ϕ) for which q_c is negative, because $\Delta U[(t - r_0/c)/(q_c/v), \phi')]$ will be nonvanishing for $t < r_0/c$.

For subsonic rupture propagation, (14.22) can be written as

$$\Omega(\mathbf{x}, t) = v^2\left(t - \frac{r_0}{c}\right)H\left(t - \frac{r_0}{c}\right)\int \frac{\Delta U\left(\dfrac{t - r_0/c}{q_c/v}, \phi'\right)}{q_c^2}\, d\phi', \quad (14.23)$$

where the integral is taken over the range of ϕ' for which $[(t - r_0/c)/(q_c/v)] < \rho_b$.

Suppose that the final slip ΔU is uniform except near the periphery, and suppose that we look at the beginning of the far-field displacement waveform, at which $t - r_0/c$ is small and the range of integration for ϕ' covers 0 to 2π. In that case, it is clear from (14.23) that the displacement waveform is a linear function of time. The linearity will hold until the rupture front reaches the periphery of the prescribed fault surface.

Thus a subsonically spreading rupture with a uniform step-function slip generates a far-field displacement waveform $(t - r_0/c)H(t - r_0/c)$ until the stopping signal arrives from the periphery of the fault. The corresponding particle-velocity waveform is a step function with a discontinuity at $t = r_0/c$. The acceleration is a δ-function, reaching infinity at $t = r_0/c$. The spectral density of acceleration, velocity, and displacement are, therefore, constant, proportional to ω^{-1} and ω^{-2}, respectively.

In order to see what happens when the rupture stops propagating, let us consider the case of a circular fault with uniform slip, in which $\rho_b = \rho_0$ (constant), $\Delta U(\rho, \phi') = \Delta U_0$ (constant), and $\Delta u(\boldsymbol{\xi}, t)$ is a function of ρ and t, as shown in Figure 14.7. The simplest result is obtained in the direction normal to the fault plane. For $\theta = 0$, $q_c = 1$, and equation (14.23) shows that the integral with respect to ϕ' is constant for $v(t - r_0/c) < \rho_0$ and vanishes for $v(t - r_0/c) > \rho_0$. In other words, the far-field displacement waveform $\Omega(\mathbf{x}, t)$ for $\theta = 0$ has a jump discontinuity at $t = r_0/c + \rho_0/v$, where $\Omega(\mathbf{x}, t)$ suddenly becomes zero. This jump discontinuity gives infinite particle velocity and acceleration. The

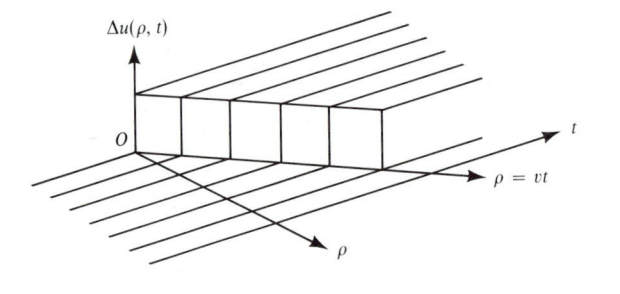

FIGURE **14.7**

Slip function for a circular fault with uniform slip.

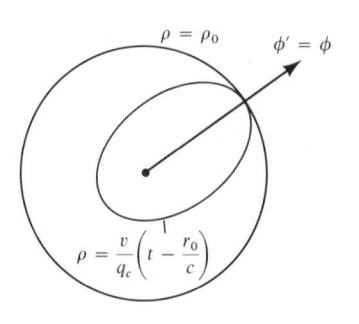

FIGURE **14.8**

spectral density of displacement will have a frequency asymptotic limit decaying as ω^{-1}. A seismic signal associated with the stopping of rupture was named a "stopping phase" by Savage.

For $\theta \neq 0$, $q_c = 1 - (v/c) \sin \theta \cos(\phi - \phi')$ is a function of ϕ', taking the minimum value in the azimuth ϕ to the station and the maximum in the opposite azimuth $\phi + \pi$. Since $\Delta U(\rho, \phi')$ is constant and q_c is a smooth function, the integral in equation (14.23) is proportional to the range of ϕ' for which $[v(t - r_0/c)]/q_c < \rho_0$. As long as the locus of $\rho = [v(t - r_0/c)]/q_c$ is contained inside the circle $\rho = \rho_0$, the integration range of ϕ' is 2π. Since the minimum of q_c is at $\phi' = \phi$, the locus of $\rho = [v(t - r_0/c)]/q_c$ will touch the circle $\rho = \rho_0$ first at $\phi' = \phi$, as shown in Figure 14.8. The growth $\Delta\phi'$ of the portion of ϕ' for which $[v(t - r_0/c)]/q_c > \rho_0$ can be found from the geometry shown in Figure 14.9.

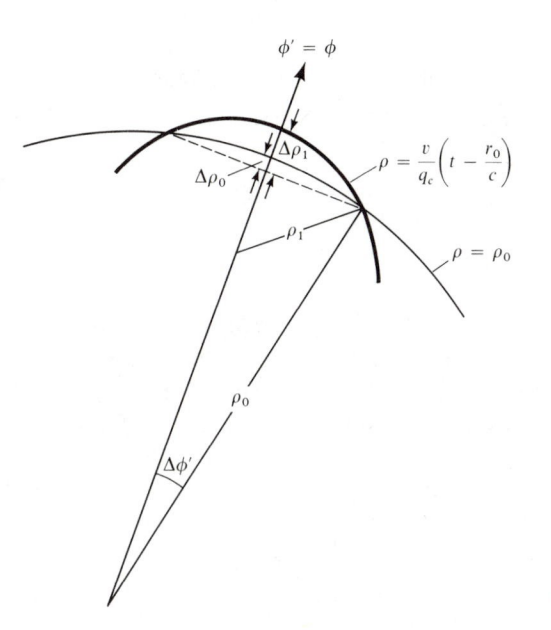

FIGURE **14.9**

Expressing ρ_1 as the radius of curvature of $\rho = [v(t - r_0/c)]/q_c$ at the point of contact, and taking into account the relation

$$\Delta\rho_0 = \rho_0 - \sqrt{\rho_0^2 - (\rho_0 \Delta\phi')^2} \sim (\rho_0/2)(\Delta\phi')^2$$

and a similar one for $\Delta\rho_1$, we get

$$\Delta\rho = \Delta\rho_1 - \Delta\rho_0 \sim \left(\frac{\rho_0^2}{2\rho_1} - \frac{\rho_0}{2}\right)(\Delta\phi')^2.$$

Since $\Delta\rho$ is proportional to the time Δt after the two curves make contact, the integration range $\Delta\phi'$ will be proportional to $\sqrt{\Delta t}$, and the far-field displacement $\Omega(\mathbf{x}, t)$ will therefore have a sudden change proportional to $\sqrt{\Delta t}$. The corresponding spectral density will have a high-frequency asymptotic limit decaying as $\omega^{-3/2}$. Both the particle velocity and acceleration will be infinite at the arrival of this stopping phase. Thus the stopping phase dominates the nucleation phase in this model.

As can be seen from the derivation given above, this $\omega^{-3/2}$ frequency dependence applies not only to the stopping of a circular crack front but to that of any smoothly curved crack front.

There are some unrealistic aspects about the above model as a shear fracture. First, it is not consistent with the known static solution; second, the stopping of slip in the interior of the fault has no causal relation with the stopping of the rupture front. The first point was partially taken into account by Savage (1966) but was more fully considered by Sato and Hirasawa (1973), who assumed the following slip function:

$$\Delta u(\rho, t) = K[(vt)^2 - \rho^2]^{1/2}H(t - \rho/v)[1 - H(\rho - \rho_0)] \quad \text{for } vt < \rho_0$$
$$= K(\rho_0^2 - \rho^2)^{1/2}[1 - H(\rho - \rho_0)] \quad \text{for } vt > \rho_0, \quad (14.24)$$

where

$$K = \left(\frac{24}{7\pi}\right)\left(\frac{\Delta\sigma}{\mu}\right).$$

This model is constructed by assuming that Eshelby's (1957) static solution holds at every successive instant of rupture formation for a circular crack under uniform shear stress (see Fig. 14.10). Putting equation (14.24) into equation (14.13) and carrying out the integration, Sato and Hirasawa obtain the following compact result:

$$\Omega(\mathbf{x}, t) = 2Kv\rho_0^2[\pi/(1 - k^2)^2]x^2 \quad \text{for } 0 < x < 1 - k$$
$$= 2Kv\rho_0^2(\pi/4)[1/k - (x^2/k)(1 + k)^{-2}] \quad \text{for } 1 - k < x < 1 + k,$$
$$(14.25)$$

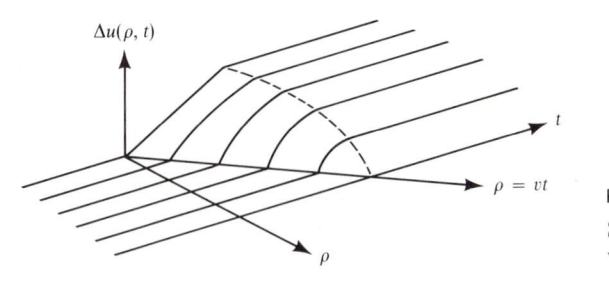

FIGURE **14.10**

Slip function for a circular fault on which the shear stress is constant.

where $k = (v/c) \sin \theta$ and $x = v(t - r_0/c)/\rho_0$. An example of $\Omega(\mathbf{x}, t)$ is shown in Figure 14.11. The initial rise of the far-field displacement is now proportional to $(t - r_0/c)^2$ instead of $(t - r_0/c)$ in the previous case of uniform step-function slip. The spectral density corresponding to this rising part will have high-frequency asymptotic decay of ω^{-3}. The spectral density for the total waveform, however, shows the high-frequency asymptotic decay of ω^{-2}, indicating that the stopping phase dominates the nucleation phase at high frequencies. The ω^{-2} decay is expected for this case because of an additional $\omega^{-1/2}$ due to the square root dependence of the slip function on distance from the crack tip, as compared with the uniform-slip case in which we obtained the $\omega^{-3/2}$ decay.

A defect of Sato and Hirasawa's model is that particle motions cease at the same instant, everywhere over the fault plane. Let us now take a look at another kinematic model of circular rupture, proposed by Molnar et al. (1973). The slip-velocity function for this model is given by

$$\Delta \dot{u}(\rho, t) = \Delta V \left[H\left(t - \frac{\rho}{v} + \frac{\rho_0}{v} \right) - H\left(t + \frac{\rho}{v} - \frac{\rho_0}{v} \right) \right] H(\rho_0 - \rho), \quad (14.26)$$

where ρ_0 is the radius of a circular ruptured area and ΔV is the relative particle velocity assumed to be constant over the area. The rupture nucleates at the center, grows radially in all directions at a constant velocity v to the radius ρ_0, and then contracts back to the center at the same velocity. This is a crude kinematic model of the spontaneous rupture process in which slip starts with the arrival of the rupture front and continues until information from the edges of the fault is radiated back to the point. The slip function of this model is shown in Figure 14.12.

Putting the Fourier transform of (14.26) into (14.14) and evaluating the integral, Molnar et al. obtained the high-frequency asymptotic decay of ω^{-2} to ω^{-3} depending on θ. The $\omega^{-5/2}$ decay is expected for a stopping phase in this case because of an additional ω^{-1} due to the linear dependence of the slip function on distance from the crack tip, as compared with the uniform-slip case in which we obtained the $\omega^{-3/2}$ decay.

As far as the initial part is concerned, this slip function is simply a time integral of the one for the uniform step-function case (equation (14.21)). Therefore, the

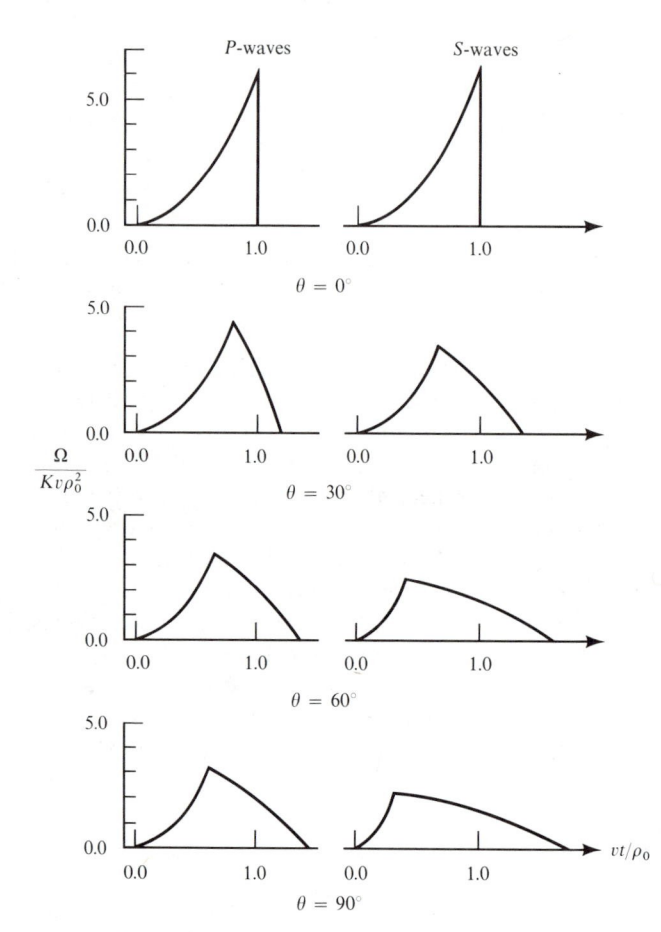

FIGURE **14.11**

Far-field wave forms according to equation (14.25); θ is defined in Figure 14.6. [From Sato and Hirasawa, 1973.]

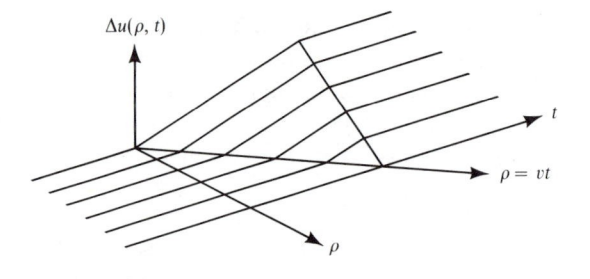

FIGURE **14.12**

Slip function for a circular fault on which a *healing front* (bringing motions to a stop) propagates inward after the fault has reached its final size.

initial rise of far-field displacement is the time-integral of linear increase. It is thus a parabolic increase, in common with the model of Sato and Hirasawa.

Dahlen (1974) extended the analysis of rupture kinematics to an elliptical crack that keeps on growing with the same shape. He used the following slip function, obtained by Burridge and Willis (1969), as an exact solution of the self-similar problem of an expanding elliptical crack:

$$\Delta u(\xi, t) = \Delta V \left(t^2 - \frac{\xi_1^2}{u^2} - \frac{\xi_2^2}{v^2} \right)^{1/2} H \left[t - \left(\frac{\xi_1^2}{u^2} + \frac{\xi_2^2}{v^2} \right)^{1/2} \right], \qquad (14.27)$$

where ΔV is the relative velocity across the center of the crack, and u and v are rupture-propagation velocities in the ξ_1- and ξ_2-directions, respectively. If $u = v$, the above equation reduces to (14.24) for the circular crack considered by Sato and Hirasawa for $t < \rho_0/v$. The slip function leads to a uniform stress drop on the fracture surface, and thus is a solution of the crack problem as long as the crack continues to grow self-similarly. The initial rise of far-field displacement is again parabolic, and the corresponding high-frequency asymptotic decay is proportional to ω^{-3}. Unlike the models discussed earlier, Dahlen's model considers the rupture as being slowly ground to a halt as the crack edge propagates into regions of either increasing friction or decreasing tectonic stress, and he concluded that the nucleation phase should dominate the stopping phase at high frequencies.

Once we neglect the contribution from the stopping phase, the high-frequency limits of the far-field spectrum can be obtained by putting the Fourier transform of (14.27) into equation (14.14). The result is given by

$$|\Omega(\mathbf{x}, \omega)| = \frac{4\pi uv\, \Delta V \omega^{-3}}{\left(1 - \frac{u^2}{c^2} \sin^2 \theta \cos^2 \phi - \frac{v^2}{c^2} \sin^2 \theta \sin^2 \phi \right)^2}, \qquad (14.28)$$

where θ and ϕ are defined in Figure 15.7. Since (14.28) does not contain parameters involving the final size of the crack, the above model predicts the same high-frequency behavior independent of the size of the earthquake. On the other hand, if the stopping phase dominates at high frequencies, the size of the earthquake will be a factor, and the high frequency values of $|\Omega|$ for larger earthquakes will be larger than for small ones. We shall come back to this point later in the discussion of scaling laws for seismic spectra.

14.1.7 Corner frequency and the high-frequency asymptote

As shown in preceding sections, the far-field displacement due to any reasonable kinematic model of an earthquake is expected to have a spectrum with a constant

value at low frequencies and proportional to a negative power of frequency at high frequencies. Following Brune (1970), we shall define a corner frequency as the frequency at the intersection of the low- and high-frequency asymptotes in the spectrum. The far-field spectrum is now characterized roughly by 3 parameters: (i) the low-frequency level, which is proportional to seismic moment; (ii) the corner frequency; and (iii) the power of the high-frequency asymptote.

Let us find the corner frequency as a function of source parameters for some of the kinematic models discussed earlier. A major disagreement arises among various models as to the relative magnitude of the P-wave corner frequency to the S-wave corner frequency.

Savage (1972) calculated the corner frequencies for P- and S-waves assuming bilateral faulting with rupture velocity v and final fault length L:

$$
\begin{aligned}
\Delta u(\xi, t) &= D_0 G(t - \xi_1/v) & 0 < \xi_1 < L/2 \\
&= D_0 G(t + \xi_1/v) & -L/2 < \xi_1 < 0 \\
&= 0 & \text{otherwise,}
\end{aligned}
\tag{14.29}
$$

where

$$
\begin{aligned}
G(t) &= 0, & t < 0 \\
&= 1 - \exp(-t/T) & t > 0.
\end{aligned}
$$

Assuming that the rise time T is equal to the travel time of rupture front over half a fault width, i.e., $T = W/2v$, Savage obtained the corner frequency as a geometric mean of two corner frequencies associated with the finite rupture propagation and the rise time. In this case, the high-frequency asymptote is proportional to ω^{-2}. Assuming further that $v = 0.9\beta$, the corner frequency averaged over all directions is obtained as

$$
2\pi\langle f_P \rangle = \sqrt{2.9} \cdot \alpha/\sqrt{LW}
\tag{14.30}
$$

for P-waves and

$$
2\pi\langle f_S \rangle = \sqrt{14.8} \cdot \beta/\sqrt{LW}
\tag{14.31}
$$

for S-waves. For a normal Poisson's ratio, the above formula shows that $\langle f_S \rangle$ is higher than $\langle f_P \rangle$.

The circular-crack model of Sato and Hirasawa also gives an asymptote like ω^{-2}, as discussed in the preceding section, but predicts higher corner frequency for P-waves than for S-waves. Their corner frequencies averaged over all directions are

$$
2\pi\langle f_P \rangle = C_p \alpha/R
\tag{14.32}
$$

for P-waves and

$$2\pi\langle f_S\rangle = C_S\beta/R \qquad (14.33)$$

for S-waves, where R is the radius of the crack and C_p and C_S are function of rupture velocity, as shown below.

Table of C_P and C_S

v/β	C_P	C_S
0.5	1.11	1.53
0.6	1.25	1.70
0.7	1.32	1.72
0.8	1.43	1.76
0.9	1.53	1.85

In this case, $\langle f_P\rangle$ is higher than $\langle f_s\rangle$, and their ratio varies from 1.26 to 1.43 as v/β increases from 0.5 to 0.9.

The model of Molnar et al., discussed in the preceding section, also predicts higher $\langle f_P\rangle$ than $\langle f_s\rangle$, which is generally supported by observation. Furuya (1969) observed that the predominant period of S-waves is 1.3 to 1.5 times greater than that of P-waves for a given magnitude, and pointed out that the simple propagating fault model cannot explain the observation. Sato and Hirasawa attribute the inadequacy of a propagating fault model to the restrictive form of $\Delta u(\xi, t)$ given in (14.15) or (14.29), where the time-dependence is assumed to be common to all the points on the fault. This assumption may be approximately valid for a long thin fault in which the slip function is determined by the width alone, but it is poor for an equidimensional fault in which the rupture nucleates at a point and spreads out to all directions on the fault plane. Both their model and the model of Molnar et al., which predict higher $\langle f_P\rangle$ than $\langle f_s\rangle$ are free from this restriction. As we shall discuss in Section 15.1.5, some of the important features of the solution to a dynamic problem of finite circular crack formation are contained in both of the above kinematic models.

On the other hand, formula (14.28) for a self-similar elliptic crack predicts lower corner frequency for P-waves than for S-waves. The formula is based on the assumption that the nucleation phase dominates the stopping phase at high frequencies. This assumption appears to contradict observations on the scale effect on seismic spectra, as discussed below.

The high-frequency asymptote given in (14.28) is determined by the rupture velocities and the particle velocity. Since the rupture velocities are given as material constants and the particle velocity is determined by rupture velocities and the initial stress, the asymptote is determined independent of the size of the final ruptured area. At a given travel distance the far-field seismic waves would have the same absolute spectrum for frequencies higher than the corner frequency, independent of earthquake magnitude, if the formula is correct.

An observed seismic spectrum is a function of source, path, and receiver. The simplest way of eliminating the path and receiver effects is to compare seismograms obtained by the same seismograph at the same station from two earthquakes with the same epicenter. Berckhemer (1962) was able to collect six such earthquake pairs from the Stuttgart records for the period 1931–1951, and found a strong frequency dependence of amplitude ratios between the pair. These data were interpreted by Aki (1967) using two kinematic models: ω-square and ω-cube models. The ω-cube model is a special case of the earthquake model suggested by Haskell (1966) and has the exact property discussed by Dahlen (1974). The far-field displacement spectrum for the ω-cube model is expressed as

$$\Omega(\omega) = \frac{\Omega(0)}{[1 + (\omega/\omega_0)^2]^{3/2}}, \tag{14.34}$$

where ω_0 is the corner frequency and $\Omega(0)$ is proportional to seismic moment. If we assume similarity between large and small earthquakes, the seismic moment will be proportional to L^3, the corner frequency to L^{-1}, and we have

$$\Omega(0) = \text{const.} \times \omega_0^{-3}. \tag{14.35}$$

Once the above constant is fixed, a family of spectral curves is determined that will describe the scaling law of seismic spectra. Neighboring curves are separated by a constant factor at frequency 0.05 Hz, so that the curves are designated by a uniform scale of M_s defined by Gutenberg and Richter (Appendix 2, Volume I) using the amplitude of surface waves at a 20-sec period. If M_s of one curve is fixed, then M_s is determined for the rest. One can then find the amplitude ratio between two earthquakes of any magnitude as a function of frequency. By trial and error, Aki (1967) found the family of spectral curves shown in Figure 14.13, which best fit Berckhemer's observed amplitude ratio. The spectral curves in Figure 14.13 share the same high-frequency asymptote in the absolute sense, independent of earthquake magnitude. This is expected when the nucleation is responsible for the high-frequency asymptote as in Dahlen's model. Then this effect may be observed for the initial portion of the seismogram in which the effect of stopping has not appeared. In fact, the body-wave magnitude m_b (Appendix 2, Volume 1) defined by the amplitude of short-period (about 1 sec) P-waves in the first 5 sec reaches a maximum value at around $m_b = 6$, and does not increase with earthquake size (Geller, 1976). This is in agreement with the spectral curves of Figure 14.13, which show that the spectral density at frequencies of 1 Hz and higher is the same for all M_S greater than 5.5.

The curves in Figure 14.13, however, do not seem to apply to the total seismogram. Since the duration of seismic signal is longer for larger earthquakes, if the spectral density is independent of magnitude, larger earthquakes should show smaller amplitudes for frequencies higher than 1 Hz. This clearly contradicts the observation on peak accelerations for $6 < M < 8$, as shown in

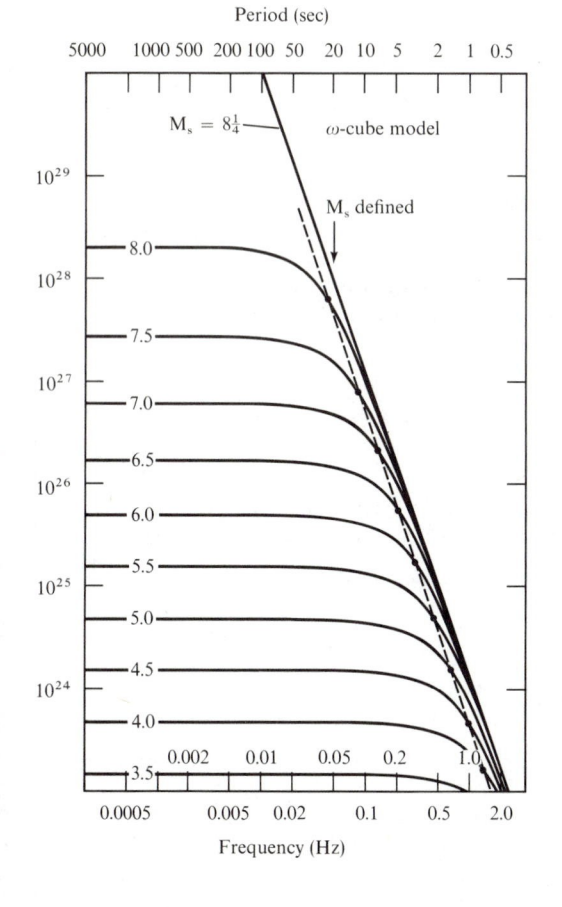

FIGURE **14.13**

Spectra of far-field body-wave displacement observed at a fixed distance from earthquakes with different M_S. The vertical coordinate shows the corresponding seismic moment. All curves share the common shape of equation (14.34), and similarity between large and small earthquakes is assumed. The broken line is the locus of ω_0. [From Aki, 1967; copyrighted by the American Geophysical Union.]

Figure 10.12. The peak accelerations observed at short distances are of frequencies higher than 1 Hz, and they are greater for larger earthquakes at a given distance.

On the other hand, the ω-square model gives more satisfactory results. This model has the far-field spectrum given by

$$\Omega(\omega) = \frac{\Omega(0)}{[1 + (\omega/\omega_0)]^2}. \tag{14.36}$$

The corresponding family of spectral curves fitting the Berckhemer data is shown in Figure 14.14. In this case, the spectral amplitude increases with magnitude M_s for all frequencies, in agreement with observation.

That the high-frequency spectrum for the ω-square model should increase with magnitude without ceiling was expected because the ω^{-2} asymptote indicates the dominance of a stopping phase at high frequencies, and the number of stopping points (or the length of stopping loop) increases with the ruptured

BOX **14.1**

Allowance for finite faulting in calculating far-field body waves within depth-dependent structures

In all our chapters on wave propagation in heterogeneous media, we have used very simple sources, usually a point source. But in order for the theories of wave propagation to be useful in explaining seismic data, it is necessary to merge them with the source theory we are developing in this chapter and the next. All that is needed, for purposes of computing the radiation from faulting, is to know the slip function $[\mathbf{u}(\boldsymbol{\xi}, \tau)]$ over the fault Σ and the Green function $\mathbf{G}(\mathbf{x}, t; \boldsymbol{\xi}, \tau)$.

In the far-field, it is often adequate to make the approximation

$$\mathbf{G}(\mathbf{x}, t; \boldsymbol{\xi}, \tau) \sim \mathbf{G}\left(\mathbf{x}, t; \mathbf{0}, \tau - \frac{\boldsymbol{\xi} \cdot \boldsymbol{\gamma}}{c}\right) \tag{1}$$

for that part of the wave field at $\mathbf{x}$ associated with waves having velocity $c(\boldsymbol{\xi})$ in the source region. Here $\boldsymbol{\gamma}$ is a unit vector at the origin of coordinates (taken on Σ), and $\boldsymbol{\gamma}$ is in the direction of waves departing from Σ.

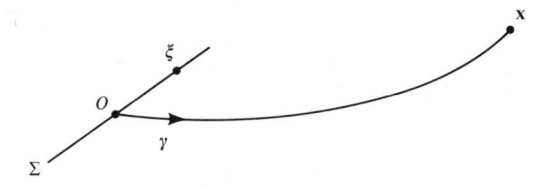

The approximation (1) above is equivalent to approximation (14.11), which was used in deriving (14.13). When the Green function is integrated over Σ, we need only make an allowance for the far-field phase correction (i.e., the travel-time difference) between $\boldsymbol{\xi}$ and $\mathbf{0}$. This is appropriate for far-field body waves and surface waves in depth-varying and laterally varying media.

For example, in Chapter 9 we obtained integrals over ray parameter p that gave far-field body-wave pulse shapes in depth-varying media. Point sources were used, and often the source was characterized by $M_0(\omega)$, together with some strike, dip, and rake (see, e.g., (9.71), (9.77)). To generalize these results to allow for finite faulting, it is clear that one must replace

$$M_0(\omega) \quad \text{by} \quad \iint_{\Sigma} \mu \, \Delta u(\boldsymbol{\xi}, \omega) \exp\left[-\frac{i\omega(\boldsymbol{\xi} \cdot \boldsymbol{\gamma})}{c(\boldsymbol{\xi})}\right] d\Sigma \tag{2}$$

(assuming slip in the same direction everywhere over Σ).

Note that $\boldsymbol{\gamma}$ itself is dependent on ray parameter, so that the expression (2) should appear within the integration over p, which characterized much of our numerical work in Chapter 9. However, for many practical purposes it is adequate to evaluate (2) at one representative value of $\boldsymbol{\gamma}$ for each $\mathbf{x}$.

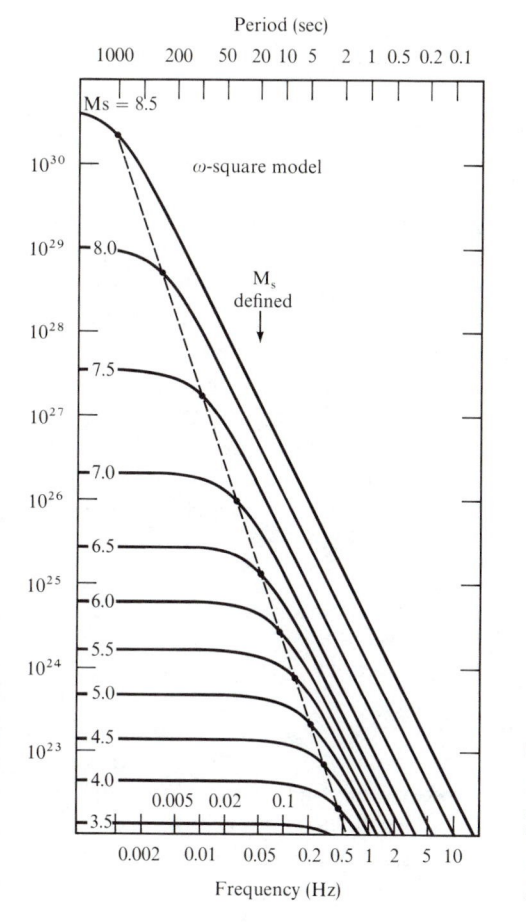

FIGURE **14.14**

Same as Figure 14.13 except that the spectra share the common shape of equation (14.36). The broken line is the locus of ω_0. [From Aki, 1967; copyrighted by the American Geophysical Union.]

area. There was a ceiling for the ω-cube model, because the nucleation point is a single point for any earthquake.

14.2 Kinematics of an Earthquake as Seen at Near Field

In the preceding section, we have studied seismic waves observed at far field, where a simple relation exists between the waveform and the fault slip function. There are, however, two major drawbacks to the study of seismic sources from far-field observation. First, as shown by (14.14), the far-field waves carry information about the source function only for that part of the space-time spectrum for which $|\omega/k| > c$, where ω is the frequency, k is the wave number in the fault plane, and c is the wave-propagation velocity. A complete determination of the slip function requires observation near the seismic source.

Second, the waves recorded at far field had to travel a long distance from the source. During the propagation, waves will suffer from attenuation, scattering, spreading, focusing, multi-path interference, and other complex path effects. One way of minimizing the path effects is to make observations at a short distance from the seismic source, again leaning upon the near-field data for more complete study of the source mechanism.

Ideally, we wish to measure the slip function $\Delta u(\xi, t)$ at various points ξ directly on the fault plane. Since such a measurement is practically impossible, we must know how the seismic motion close to but at some distance from the fault is related to the slip on the fault. This relation is not very simple, because the seismograms at short distances are composed of far-field and near-field terms of P- and S-waves coming from each element of the fault, as discussed in Section 4.2. These terms cannot be isolated on the records, and therefore the total seismogram must be computed for comparison with observations. Such a computation is useful also for predicting seismic effects at the site of an engineering structure due to a nearby earthquake fault on which the kinematic motions have been prescribed.

14.2.1   Synthesis of near-field seismograms for a finite dislocation source in a homogeneous, unbounded, isotropic medium

The near-field seismic motion for a finite dislocation source buried in a homogeneous, isotropic, unbounded medium can be calculated by integrating the solution obtained in Section 4.2 for seismic displacement due to an infinitesimal fault $d\Sigma$ across which the slip is given as $[\mathbf{u}(t)]$. The solution for an arbitrary slip function over a finite fault surface $\Sigma(\xi)$ can be obtained using equation (4.30), or equations (14.1) and (14.2), as

$$
\begin{aligned}
u_i(\mathbf{x}, t) = \iint_\Sigma \mu & \left[\left(\frac{30\gamma_i n_p \gamma_p \gamma_q v_q - 6v_i n_p \gamma_p - 6n_i \gamma_q v_q}{4\pi\rho r^4} \right) \right. \\
& \times \left(F\left(t - \frac{r}{\alpha}\right) - F\left(t - \frac{r}{\beta}\right) + \frac{r}{\alpha}\dot{F}\left(t - \frac{r}{\alpha}\right) - \frac{r}{\beta}\dot{F}\left(t - \frac{r}{\beta}\right) \right) \\
& + \left(\frac{12\gamma_i n_p \gamma_p \gamma_q v_q - 2v_i n_p \gamma_p - 2n_i \gamma_q v_q}{4\pi\rho\alpha^2 r^2} \right) \Delta u\left(\xi, t - \frac{r}{\alpha}\right) \\
& - \left(\frac{12\gamma_i n_p \gamma_p \gamma_q v_q - 3v_i n_p \gamma_p - 3n_i \gamma_q v_q}{4\pi\rho\beta^2 r^2} \right) \Delta u\left(\xi, t - \frac{r}{\beta}\right) \\
& + \frac{2\gamma_i n_p \gamma_p \gamma_q v_q}{4\pi\rho\alpha^3 r} \Delta\dot{u}\left(\xi, t - \frac{r}{\alpha}\right) \\
& \left. - \left(\frac{2\gamma_i n_p \gamma_p \gamma_q v_q - v_i n_p \gamma_p - n_i \gamma_q v_q}{4\pi\rho\beta^3 r} \right) \Delta\dot{u}\left(\xi, t - \frac{r}{\beta}\right) \right] d\Sigma(\xi), \quad (14.37)
\end{aligned}
$$

where $F(t) = \int_0^t dt' \int_0^{t'} \Delta u(\xi, t'') \, dt''$, $\mathbf{n} \, \Delta u(\xi, t) = [\mathbf{u}]$, v is the fault normal, $r = |\mathbf{x} - \xi|$, and $\gamma = (\mathbf{x} - \xi)/r$. Each of the terms under the surface integral has a simple form identifiable as waves propagating either as P or S, attenuating as a certain negative power of distance from the source. The waveform of each term can be easily calculated for a given slip function $\Delta u(\xi, t)$. It is difficult, however, to make a general statement on the total displacement because, at short distances, those terms arrive almost simultaneously, often canceling each other, and the behavior of the sum of all the terms is quite unpredictable from separate consideration of each individual term. This is especially true for motion close to the fault, because each term tends to infinity as $r \to 0$, although physically we expect the sum of all the terms to be finite.

Work done in this area before the mid-1970's was based mostly on the direct numerical integration of (14.37) with respect to ξ. For a numerical integration of (14.37), we replace the integral by a summation over grid points, assuming that the integrand varies smoothly in the grid interval. Each term of the integrand of Equation (14.37) contains two distinct factors: one is a negative power of $r = |\mathbf{x} - \xi|$, and the other is a function directly derivable from $\Delta u(\xi, t - r/c)$, where c is the wave velocity.

For r^{-n} to be smooth over $(r_0, r_0 + l)$, nl/r_0 must be negligible as compared to 1. Therefore, the grid interval l must be taken to be much smaller than $r_{\min}/n$, where $r_{\min}$ is the minimum distance from the observation point $\mathbf{x}$ to the fault plane $\Sigma(\xi)$.

The smoothness of the other factor is determined by the slip-time function. Since $\Delta u(\xi, t - r/c)$ contains $t - r/c$ in place of the time variable of the slip function, it varies rapidly as a function of ξ over a distance l if the slip function varies rapidly over a time interval l/c. For this factor to be smooth over the grid interval l, l/c must be much smaller than the minimum period $T_{\min}$ contained in the slip function. For example, if the slip function is characterized by the rise time T, then l must be much smaller than cT. The choice of l is restricted by the above two conditions to the shaded region of Figure 14.15, where $\lambda_{\min}$ is the wavelength corresponding to $T_{\min}$, and ε and ε' are small fractions.

For a relatively large $r_{\min}$, we can relax the above restriction on l to some degree by the same approximate method used for the far-field calculation in Section 14.1.3. If l^2 is much less than $r_{\min} \cdot \lambda_{\min}$, then from (14.11) we can put

$$r = r_0 - (\gamma \cdot \xi)$$

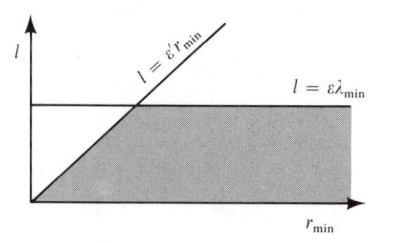

FIGURE **14.15**

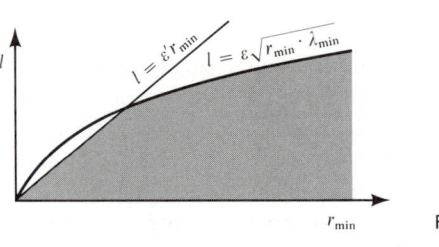

FIGURE **14.16**

in each grid interval, where γ is the unit vector directed from a grid point to the observation point. Then we can get a compact result for the integral over each grid interval, assuming that the factor r^{-n} is constant. For example, if we set the time dependence of $\Delta u(\xi, t)$ as $\exp(-i\omega t)$, the integral over the grid interval will have a factor $(x^{-1} \sin x)e^{ix}$, where $x = \omega l/2(1/v - c^{-1} \cos \Psi)$, as defined in Section 14.1.5. We can then sum these integrated terms over all the grid points. The time domain solution for a given slip function can be synthesized from the solutions for various ω. The appropriate choice of l for this method will be in the shaded region of Figure 14.16.

Both methods have been used in interpreting the records of strong-motion (or low-magnification) seismographs located short distances from earthquakes, and several interesting results have been obtained from comparison with

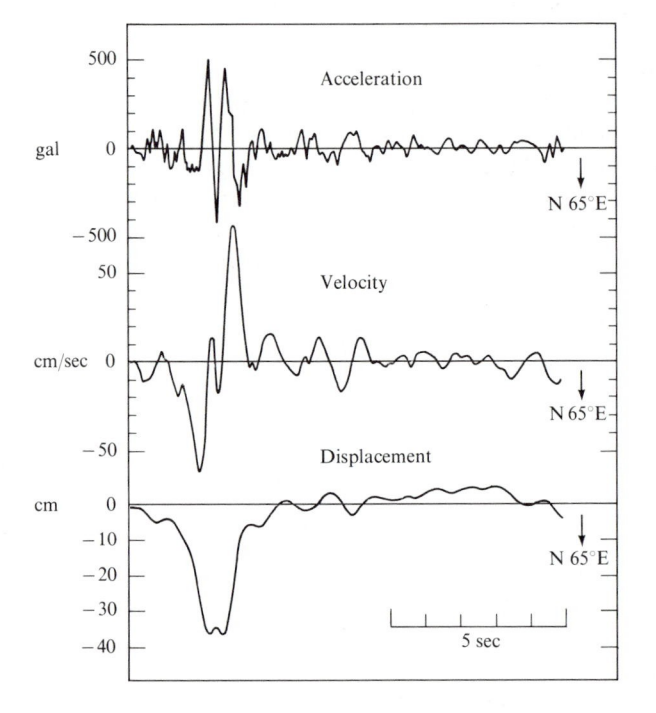

FIGURE **14.17**

Acceleration, velocity, and displacement observed at 80 meters from the San Andreas fault in the direction perpendicular to the fault trace during the 1966 Parkfield earthquake. [From Aki, 1968; copyrighted by the American Geophysical Union.]

observations. For example, for a Haskell-type (14.19) moving dislocation with slip motion parallel to the direction of rupture propagation, it was predicted that the displacement near the fault in the direction perpendicular to the fault plane should have an impulsive form with amplitude being a significant fraction of the amount of slip and with width being nearly equal to the rise time. Such an impulsive displacement with the expected sense of motion was actually observed by a strong-motion seismograph located only 80 meters from the San Andreas fault during the Parkfield earthquake of 1966 June 28. Figure 14.17 shows the perpendicular component of the acceleration, velocity, and displacement. A slightly different displacement record was published by Housner and Trifunac (1967), who used an integration technique different from the one used to obtain Figure 14.17. Figure 14.18 shows the theoretical displacement seismogram synthesized for a unilaterally propagating fault. The rise time and slip for the models that fit the observation are about 0.4–0.9 sec and 60–100 cm, respectively. Although these estimates of source parameters had to be revised by later work of Bouchon (1979), who also included the effect of a low-velocity sedimentary layer, the successful comparison between the theoretical and experimental results encouraged seismologists to pursue further the synthesis of strong motion near an earthquake fault. Results on source parameters determined by the numerical-synthesis method are summarized in Geller (1976).

The numerical integration methods were also used by Anderson and Richards (1975) in a comparative study of the near-field motion for Haskell's model with that calculated for several different kinematic models of faulting; they found that it is often difficult in practice to determine the slip function from kinematic modeling, even when several records of ground motion are available within one fault length from the source region.

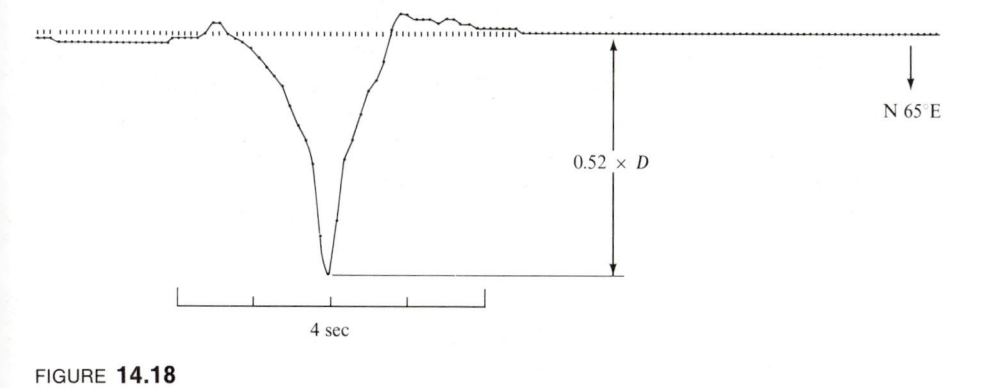

$0.52 \times D$

N 65°E

4 sec

FIGURE **14.18**

Synthetic displacement corresponding to the observation shown in Figure 14.17 for a right-lateral strike-slip fault propagating with rupture velocity 2.2 km/sec. [From Aki, 1968; copyrighted by the American Geophysical Union.]

The numerical-integration methods described above have several limitations. They are time consuming and do not give good physical insight. Thus one cannot generalize the behavior of seismic motion, but must calculate it for each specific case. This limitation is especially severe for high-frequency waves.

To overcome this limitation, compact, exact analytic solutions have been sought for simplified source models. For example, Boatwright and Boore (1975) and Sato (1975) showed that analytic solutions may be obtained for a Haskell's model in the case when the fault width W becomes zero. The other extreme model is the case when W becomes infinity. In this case, the problem is reduced to two dimensions. This reduction may be justified for frequencies higher than a certain critical frequency f_c determined by the minimum distance r_{min} to the fault from the station and by the width W of the fault plane. If the station is close to the fault plane, the seismic motion will be independent of W for frequencies higher than f_c. Then we can choose W to be infinite and reduce the problem to two dimensions. For two-dimensional problems, we can find compact and exact solutions more easily, as shown in Sections 14.2.3 and 14.2.4. The result will be useful for understanding the high-frequency motions, especially the nature of ground accelerations near the fault, which could not effectively be studied by the numerical method described above.

For a general layered medium, superposition of point-source solutions may be useful for certain cases. The point-source solution can be calculated by the Cagniard method (Section 9.1), as in Heaton and Helmberger (1977) for the 1968 Borego Mountain earthquake, or by the surface-wave method (Section 7.5), as in Swanger and Boore (1978) for the same earthquake. For more exact calculation of the near field of a finite propagating fault, however, it is advantageous to express the elastic-wave field as a superposition of plane waves. This is similar to the reflectivity method (Section 9.2), but requires inclusion of inhomogeneous plane waves because of the proximity to seismic source. The boundary and interface conditions may be satisfied separately for each plane wave, which will be superposed in accordance with the source field. This approach also eliminates the discretizing of fault plane needed for other methods, because the integration over the fault plane can be carried out analytically for each source plane wave before superposition. The method was applied by Niazy (1973, 1975) and Bouchon and Aki (1977b) to two-dimensional problems with special reference to the 1971 San Fernando earthquake. It was successfully extended to the case of three dimensions by Bouchon (1979), who applied the method to the 1966 Parkfield earthquake.

14.2.2   High-frequency motions near a propagating fault

In order to gain some physical insight into the near field of a propagating fault, we shall consider greatly simplified models. Let us put the fault plane in the xy-plane, with the rupture front parallel to the z-axis and propagating in the

x-direction, as shown in Figure 14.19. For a nearby observation point P on the xy-plane, the effect of the width of the fault on high-frequency motions may be neglected. In order to demonstrate this, we shall take a simple case of scalar spherical wave sources distributed along the z-axis. Writing the contribution from an infinitesimal line segment as $\exp[i\omega(t - r/c)]/r\, dz$, and assuming that the source function is simultaneous along the width of the fault, we get the total contribution from the line segment $(-z_0, z_0)$ as

$$\phi(P) = \int_{-z_0}^{z_0} \frac{\exp[-i\omega(t - r/c)]}{r}\, dz.$$

Changing variable z to $\zeta = \tan(\pi z/2z_0)$ (to make the integration limits recede to infinity), the above integral can be rewritten as

$$\phi(P) = \frac{2z_0}{\pi} \exp(-i\omega t) \int_{-\infty}^{\infty} \frac{\exp(i\omega r/c)}{r} \frac{d\zeta}{1 + \zeta^2}. \tag{14.38}$$

This integral can be evaluated by the method of steepest descents (Box 6.3). We first obtain the saddle point at which the exponent factor $f(\zeta) = ir/c$ is stationary by

$$\frac{df}{d\zeta} = \frac{i}{c}\frac{dr}{d\zeta} = \frac{i}{c} \cdot \frac{z}{r} \cdot \frac{1}{1 + \zeta^2} \cdot \frac{2z_0}{\pi} = 0$$

so that

$$z = 0 \quad \text{or} \quad \zeta = 0.$$

Putting $r = r_0$ for $z = 0$, $f(\zeta)$ is expanded in a Taylor series as

$$f(\zeta) = \frac{ir_0}{c} + \frac{i}{2cr_0} \zeta^2 \left(\frac{2z_0}{\pi}\right)^2 + \cdots.$$

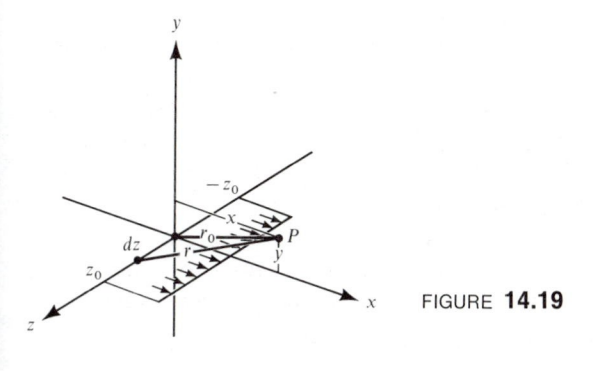

FIGURE **14.19**

The steepest-descents path in the ζ-plane is given by

$$\omega f(\zeta) = i\omega r_0/c - s^2/2,$$

where s varies from $-\infty$ to ∞ along its real axis. For small ζ, the steepest-descents path is given as

$$\frac{-i\omega}{2cr_0} \zeta^2 \left(\frac{2z_0}{\pi}\right)^2 = \frac{s^2}{2}$$

or

$$\zeta = \frac{\pi}{2z_0} \sqrt{\frac{cr_0}{\omega}}\, e^{i\pi/4} s. \tag{14.39}$$

This is a straight path through the origin of the ζ-plane, making an angle of $45°$ with the real axis. If $(\pi/2z_0)\sqrt{cr_0/\omega}$ is very small, the contribution to the integrand will come mostly from the segment of the path in the vicinity of the origin. In that case, it won't matter how the steepest-descents path would deviate from the straight line for large $|\zeta|$. Thus, if

$$\frac{\pi}{2z_0} \sqrt{\frac{cr_0}{\omega}} \ll 1, \tag{14.40}$$

we can rewrite the integral (14.38) using (14.39) as

$$\phi(P) \sim \frac{2z_0}{\pi} \exp(-i\omega t) \cdot \frac{\pi}{2z_0} \cdot \left(\frac{cr_0}{\omega}\right)^{1/2} \frac{\exp\left(\frac{i\omega r_0}{c} + \frac{i\pi}{4}\right)}{r_0} \int_{-\infty}^{\infty} e^{-s^2/2}\, ds$$

$$= \sqrt{2\pi} \cdot \left(\frac{c}{\omega r_0}\right)^{1/2} \exp\left[-i\omega\left(t - \frac{r_0}{c}\right) + \frac{i\pi}{4}\right]. \tag{14.41}$$

This expression does not contain z_0, showing the independence of high-frequency waves, which satisfy condition (14.40) on the length of the line source. This is a cylindrical wave attenuating as the inverse square root of the distance from the z-axis. A high-frequency attenuation by a factor of $\omega^{-1/2}$ and a $\pi/4$ phase delay are introduced relative to the spherical wave from the origin. Condition (14.40) is satisfied if the product of the wavelength λ and the distance r_0 is less than z_0^2 times some factor. Numerical experiments for a propagating dislocation source indicate that this factor is about 1/5.

For high-frequency motions near a fault, we can use the solution for two-dimensional problems in which the fault width is set equal to infinity. This simplifies the analysis greatly, because every quantity becomes independent of z. We shall consider two basic types of propagating faults: anti-plane and

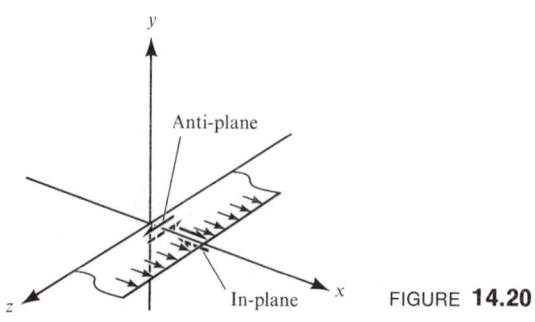

FIGURE **14.20**

in-plane. For the anti-plane type, the slip is in the z-direction, as shown in Figure 14.20, and the resultant displacement has a component only in the z-direction. In crystal-dislocation theory, this type is called a *screw dislocation*, in which the slip direction (Burger's vector) is parallel to the dislocation line. For the in-plane type, the slip is in the x-direction, and the resultant displacement has x- and y-components. In crystal-dislocation theory, this type is called an *edge dislocation*, in which the slip direction is perpendicular to the dislocation line. When the edge-dislocation line moves in the direction parallel to the slip, the movement is called *gliding*.

14.2.3   Anti-plane problems

As the simplest anti-plane problem, we shall consider the case in which phenomena appear to be stationary if looked at in coordinates

$$x' = x - vt,$$

$$y' = y,$$

$$t' = t,$$

which move with a constant velocity v. Such a case is possible for a semi-infinite fault plane propagating with velocity v from the time $-\infty$. The condition on the discontinuity in the displacement $w(x, y, t)$ across the fault plane is given by

$$w(x, +0, t) - w(x, -0, t) = \Delta w H(-x'), \tag{14.42}$$

where $H(x)$ is the unit step function. The stress is assumed to be continuous, so that

$$\mu \frac{\partial w}{\partial y}(x, +0, t) = \mu \frac{\partial w}{\partial y}(x, -0, t). \tag{14.43}$$

The equation of motion for an isotropic homogeneous body reduces to a wave equation for w:

$$\frac{1}{\beta^2}\frac{\partial^2 w}{\partial t^2} = \frac{\partial^2 w}{\partial x^2} + \frac{\partial^2 w}{\partial y^2}. \tag{14.44}$$

Using the new coordinates, and applying the stationarity condition $\partial/\partial t' = 0$, i.e.,

$$\frac{\partial}{\partial t} = \frac{\partial}{\partial t'} - v\frac{\partial}{\partial x'} = -v\frac{\partial}{\partial x'},$$

$$\frac{\partial}{\partial x} = \frac{\partial}{\partial x'},$$

$$\frac{\partial}{\partial y} = \frac{\partial}{\partial y'},$$

equation (14.44) can be rewritten as

$$\left(1 - \frac{v^2}{\beta^2}\right)\frac{\partial^2 w}{\partial x'^2} + \frac{\partial^2 w}{\partial y'^2} = 0. \tag{14.45}$$

This reduces to the Laplace equation for $v = 0$, and its solution for boundary conditions (14.42) and (14.43) is well known. The solution is

$$w(x', y') = \frac{\Delta w}{2\pi}\tan^{-1}\frac{y'}{x'},$$

which can be obtained by considering that the imaginary part of $\log(x' + iy')$ is harmonic and satisfies the Laplace equation. The function $\tan^{-1}$ here is taken to lie in the range $(-\pi, \pi)$. If $w(x', y')$ is the solution of the Laplace equation, then $w(x'/\sqrt{1 - v^2/\beta^2}, y')$ will be the solution of (14.45). Here we have assumed a subsonic rupture propagation, $v < \beta$. Our solution of (14.44) is therefore given by

$$w(x, y, t) = \frac{\Delta w}{2\pi}\tan^{-1}\frac{y\sqrt{1 - v^2/\beta^2}}{x - vt}. \tag{14.46}$$

This satisfies the boundary conditions (14.42), requiring the continuity of w everywhere except at $y = 0$, $x < vt$, because (14.46) does correctly give

$$w(x, y, t) = 0 \qquad \text{at } y = 0, x > vt$$

$$= \frac{\Delta w}{2} \qquad \text{at } y = +0, x < vt$$

$$= -\frac{\Delta w}{2} \qquad \text{at } y = -0, x < vt.$$

Since the stress components are

$$\tau_{xz} = -\frac{\mu\,\Delta w}{2\pi}\frac{\sqrt{1 - v^2/\beta^2}\cdot y}{(x - vt)^2 + (1 - v^2/\beta^2)y^2},$$

$$\tau_{yz} = \frac{\mu\,\Delta w}{2\pi}\frac{\sqrt{1 - v^2/\beta^2}(x - vt)}{(x - vt)^2 + (1 - v^2/\beta^2)y^2}, \tag{14.47}$$

τ_{xz} vanishes at $y = 0$, and the condition (14.43) is satisfied at $y = 0$, where

$$\tau_{yz} = \mu\frac{\partial w}{\partial y} = \mu\frac{\Delta w}{2\pi}\frac{\sqrt{1 - v^2/\beta^2}}{x - vt}. \tag{14.48}$$

The stress τ_{yz} on the plane $y = 0$ is an odd function of $x - vt$, and becomes $-\infty$ behind the tip and $+\infty$ ahead of the tip. It is also important to remember that τ_{yz} vanishes when the rupture velocity is β. Equation (14.46) was obtained by Frank (1949), Liebfried and Dietze (1949) and Eshelby (1949).

The particle velocity can be computed from (14.46) as

$$\frac{\partial w}{\partial t} = \frac{\Delta w}{2\pi}\frac{yv\sqrt{1 - v^2/\beta^2}}{(x - vt)^2 + y^2(1 - v^2/\beta^2)}. \tag{14.49}$$

The peak velocity occurs at $x = vt$, and the peak value is $(\Delta w/2\pi)\cdot[v/(y\sqrt{1 - v^2/\beta^2})]$. The peak value tends to infinity as the rupture velocity approaches β for this semi-infinite crack. It decays as the inverse of distance from the fault.

The acceleration can be computed from (14.49) as

$$\frac{\partial^2 w}{\partial t^2} = \frac{\Delta w}{2\pi}\cdot\frac{2y(x - vt)v^2\sqrt{1 - v^2/\beta^2}}{[(x - vt)^2 + y^2(1 - v^2/\beta^2)]^2}.$$

The peak acceleration occurs at $x - vt = (y/\sqrt{3})\sqrt{1 - v^2/\beta^2}$, and is

$$\frac{\Delta w}{2\pi}\frac{9}{8\sqrt{3}}\frac{v^2}{y^2(1 - v^2/\beta^2)}.$$

The peak acceleration also tends to infinity as the rupture velocity approaches β. It decays as the inverse square of distance from the fault.

The nature of motions in the near field of a propagating dislocation may be better exposed in their spectrum. The Fourier transform of $\partial w/\partial t$ at $x' = 0$ can be obtained by residue evaluation at poles $t = \pm iy(1 - v^2/\beta^2)/v$, and is equal to

$$\frac{\Delta w}{2}\exp\left(-|y|\cdot|\omega|\cdot\frac{\sqrt{1 - v^2/\beta^2}}{v}\right).$$

This shows an exponential decay with both y and ω, indicating that they are composed of the inhomogeneous plane waves trapped near the fault plane. Thus, for a subsonic rupture propagation with uniform velocity that starts at time $-\infty$ and continues to time $+\infty$, the near-field motion decays quickly with the distance y from the fault plane; the peak velocity decays as y^{-1}, and the peak acceleration as y^{-2}. The spectrum decays exponentially with increasing frequency, as expected for inhomogeneous plane waves.

Next let us introduce the starting and stopping in the fault propagation and see what will happen in the near field. To study the effect of starting, we shall apply the following boundary condition: instead of (14.42), we have

$$w(x, +0, t) - w(x, -0, t) = \Delta w H\left(t - \frac{x}{v}\right) H(x), \qquad (14.50)$$

which corresponds to a step-function slip starting from $x = 0$ at $t = 0$ and propagating in the $+x$-direction with velocity v, as shown in Figure 14.20. Since the motion is not stationary in the moving coordinates, equation (14.45) no longer applies. Expressing the Laplace transform of $w(x, y, t)$ as

$$w(x, y, s) = \int_0^\infty w(x, y, t)e^{-st}\, dt,$$

we rewrite the equation of motion (14.44) as

$$\frac{s^2}{\beta^2} w = \frac{\partial^2 w}{\partial x^2} + \frac{\partial^2 w}{\partial y^2}. \qquad (14.51)$$

Taking the Laplace transform of the boundary condition (14.50), we have

$$w(x, +0, s) - w(x, -0, s) = \frac{\Delta w \cdot e^{-sx/v}}{s} H(x). \qquad (14.52)$$

Since

$$e^{-sx/v} H(x) = \frac{1}{2\pi} \int_{-\infty}^{\infty} \frac{e^{ikx}}{i(k - is/v)}\, dk,$$

which can be obtained easily by residue evaluation of the pole at $k = is/v$, the boundary condition (14.52) can be rewritten as

$$w(x, +0, s) = -w(x, -0, s)$$

$$= \frac{\Delta w}{4\pi s} \int_{-\infty}^{\infty} \frac{e^{ikx}}{i(k - is/v)}\, dk, \qquad (14.53)$$

where we used the symmetry of w (an odd function of y).

To meet the above boundary condition, we assume the solution of (14.51) has the following form:

$$w(x, y, s) = \int_{-\infty}^{\infty} Q(k)e^{ikx - vy} \, dk,$$

where $v^2 = k^2 + s^2/\beta^2$. $Q(k)$ is determined by putting $y = \pm 0$ and comparing with (14.53). The result is

$$w(x, y, s) = \frac{\Delta w}{4\pi s} \int_{-\infty}^{\infty} \frac{e^{ikx - vy}}{i(k - is/v)} \, dk, \qquad y > 0$$

$$= -\frac{\Delta w}{4\pi s} \int_{-\infty}^{\infty} \frac{e^{ikx + vy}}{i(k - is/v)} \, dk, \qquad y < 0. \qquad (14.54)$$

The stress $\mu(\partial w/\partial y)$ is continuous at $y = 0$.

Equation (14.54) has a familiar form to which the Cagniard method is applicable (Mitra, 1966; Boore and Zoback, 1974). Transforming the variable k to τ by the relation

$$-s\tau = ikx - \sqrt{k^2 + s^2/\beta^2}\, y,$$

(14.54) is reduced to

$$w(x, y, s) = -\frac{\Delta w}{2\pi} \int_0^\infty \text{Re} \frac{(\tau^2 - r^2/\beta^2)^{1/2} \cos \theta - i\tau \sin \theta}{(\tau^2 - r^2/\beta^2)^{1/2} \sin \theta + i(\tau \cos \theta - r/v)}$$

$$\times \frac{H(\tau - r/\beta)}{(\tau^2 - r^2/\beta^2)^{1/2}} \frac{e^{-s\tau}}{s} \, d\tau,$$

where $x = r \cos \theta$ and $y = r \sin \theta$. Since the above equation has the form of the Laplace transform for $sw(x, y, s)$, the corresponding time-domain solution $\partial w(x, y, t)/\partial t$ can be identified as

$$\frac{\partial w(x, y, t)}{\partial t} = \frac{\Delta w}{2\pi} \frac{(t^2 - r^2/\beta^2) \sin \theta \cos \theta - t \sin \theta (t \cos \theta - r/v)}{(t^2 - r^2/\beta^2) \sin^2 \theta + (t \cos \theta - r/v)^2}$$

$$\times \frac{H(t - r/\beta)}{(t^2 - r^2/\beta^2)^{1/2}}. \qquad (14.55)$$

This equation reduces to (14.49) for the dislocation propagating from $t = -\infty$ if we make θ small and $x \to vt$. In other words, the near-field motion at the time of arrival of a rupture front is approximately explained by the simple solution given in equation (14.49).

The new solution, however, contains an additional sharp waveform originating from the starting point at $t = 0$ and propagating with the velocity of
shear waves. The particle velocity becomes infinity at $t = r/\beta$, where it has a
square-root singularity. Near $t = r/\beta$, (14.55) becomes approximately

$$\frac{\partial w}{\partial t} = \frac{\Delta w}{2\pi} \frac{\sin \theta}{(\beta/v - \cos \theta)} \frac{1}{(2r/\beta)^{1/2}} \frac{H(t - r/\beta)}{(t - r/\beta)^{1/2}}. \qquad (14.56)$$

This wave attenuates with distance as $r^{-1/2}$, representing a cylindrical wave
that originates at the starting point of the fault propagation, and shows a
radiation pattern given by $\sin \theta/(\beta/v - \cos \theta)$. For example, the x-direction
along which the fault is propagating is a node. The spectrum of particle velocity
$\partial w/\partial t$ given in (14.56) has a high-frequency asymptote of $\omega^{-1/2}$.

The acceleration associated with this "starting phase" is also infinity at
$t = r/\beta$. The singularity is of the form $(t - r/\beta)^{-3/2} H(t - r/\beta)$, and the corresponding spectrum has a high-frequency asymptote of $\omega^{1/2}$.

If the slip function is a ramp function (equation (14.19)) instead of a step
function, the peak particle velocity will be finite, but the peak acceleration will
have a square-root singularity at $t = r/\beta$.

The effect of stopping can be studied by a superposition of another moving
dislocation starting at, say, $x = L$ at $t = L/v$, propagating with the same velocity v, but with the opposite sign of slip. This will annihilate the fault ahead
of $x = L$, and gives the solution for a finite fault that started at $x = 0$ and
stopped at $x = L$. We then obtain another singularity propagating from the
stopping point as cylindrical waves. The nature of the singularity of this
"stopping" phase is nothing but the "starting" phase of the superposed second
fault.

The above equivalency of stopping and starting is due to the unidirectional
nature of fault propagation. If the rupture starts from a point and grows over an
expanding area, the two effects will be quite different, as discussed in Section
14.1.6 for the far field.

14.2.4   In-plane problems

The simplest in-plane problem (Fig. 14.20) is the semi-infinite fault plane
moving with a constant velocity v from time $-\infty$. We shall consider the case
of a step-function slip, in which the boundary condition is written as

$$u(x, +0, t) - u(x, -0, t) = \Delta u \cdot H(-x'), \qquad (14.57)$$

where $x' = x - vt$. The y-component displacement $v(x, y, t)$ and the traction
on the fault plane are continuous across $y = 0$.

The displacement components that satisfy the equation of motion in the in-plane problem can be written in terms of two scalar potentials as

$$u(x, y, t) = \frac{\partial \phi}{\partial x} - \frac{\partial \psi}{\partial y},$$

$$v(x, y, t) = \frac{\partial \phi}{\partial y} + \frac{\partial \psi}{\partial x}$$

(14.58)

(see Section 5.1 for development of two scalar potentials for two-dimensional $P - SV$ problems). The potentials ϕ and ψ satisfy the following wave equations:

$$\frac{1}{\alpha^2} \frac{\partial^2 \phi}{\partial t^2} = \frac{\partial^2 \phi}{\partial x^2} + \frac{\partial^2 \phi}{\partial y^2},$$

$$\frac{1}{\beta^2} \frac{\partial^2 \psi}{\partial t^2} = \frac{\partial^2 \psi}{\partial x^2} + \frac{\partial^2 \psi}{\partial y^2},$$

where $\alpha = \sqrt{(\lambda + 2\mu)/\rho}$ and $\beta = \sqrt{\mu/\rho}$ are P and S velocities, respectively. Using the moving-coordinate system (compare with the development of (14.45)), the wave equation can be rewritten as

$$\left(1 - \frac{v^2}{\alpha^2}\right) \frac{\partial^2 \phi}{\partial x'^2} + \frac{\partial^2 \phi}{\partial y^2} = 0,$$

$$\left(1 - \frac{v^2}{\beta^2}\right) \frac{\partial^2 \psi}{\partial x'^2} + \frac{\partial^2 \psi}{\partial y^2} = 0.$$

We consider here only subsonic motion, i.e., $v < \alpha$ and $v < \beta$.

The above equations have solutions of the form

$$\phi = \exp(ikx' \pm \sqrt{1 - v^2/\alpha^2}\, ky).$$

and

$$\psi = \exp(ikx' \pm \sqrt{1 - v^2/\beta^2}\, ky).$$

Since there are no sources of waves at infinity, we require that ϕ and ψ propagate away from $y = 0$. For $y > 0$, then, we have

$$\phi = \phi^+ \exp(ikx' - \sqrt{1 - v^2/\alpha^2}|k|y),$$

$$\psi = \psi^+ \exp(ikx' - \sqrt{1 - v^2/\beta^2}|k|y),$$

(14.59)

and for $y < 0$,

$$\phi = \phi^- \exp(ikx' + \sqrt{1 - v^2/\alpha^2}|k|y),$$

$$\psi = \psi^- \exp(ikx' + \sqrt{1 - v^2/\beta^2}|k|y).$$

As usual, we express the displacement u, v in terms of ϕ^+, ψ^+, ϕ^-, and ψ^- using (14.58) and also the stress components τ_{xy} and τ_{yy} by

$$\tau_{yy} = \lambda\left(\frac{\partial u}{\partial x} + \frac{\partial v}{\partial y}\right) + 2\mu\frac{\partial v}{\partial y},$$

$$\tau_{xy} = \mu\left(\frac{\partial u}{\partial y} + \frac{\partial v}{\partial x}\right).$$

$$(14.60)$$

The condition of discontinuity for u (see (14.57)) and the condition of continuity for v, τ_{xy}, τ_{yy} give four equations to determine four unknowns, ϕ^+, ψ^+, ϕ^-, and ψ^-. The continuity of v and τ_{xy} imposes the following constraint on $(\phi^+ + \phi^-)$ and $(\psi^+ - \psi^-)$:

$$-\sqrt{1 - v^2/\alpha^2}(\phi^+ + \phi^-) + i(\psi^+ - \psi^-) = 0,$$

$$2i\sqrt{1 - v^2/\alpha^2}(\phi^+ + \phi^-) + (2 - v^2/\beta^2)(\psi^+ - \psi^-) = 0.$$

The determinant of coefficients here is

$$\begin{vmatrix} -\sqrt{1 - v^2/\alpha^2} & i \\ 2ik\sqrt{1 - v^2/\alpha^2} & 2 - v^2/\beta^2 \end{vmatrix} = -\sqrt{1 - v^2/\alpha^2} \cdot v^2/\beta^2,$$

and it does not vanish as long as $v \neq \alpha$, so we must have

$$\phi^+ + \phi^- = 0,$$

$$\psi^+ - \psi^- = 0.$$

From this, we find that $u(x, y, t)$ and $\tau_{yy}(x, y, t)$ are odd functions of y, and $v(x, y, t)$ and $\tau_{xy}(x, y, t)$ are even functions of y. Since the function τ_{yy} must be continuous at $y = 0$, it must vanish at $y = 0$:

$$\tau_{yy} = 0 \quad \text{at } y = 0. \qquad (14.61)$$

Since u is an odd function, the discontinuity condition (14.57) can be rewritten as

$$u(x, +0, t) = -u(x, -0, t)$$

$$= \frac{\Delta u}{2} H(-x')$$

$$= -\frac{\Delta u}{4\pi} \int_{-\infty + i\varepsilon}^{\infty + i\varepsilon} \frac{e^{ikx'}}{ik} dk, \qquad (14.62)$$

where ε is a small positive number.

From the conditions (14.61) and (14.62), we can determine ϕ_+ and ψ_+ for each k and express u and v as an integral of (14.59) with respect to k. The result for $y > 0$ is

$$u(x, y, t) = -\frac{\Delta u}{2\pi} \int_{-\infty + i\varepsilon}^{+\infty + i\varepsilon} \left[\frac{\beta^2}{v^2} \exp(-\sqrt{1 - v^2/\alpha^2}|k|y) \right.$$

$$\left. - \frac{(\beta^2 - v^2/2)}{v^2} \exp(-\sqrt{1 - v^2/\beta^2}|k|y) \right] e^{ikx'} \cdot \frac{dk}{ik},$$

$$v(x, y, t) = -\frac{\Delta u}{2\pi} \int_{-\infty + i\varepsilon}^{+\infty + i\varepsilon} \left[\frac{\beta^2}{v^2} \sqrt{1 - v^2/\alpha^2} \exp(-\sqrt{1 - v^2/\alpha^2}|k|y) \right.$$

$$\left. - \frac{(\beta^2 - v^2/2)}{v^2 \sqrt{1 - v^2/\beta^2}} \exp(-\sqrt{1 - v^2/\beta^2}|k|y) \right]$$

$$\times i \, \text{sign(Re } k) \cdot e^{ikx'} \frac{dk}{ik},$$

where sign(Re k) means the sign of the real part of k. Putting $y = 0^+$, we get, of course, $u(x, 0, t) = (\Delta u/2) \cdot H(-x')$. The integrand decays exponentially with k and y, showing that they are superpositions of inhomogeneous plane waves, as in the anti-plane case. We expect, therefore, that they are trapped near the fault plane and attenuate quickly with distance from the fault. Both components show a similar amplitude spectrum, but there is a $\pi/2$ phase shift between the two components indicated by the factor i in the integrand for $v(x, y, t)$. If $u(x, y, t)$ is antisymmetric with respect to $x' = 0$, then $v(x, y, t)$ will be symmetric. We shall see shortly that for the step-function slip, the transverse component displacement $v(x, y, t)$ shows an impulsive form with a logarithmic singularity at $x' = 0$.

To avoid the singularity at $k = 0$, we shall first evaluate the particle velocities $\partial u/\partial t$ and $\partial v/\partial t$. Since $\partial/\partial t$ introduces a factor $-ikv$ in the integrand that

removes the singularity, we can put $\varepsilon = 0$. Then the integrals will be either of the form

$$\int_{-\infty}^{\infty} \exp(ikx' - \sqrt{1 - v^2/\alpha^2}|k|y) \, dk$$

$$= \int_{0}^{\infty} \exp(ikx' - \sqrt{1 - v^2/\alpha^2}ky) \, dk + \int_{-\infty}^{0} \exp(ikx' + \sqrt{1 - v^2/\alpha^2}ky) \, dk$$

$$= \frac{2\sqrt{1 - v^2/\alpha^2}\,y}{x'^2 + (1 - v^2/\alpha^2)y^2} \qquad (14.63)$$

or of the form

$$\int_{-\infty}^{\infty} i\,\text{sign}(k)\,\exp(ikx' - \sqrt{1 - v^2/\alpha^2}|k|y) \, dk$$

$$= \int_{0}^{\infty} i\exp(ikx' - \sqrt{1 - v^2/\alpha^2}|k|y) \, dk - \int_{-\infty}^{0} i\exp(ikx' + \sqrt{1 - v^2/\alpha^2}|k|y) \, dk$$

$$= \frac{-2x'}{x'^2 + (1 - v^2/\alpha^2)y^2}. \qquad (14.64)$$

We can also immediately recognize that (14.63) is the derivative of

$$-2\tan^{-1}(\sqrt{1 - v^2/\alpha^2}\,y/x')$$

and that (14.64) is the derivative of $\ln[x'^2 + (1 - v^2/\alpha^2)y^2]$ with respect to x'. Using these relations, we obtain the particle velocity for $y > 0$ as

$$\frac{\partial u(x, y, t)}{\partial t} = -v\frac{\partial u(x, y, t)}{\partial x'}$$

$$= \frac{v\,\Delta u}{\pi}\left[\frac{\beta^2}{v^2}\frac{\sqrt{1 - v^2/\alpha^2}\cdot y}{x'^2 + (1 - v^2/\alpha^2)y^2}\right.$$

$$\left. - \frac{(\beta^2 - v^2/2)}{v^2}\frac{\sqrt{1 - v^2/\beta^2}\,y}{x'^2 + (1 - v^2/\beta^2)y^2}\right], \qquad (14.65)$$

$$\frac{\partial v(x, y, t)}{\partial t} = -v\frac{\partial v(x, y, t)}{\partial x'}$$

$$= -\frac{v\,\Delta u}{\pi}\left[\frac{\beta^2}{v^2}\sqrt{1 - v^2/\alpha^2}\frac{x'}{x'^2 + (1 - v^2/\alpha^2)y^2}\right.$$

$$\left. - \frac{(\beta^2 - v^2/2)}{v^2\sqrt{1 - v^2/\beta^2}}\frac{x'}{x'^2 + (1 - v^2/\beta^2)y^2}\right]$$

and the displacement for $y > 0$ as

$$u(x, y, t) = \frac{\Delta u}{\pi} \left[\frac{\beta^2}{v^2} \tan^{-1} \frac{\sqrt{1 - v^2/\alpha^2} \cdot y}{x'} \right.$$
$$\left. - \frac{(\beta^2 - v^2/2)}{v^2} \tan^{-1} \frac{\sqrt{1 - v^2/\beta^2} \cdot y}{x'} \right], \qquad (14.66)$$

$$v(x, y, t) = \frac{\Delta u}{\pi} \left\{ \frac{\beta^2}{v^2} \sqrt{1 - v^2/\alpha^2} \ln[x'^2 + (1 - v^2/\alpha^2)y^2]^{1/2} \right.$$
$$\left. - \frac{(\beta^2 - v^2/2)}{v^2 \sqrt{1 - v^2/\beta^2}} \ln[x'^2 + (1 - v^2/\beta^2)y^2]^{1/2} \right\}.$$

The above formula was first obtained by Eshelby (1949), and is in fact valid for $y > 0$ and $y < 0$. Values of $\tan^{-1}$ lie in the range $(-\pi, \pi)$, so that there is a step discontinuity in u of amount Δu across $y = 0$ for $x' < 0$. This is the fault plane, and (14.66) correctly reproduces the discontinuity (14.62). The transverse component $v(x, y, t)$ shows an impulsive, symmetric form with a logarithmic singularity $\ln|x'|$ at $x' = 0$ or $x = vt$. This result qualitatively agrees with the result of the numerical solution discussed in Section 14.2.1 in relation to the record of the Parkfield earthquake.

The stress components may be obtained from (14.65) and (14.59) as

$$\tau_{xx} = \frac{2\mu \, \Delta u \beta^2}{\pi v^2} \left[\frac{(1 - v^2/\alpha^2)^{1/2}(v^2/\alpha^2 - v^2/2\beta^2)y}{x'^2 + (1 - v^2/\alpha^2)y^2} \right.$$
$$\left. + \frac{(1 - v^2/2\beta^2)(1 - v^2/\beta^2)^{1/2}y}{x'^2 + (1 - v^2/\beta^2)y^2} \right],$$

$$\tau_{yy} = \frac{2\mu \, \Delta u \beta^2}{\pi v^2} \left[\frac{(1 - v^2/2\beta^2)(1 - v^2/\alpha^2)^{1/2}y}{x'^2 + (1 - v^2/\alpha^2)y^2} \right.$$
$$\left. - \frac{(1 - v^2/2\beta^2)(1 - v^2/\beta^2)^{1/2}y}{x'^2 + (1 - v^2/\beta^2)y^2} \right],$$

$$\tau_{xy} = \frac{2\mu \, \Delta u \beta^2}{\pi v^2} \left[\frac{(1 - v^2/\alpha^2)^{1/2}x'}{x'^2 + (1 - v^2/\alpha^2)y^2} \right.$$
$$\left. - \frac{(1 - v^2/2\beta^2)^2 x'}{(1 - v^2/\beta^2)^{1/2}[x'^2 + (1 - v^2/\beta^2)y^2]} \right].$$

As imposed by the boundary condition (14.61), $\tau_{yy} = 0$ on the fault plane $y = 0$, and τ_{xy} is continuous across $y = 0$, where

$$\tau_{xy} = \frac{2\mu \, \Delta u \beta^2}{\pi v^2 x'} [(1 - v^2/\alpha^2)^{1/2} - (1 - v^2/2\beta^2)^2(1 - v^2/\beta^2)^{-1/2}]. \quad (14.67)$$

Thus, τ_{xy} has an $(x')^{-1}$ singularity at the crack tip, reaching $-\infty$ behind the tip and $+\infty$ ahead of the tip. The function in the bracket [ ] is the familiar form that appeared in equation (5.54) for determining phase velocity of Rayleigh waves in a homogeneous half-space. When v is equal to the Rayleigh velocity, it vanishes. Thus the in-plane shear stress across the fault plane vanishes when the crack tip propagates with the Rayleigh velocity.

The result for the particle velocity is quite similar to that obtained earlier for the anti-plane problem with respect to the spectral contents and attenuation with distance. For example, the peak velocity decays inversely in proportion to the distance from the fault for both components. On the fault, the particle velocity is a δ-function for the parallel component and proportional to $(x - vt)^{-1}$ for the transverse component. Both functions have the same constant spectral density, but they differ by $\pi/2$ in phase for all frequencies. Off the fault, the high-frequency asymptote has an exponential decay, as expected for inhomogeneous plane waves.

Let us now consider the starting effect of in-plane faulting by solving a problem similar to the one studied in the anti-plane case (Fig. 11.20). The faulting starts at $x = 0$ and propagates in the x-direction with velocity v. The boundary conditions are

$$u(x, +0, t) - u(x, -0, t) = \Delta u \cdot H(t - x/v)H(x), \qquad (14.68)$$

with continuity for v, τ_{xy}, and τ_{yy} as before. Again u and τ_{yy} are odd functions of y, and v and τ_{xy} are even functions. It follows, then, that τ_{yy} must vanish at $y = 0$, as in the case of the fault propagated from $t = -\infty$ (see Problem 14.2, for a general result in three dimensions).

Working with potentials (14.58) and the Laplace transform, e.g.,

$$\phi(x, y, s) = \int \phi(x, y, t)e^{-st} \, dt,$$

the equation of motion is satisfied if the potentials satisfy the following wave equations:

$$\frac{\partial^2 \phi}{\partial x^2} + \frac{\partial^2 \phi}{\partial y^2} = \frac{s^2}{\alpha^2} \phi,$$

$$\frac{\partial^2 \psi}{\partial x^2} + \frac{\partial^2 \psi}{\partial y^2} = \frac{s^2}{\beta^2} \psi.$$

The solutions of these equations are of the forms $e^{ikx \pm \gamma y}$ and $e^{ikx \pm vy}$, where

$$\gamma^2 = k^2 + s^2/\alpha^2,$$

$$v^2 = k^2 + s^2/\beta^2.$$

The boundary condition for $u(x, +0, s)$ can be obtained by taking the Laplace transform of (14.68):

$$u(x, +0, s) = -u(x, -0, s)$$

$$= \frac{\Delta u}{2} \frac{e^{-sx/v}}{s} H(x)$$

$$= \frac{\Delta u}{4\pi s} \int_{-\infty}^{\infty} \frac{e^{ikx} \, dk}{i(k - is/v)}. \tag{14.69}$$

This condition and the vanishing τ_{yy} at $y = 0$ determine the solution as

$$u(x, y, s) = -\frac{\Delta u}{2\pi} \int_{-\infty}^{\infty} \left(\frac{\beta^2 k^2}{s^2} e^{-\gamma y} - \frac{2\beta^2 k^2 + s^2}{2s^2} e^{-vy} \right) \frac{e^{ikx} \, dk}{i(k - is/v)}$$

$$v(x, y, s) = -\frac{\Delta u}{2\pi} \int_{-\infty}^{\infty} \left[\frac{\beta^2}{s^2} ik\gamma e^{-\gamma y} - \frac{(2\beta^2 k^2 + s^2)}{2vs^2} ike^{-vy} \right] \frac{e^{ikx} \, dk}{i(k - is/v)}.$$

To each term of the above integral, we can apply Cagniard's method. By transforming the variable k to either

$$\tau = \frac{1}{s}(-ikx + \gamma y)$$

or

$$\tau = \frac{1}{s}(-ikx + vy)$$

and identifying the resulting integral as a Laplace transform, we obtain the following result:

$$\frac{\partial u(x, y, t)}{\partial t} = \frac{\beta^2 \Delta u}{\pi} \left\{ \mathrm{Im} \left[\frac{p_1^2 [t \sin \theta + i \cos \theta (t^2 - r^2/\alpha^2)^{1/2}]}{i/v - p_1} \right] \right.$$

$$\times \frac{H(t - r/\alpha)}{r(t^2 - r^2/\alpha^2)^{1/2}}$$

$$- \mathrm{Im} \left[\frac{(\frac{1}{2}\beta^2 + p_2^2)[t \sin \theta + i \cos \theta (t^2 - r^2/\beta^2)^{1/2}]}{i/v - p_2} \right]$$

$$\left. \times \frac{H(t - r/\beta)}{r(t^2 - r^2/\beta^2)^{1/2}} \right\} \tag{14.70}$$

$$\frac{\partial v(x, y, t)}{\partial t} = \frac{\beta^2 \Delta u}{\pi} \left\{ \mathrm{Re} \left[\frac{(1/\alpha^2 + p_1^2)p_1}{i/v - p_1} \right] \frac{H(t - r/\alpha)}{(t^2 - r^2/\alpha^2)^{1/2}} \right.$$

$$\left. - \mathrm{Re} \left[\frac{(\frac{1}{2}\beta^2 + p_2^2)p_2}{i/v - p_2} \right] \frac{H(t - r/\beta)}{(t^2 - r^2/\beta^2)^{1/2}} \right\},$$

where

$$x = r \cos \theta,$$

$$y = r \sin \theta,$$

$$p_1 = \frac{1}{r}(t^2 - r^2/\alpha^2)^{1/2} \sin \theta + i \frac{t}{r} \cos \theta,$$

$$p_2 = \frac{1}{r}(t^2 - r^2/\beta^2)^{1/2} \sin \theta + i \frac{t}{r} \cos \theta.$$

The above formulas were obtained by Ang and Williams (1959) and were used in the interpretation of accelerograms recorded at Pacoima Dam during the San Fernando earthquake of 1971 February 9 by Boore and Zoback (1974).

At a long distance from the origin, the motion near the fault should look like the one obtained earlier in equations (14.65). Actually, if we make both $y/r (= \sin \theta)$ and $r - vt$ small, equations (14.70) reduce to (14.65). That is, the near-field motion at the time of arrival of the rupture front is approximately explained by the simple forms given in (14.65).

Equation (14.70) contains additional arrivals propagated as P- and S-waves from the starting point of rupture. Making $t - r/\alpha$ small, we find P-waves of the form

$$\begin{matrix} \dfrac{\partial u}{\partial t} \\[2mm] \dfrac{\partial v}{\partial t} \end{matrix} = \begin{matrix} \cos \theta \\[2mm] \sin \theta \end{matrix} \times \frac{\Delta u}{\pi} \frac{\beta^2}{\alpha^2} \frac{\sin 2\theta}{2(\alpha/v - \cos \theta)} \frac{H(t - r/\alpha)}{(t - r/\alpha)^{1/2}(2r/\alpha)^{1/2}}, \quad (14.71)$$

and making $t - r/\beta$ small, we find S-waves of the form

$$\begin{matrix} \dfrac{\partial u}{\partial t} \\[2mm] \dfrac{\partial v}{\partial t} \end{matrix} = \begin{matrix} -\sin \theta \\[2mm] \cos \theta \end{matrix} \times \frac{\Delta u}{\pi} \frac{\cos 2\theta}{2(\beta/v - \cos \theta)} \frac{H(t - r/\beta)}{(t - r/\beta)^{1/2}(2r/\beta)^{1/2}}. \quad (14.72)$$

As shown schematically in Figure 14.21, the radiation patterns of these waves have a double-couple symmetry modified by the factor $(\alpha/v - \cos \theta)^{-1}$ for P and $(\beta/v - \cos \theta)^{-1}$ for S. They are cylindrical waves, attenuating as $r^{-1/2}$. As in the case of the anti-plane problem, the particle velocity has a square-root singularity at the onset. The accelerations associated with these "starting phases" are also infinity at the onset, where they have 3/2 power singularities. If the slip function is a ramp function, the peak particle velocity will be finite, but the peak acceleration will have a square-root singularity at the onset.

As discussed in the preceding section, the effect of stopping of the fault propagation is similar to that of starting in the case of unidirectional fault propagation.

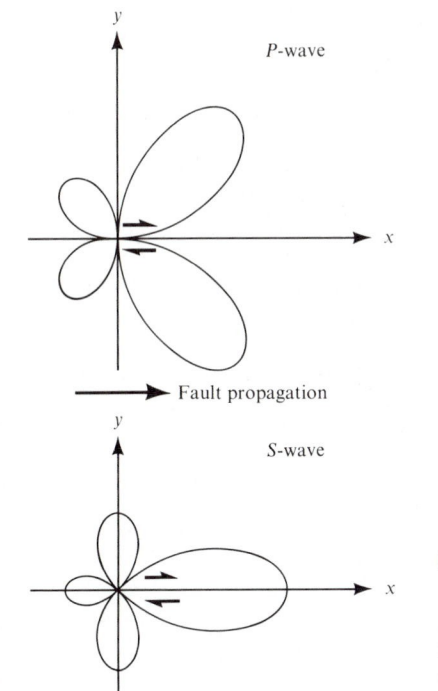

FIGURE **14.21**

Radiation patterns for the body waves radiating from the point of nucleation of a propagating in-plane shear fault. Compare with the usual double-couple radiation patterns (Figs. 4.5a and 4.6a).

For the Haskell model with a uniform slip function over a rectangular fault, Madariaga (1978) obtained an exact analytic solution for motions at any point in an unbounded, elastic, homogeneous medium. The solution consists of (*i*) cylindrical waves from the suddenly appearing initial dislocation line of length W and from the sudden arrest of rupture and (*ii*) spherical waves radiated from the corners of the rectangular fault. The cylindrical waves dominate in the region of a slab normal to the dislocation line containing the fault plane, and have the same characteristics as the cylindrical waves from the moving dislocation line given in equations (14.56), (14.71), and (14.72).

SUGGESTIONS FOR FURTHER READING

Aki, K. Earthquake mechanism. *Tectonophysics*, **13**, 423–446, 1972.

Haskell, N. Total energy and energy spectral density of elastic wave radiation from propagating faults, *Bulletin of the Seismological Society of America*, **54**, 1811–1842, 1964; **56**, 125–140, 1966.

Honda, H. Earthquake mechanism and seismic waves. *Journal of Physics of the Earth*, **10**, 1–98, 1962.

Jeffreys, H. On the mechanics of faulting. *Geological Magazine*, **79**, 291–295, 1942.

Kanamori, H. Great earthquakes at island arcs and the lithosphere. *Tectonophysics*, **12**, 187–198, 1971.

Mansinha, L., D. E. Smylie, and A. E. Beck (editors). *Earthquake Displacement Fields and the Rotation of the Earth*. New York: Springer-Verlag, 1970.

Savage, J. C. Corner frequency and fault dimensions. *Journal of Geophysical Research*, **77**, 3788–3795, 1972.

Trifunac, M. D. A three-dimensional dislocation model for the San Fernando, California, earthquake of February 9, 1971. *Bulletin of the Seismological Society of America*, **64**, 149–172, 1974.

Weertman, J. Dislocations in uniform motion on slip or climb planes having periodic force laws. *In* T. Mura (editor), *Mathematical Theory of Dislocations*. New York: American Society of Mechanical Engineers, 1969.

PROBLEMS

14.1   One of the most powerful methods for discriminating an underground nuclear explosion from an earthquake is based on the excitation of short-period P-waves relative to long-period surface waves. If we take an explosion and a shallow earthquake that generate comparable P-waves with period around 1 sec, it is observed that the Rayleigh waves generated by the explosion are an order of magnitude smaller (with period around 20 sec) than those generated by the earthquake. Assume a double-couple point source for the earthquake and a point source with isotropic moment tensor for the explosion, both buried in a homogeneous half-space. Find out if the difference in source type and focal depth (the depth of an explosion cannot be greater than a few kilometers) can cause an order of magnitude difference in Rayleigh vs P-wave excitation. If not, what other effects can account for this observation?

14.2   Some important symmetry properties for the radiation from general shear faulting on a plane surface within an infinite, homogeneous, isotropic medium can be inferred from (14.37). For shear faulting, show that displacement components *parallel* to the fault plane are *odd* functions of distance x_3 from the fault and that the displacement *normal* to the fault is an *even* function of x_3. Hence, for traction on planes parallel to Σ at distance x_3, show that the normal component is an odd function of x_3 and that the shear components are even. Hence show that the normal component of traction on a planar fault (in an infinite, homogeneous, isotropic medium) does not change at any time for any shearing event on the fault.

14.3   Equation (14.43) amounts to a dynamic boundary condition for tractions on the fault plane. Where do we take this condition into account in setting up a representation of the solution, such as (14.37)? Verify that this representation of the radiated

field does indeed have continuity of shear stress across the fault (use results of Problem 14.2).

14.4 The opening of a crack may be represented by a displacement discontinuity $[\mathbf{u}]$ that is parallel to $\mathbf{v}$, the fault normal. Obtain the equivalent body force in an isotropic elastic body, and find the far-field body waves (P and S) in an infinite homogeneous medium (cf. (equation 14.6)).

14.5 The "finiteness factor" $X^{-1} \sin X$ that appears in equations (14.18)–(14.20) is very simple, because (i) the rupture is unilateral (i.e., it proceeds from one end of the fault to the other); (ii) it has constant rupture velocity; (iii) the fault width W is very small; and (iv) the slip function at each point of the fault plane is the same, apart from a delay due to the time taken for rupture to initiate.

a) Suppose that we drop assumptions (i), (ii), and (iii), but retain (iv). Show that the far-field pulse shape is given by

$$\Omega(\mathbf{x}, \omega) = \Omega_0(\mathbf{x}, \omega) F(\gamma, \omega),$$

where $\Omega_0(\mathbf{x}, \omega)$ is the pulse shape radiated by a point shear dislocation of strength $A \times \Delta u(\omega)$, and the finiteness factor in this more general case is

$$F(\gamma, \omega) = \frac{1}{A} \iint_{\Sigma} \exp i\omega \left[\tau(\xi) - \frac{\xi \cdot \gamma}{c} \right] d\Sigma.$$

Here A is the fault area, $\tau(\xi)$ is the time taken for the rupture to reach ξ on the fault plane, and γ is the ray direction from source to receiver.

b) In the time domain, show that $\Omega(\mathbf{x}, t)$ is given by convolving $\Omega_0(\mathbf{x}, t)$, with a pulse shape having unit "area," i.e., that $\int_{-\infty}^{\infty} F(\gamma, t) \, dt = 1$.

c) Now drop assumption (iv) also, and show that the corresponding finiteness factor, with unit area in the time domain, is

$$F(\gamma, \omega) = \frac{1}{M_0(\omega)} \iint_{\Sigma} \mu(\xi) \, \Delta u(\xi, \omega) \exp i\omega \left[\frac{-\xi \cdot \gamma}{c} \right] d\Sigma,$$

where $M_0(\omega) = \iint \mu(\xi) \, \Delta u(\xi, \omega) \, d\Sigma$.

CHAPTER 15

The Seismic Source: Dynamics

So far, we have studied the seismic motion at near and far field for a propagating dislocation with a given slip function. The form of the slip function was adopted intuitively to simulate geologic faulting with the least number of parameters. As such, some of the slip functions analyzed in Chapter 14 have consequences that are physically unacceptable.

Take the simplest case of an anti-plane problem in which a semi-infinite fault is propagating with uniform velocity (Section 14.2.3). When the slip function is a step function, the shear stress acting on the fault plane is given by equation (14.48),

$$\tau_{yz} = \frac{\mu \, \Delta w \sqrt{1 - v^2/\beta^2}}{2\pi x'} \qquad \text{(putting } x' = x - vt\text{)}, \qquad (15.1)$$

which shows that the shear stress has nonzero values inside the fault plane ($x' < 0$) and amounts to infinity just behind the crack tip ($x' = 0$).

Since faulting is a failure along the fault plane, we expect that the fault plane, once ruptured, cannot sustain stress beyond the frictional stress. Obviously, the step-function slip is a gross violation of this expectation. In this chapter we shall develop a variety of better alternatives, with the principal aim of finding

fault motions that not only are kinematically satisfactory for shear failure, but are also associated with plausible stresses on the fault plane.

This chapter divides conveniently into two main sections. In the first, we shall suppose that the rupture velocity is prescribed (usually, we shall assume it has some constant value). We obtain a simple relation between slip and shear stress on a fault plane for anti-plane problems. Then we describe the energy balance at the rupture front for anti-plane and in-plane faulting and introduce the concept of cohesive force. As a useful illustration of rupture propagation that originates from a point (and therefore involves both anti-plane and in-plane motions), we look at the case of a growing elliptical fault, for which the radiated motions are known in detail. As an example of a fault that grows steadily (from a point) and then suddenly stops, we describe a growing circular fault with known final radius and use an important numerical procedure to obtain the far-field motions. In the second main section, we recognize that shear failure is a spontaneous process and that the velocity of rupture is itself an unknown (probably varying) quantity, to be determined as part of the solution to the problem in hand. The rupture-velocity history is known for a variety of anti-plane problems and for certain in-plane problems.

15.1 Dynamics of a Crack Propagating with Prescribed Velocity

15.1.1 *Relations between stress and slip for a propagating crack*

In order to find an appropriate slip function for a crack propagating with a constant velocity v, we shall first find a relation between the stress and slip on the fault plane for a propagating anti-plane dislocation.

Let the slip function $\Delta w(x')$ be an arbitrary function. We shall express $\Delta w(x')$ by a superposition of step functions, as shown in Figure 15.1. An arbitrary $\Delta w(x')$ can be written as

$$\Delta w(x') = -\int_{x'}^{0} \frac{\partial}{\partial \xi} \Delta w(\xi)\, d\xi$$

$$= -\int_{-\infty}^{0} \frac{\partial \Delta w}{\partial \xi} H(\xi - x')\, d\xi. \tag{15.2}$$

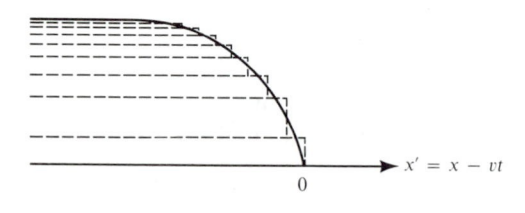

$x' = x - vt$

0

FIGURE **15.1**

Approximating an arbitrary slip function $\Delta w(x')$ by a superposition of step functions.

The step-function slip $H(\xi - x')$ with tip at $x' = \xi$ will generate a stress component τ_{yz} according to equation (15.1) given by

$$\tau_{yz}(x', \xi) = \frac{\mu}{2\pi} \cdot \frac{\sqrt{1 - v^2/\beta^2}}{x' - \xi}.$$

Multiplying by the step-height $-(\partial \Delta w/\partial \xi)\, d\xi$ and integrating over ξ from $-\infty$ to 0, we find that the stress due to the slip function $\Delta w(x')$ is

$$\tau_{yz}(x') = -\frac{\mu}{2\pi} \sqrt{1 - v^2/\beta^2} \int_{-\infty}^{0} \frac{\partial \Delta w/\partial \xi}{x' - \xi}\, d\xi$$

$$= -\frac{\mu}{2\pi} \frac{\sqrt{1 - v^2/\beta^2}}{v} \int_{-\infty}^{0} \frac{\Delta \dot{w}}{\xi - x'}\, d\xi, \tag{15.3}$$

where $\Delta \dot{w}$ is the slip velocity.

For the in-plane problem, a similar relation is found between the shear stress τ_{xy} on the fault plane and slip velocity $\Delta \dot{u}$. Applying the same superposition to equation (14.67), we get

$$\tau_{xy}(x') = -\frac{2\mu\beta^2}{\pi v^3} \left[(1 - v^2/\alpha^2)^{1/2} - (1 - v^2/2\beta^2)^2 \right.$$

$$\left. \times (1 - v^2/\beta^2)^{-1/2} \right] \int_{-\infty}^{0} \frac{\Delta \dot{u}}{\xi - x'}\, d\xi. \tag{15.4}$$

In both (15.3) and (15.4), the shear stress on the fault plane is a constant times the Hilbert transform of slip velocity.

A function and its Hilbert transform are very closely related. From Cauchy's theorem, if $f(\xi)$ is an analytic function on and inside a closed circuit C on the complex ξ-plane, then for $\xi = x$ on C, we have

$$f(x) = \frac{1}{\pi i} \int_C \frac{f(\xi)}{\xi - x}\, d\xi.$$

For $f(\xi) = e^{ik\xi}$ ($k > 0$), choosing the entire real axis and upper semicircle as the circuit C, we find

$$ie^{ikx} = \frac{1}{\pi} \int_{-\infty}^{\infty} \frac{e^{ik\xi}}{\xi - x}\, d\xi \qquad k > 0.$$

Equating the real and imaginary parts separately, we have

$$\cos kx = \frac{1}{\pi} \int_{-\infty}^{\infty} \frac{\sin k\xi}{\xi - x}\, d\xi \qquad k > 0,$$

$$\sin kx = -\frac{1}{\pi} \int_{-\infty}^{\infty} \frac{\cos k\xi}{\xi - x}\, d\xi \qquad k > 0.$$

Thus $\cos kx$ is the Hilbert transform of $\sin kx$, and $\sin kx$ is the Hilbert transform of $-\cos kx$ for $k > 0$. Combining the result for $k < 0$, we find in general that the Hilbert transform of

$$f(x) = \int_{-\infty}^{\infty} F(k)e^{ikx}\, dk$$

is

$$g(x) = \int_{0}^{\infty} iF(k)e^{ikx}\, dk - \int_{-\infty}^{0} iF(k)e^{ikx}\, dk.$$

They share a common amplitude spectral density, and their phases differ by $\pi/2$.

Thus the shear stress and the slip velocity on the plane $y = 0$ must share a common amplitude spectral density, with a phase difference of $\pi/2$. Furthermore, the slip velocity must vanish outside the crack (because no slip occurs there yet), and the shear stress must vanish inside the crack (assuming no frictional stress for simplicity). In other words, we must find a Hilbert transform pair $f(x)$ and $g(x)$ that satisfies

$$f(x) = 0 \qquad x > 0,$$

$$g(x) = 0 \qquad x < 0,$$

and

$$g(x) = \frac{1}{\pi} \int_{-\infty}^{\infty} \frac{f(\xi)}{\xi - x}\, d\xi. \qquad (15.5)$$

From tables of Hilbert transforms, we find the following results for $f(x)$ and $g(x)$ in the undefined regions, satisfy the above three conditions:

$$f(x) = \frac{1}{\sqrt{-x}} \qquad x < 0,$$

$$g(x) = \frac{-1}{\sqrt{x}} \qquad x > 0.$$

It is easy to show that they satisfy equation (15.5) by extending ξ to a complex plane and making a branch cut along the negative real axis (Fig. 15.2). The integral along AO will be equal to the one along OB because of the opposite signs of $\sqrt{-\xi}$ on the two paths. For $x > 0$, the residue evaluation at $\xi = x$ gives $g(x) = 1/\sqrt{x}$, and for $x < 0$ the integral vanishes. Thus we find that a square-root singularity in stress ahead of the crack tip and another square-root singularity in slip velocity behind the crack tip are needed to satisfy the bound-

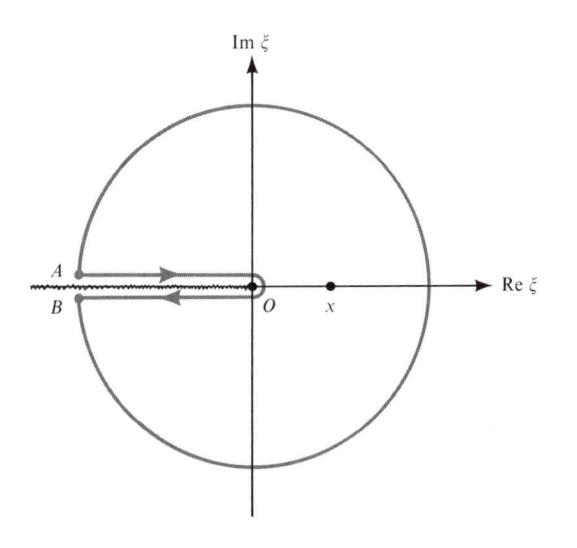

FIGURE **15.2**
Integration path for (15.5) when
$f(\xi) = 1/\sqrt{-\xi}[H(-\xi)]$.

ary conditions for a moving crack. The square-root singularity in stress is well
known for a static crack.

By integrating the slip velocity, we get the slip function proportional to
$\sqrt{-x'}$ for $x' < 0$. We can now summarize results for the anti-plane case as

$$\Delta w = A\sqrt{-x'}H(-x'),$$

$$\dot{w} = \frac{Av}{2\sqrt{-x'}}H(-x'),$$

$$\tau_{yz} = \frac{K}{\sqrt{2\pi x'}}H(x'),$$

(15.6)

where

$$K = \sqrt{2\pi}\,\frac{\mu A\sqrt{1 - v^2/\beta^2}}{4},$$

and the in-plane case as

$$\Delta u = A'\sqrt{-x'}H(-x'),$$

$$\dot{u} = \frac{A'v}{2\sqrt{-x'}}H(-x'),$$

$$\tau_{xy} = \frac{K'}{\sqrt{2\pi x'}}H(x'),$$

(15.7)

BOX **15.1**

Stress singularities for static, in-plane and anti-plane shear cracks of finite width 2a.

The equilibrium equation for the anti-plane displacement $w(x, y)$ is given by

$$\frac{\partial^2 w}{\partial x^2} + \frac{\partial^2 w}{\partial y^2} = 0. \tag{1}$$

For a crack plane defined by $|x| < a$, $y = 0$, and a uniform stress τ_∞ acting at $x, y \to \infty$, the boundary conditions for a stress-free crack are

$$\frac{\partial w}{\partial y} = 0 \qquad |x| < a, y = 0, \tag{2}$$

and

$$w \to \frac{\tau_\infty}{\mu} y \qquad \text{as } x, y \to \infty. \tag{3}$$

(The reference state for displacement is here taken as the stress-free state, in contrast with many of the dynamic solutions in this chapter and the previous one, where the reference state is the static strained state just prior to crack growth).

Equation (1) can be satisfied by the real or imaginary part of an analytic function of $z = x + iy$. It is easy to show that the imaginary part of

$$f(x + iy) = \frac{\tau_\infty}{\mu} [(x + iy)^2 - a^2]^{1/2} \tag{4}$$

satisfies equations (1), (2), and (3).

$$w = \text{Im } f(x + iy)$$

$$= \frac{\tau_\infty}{\mu} \text{Im}\{[(x^2 - y^2 - a^2)^2 + 4x^2 y^2]^{1/4} e^{i\theta/2}\}, \tag{5}$$

where $\sin \theta = 2xy/[(x^2 - y^2 - a^2)^2 + 4x^2 y^2]^{1/2}$. We then have

$$w = \frac{\tau_\infty}{\mu} \sqrt{a^2 - x^2} \qquad y = +0, |x| < a$$

$$= -\frac{\tau_\infty}{\mu} \sqrt{a^2 - x^2} \qquad y = -0, |x| < a. \tag{6}$$

The stress on the plane $y = 0$, but outside of the crack, is

$$\mu \frac{\partial w}{\partial y}\bigg|_{y=0} = \tau_\infty \frac{x}{(x^2 - a^2)^{1/2}} \qquad |x| > a. \tag{7}$$

Here we find the square-root singularity of stress at both ends of the crack $x = a$. The stress intensity factor is $\tau_\infty \sqrt{\pi a}$, and grows proportionally to the square root of the crack length a. The above solution was given by Knopoff (1958). A solution for an in-plane shear crack was given by Starr (1928), with the following results:

$$u = \frac{\tau_\infty}{2} \frac{\lambda + 2\mu}{\mu(\lambda + \mu)} \sqrt{a^2 - x^2} \qquad y = +0, |x| < a$$

$$= -\frac{\tau_\infty}{2} \frac{\lambda + 2\mu}{\mu(\lambda + \mu)} \sqrt{a^2 - x^2} \qquad y = -0, |x| < a \qquad (8)$$

and

$$\tau_{xy} = \tau_\infty \frac{x}{(x^2 - a^2)^{1/2}} \qquad y = 0, |x| > a. \qquad (9)$$

where

$$K' = \sqrt{2\pi} \frac{\mu A' \beta^2 [(1 - v^2/\alpha^2)^{1/2} - (1 - v^2/2\beta^2)^2 (1 - v^2/\beta^2)^{-1/2}]}{v^2}.$$

K and K' are called *stress-intensity factors* in fracture mechanics.

Note that K vanishes for $v = \beta$. Since

$$K' = -\tfrac{1}{4}\sqrt{2\pi}\mu A' \beta^2 v^2 (1 - v^2/\beta^2)^{-1/2} R(1/v),$$

where R is the Rayleigh function first introduced in (5.54), we find that the stress singularity ahead of the in-plane crack vanishes for rupture speed $v = c_R$, the Rayleigh wave speed.

In the preceding chapter, we studied seismic motion from a propagating dislocation with step-function slip. Now that we have found a more appropriate slip function for the crack in the form $\sqrt{-x'}H(-x')$ (instead of $H(-x')$), we shall re-examine the motion in the vicinity of the fault. Using equation (15.2), we can express the slip function for the moving crack as a superposition of step functions:

$$A\sqrt{-x'}H(-x') = \frac{A}{2} \int_{-\infty}^{0} \frac{1}{\sqrt{-\xi}} H(\xi - x') \, d\xi.$$

Since our system is linear, if the seismic motion corresponding to unit step-function slip $H(-x')$ was $f(x', y)$, then the motion $g(x', y)$ for the moving crack

will be

$$g(x', y) = \frac{A}{2} \int_{-\infty}^{0} \frac{1}{\sqrt{-\xi}} f(x' - \xi, y) \, d\xi. \tag{15.8}$$

Using this relation, let us obtain the motion and stress around the tip of the anti-plane crack from the results previously obtained for a step-function dislocation. Putting equations (14.49) and (14.47) into $f(x', y)$ of equation (15.8), the particle velocity $\dot{w}$ and stress component τ_{yz} for the moving crack can be written as

$$\dot{w} = \frac{Av}{4\pi} \int_{-\infty}^{0} \frac{1}{\sqrt{-\xi}} \frac{\gamma y \, d\xi}{(x' - \xi)^2 + \gamma^2 y^2},$$

$$\tau_{yz} = \frac{A\mu}{4\pi} \int_{-\infty}^{0} \frac{1}{\sqrt{-\xi}} \frac{\gamma(x' - \xi) \, d\xi}{(x' - \xi)^2 + \gamma^2 y^2}$$

where $\gamma = \sqrt{1 - v^2/\beta^2}$. Both integrals can be evaluated easily by using the same contour used earlier in Figure 15.2. Now poles are located at $\xi = x' \pm i\gamma y$, and the evaluation of residues at these poles gives

$$\dot{w} = \frac{Av}{4\pi} \pi \left(\frac{1}{2i\sqrt{x' - i\gamma y}} - \frac{1}{2i\sqrt{x' + i\gamma y}} \right)$$

$$= \frac{Av}{4\sqrt{2}} \frac{(\sqrt{x'^2 + \gamma^2 y^2} - x')^{1/2}}{(x'^2 + \gamma^2 y^2)^{1/2}}, \tag{15.9}$$

$$\tau_{yz} = \frac{A\mu\gamma}{4\pi} \pi \left(\frac{1}{2\sqrt{x' + i\gamma y}} + \frac{1}{2\sqrt{x' - i\gamma y}} \right)$$

$$= \frac{A\mu\gamma}{4\sqrt{2}} \frac{(\sqrt{x'^2 + \gamma^2 y^2} + x')^{1/2}}{(x'^2 + \gamma^2 y^2)^{1/2}}. \tag{15.10}$$

In contrast to the case of a step-function dislocation, the peak amplitude of particle velocity decays with distance from the fault as $y^{-1/2}$. The motion for the crack is smoother than for a dislocation. Before discussing the difference in their spectra, we shall point out a drastic difference in the transverse component of particle motion between the in-plane crack and the in-plane step-function dislocation.

The transverse component of particle velocity for the in-plane step-function dislocation is of the form $f(x', y) = 1/x'$ along $y = 0$ ($\dot{v}(x, 0, t)$ in (14.65)).

Using (15.8), the corresponding solution for the crack is (Fig. 15.3)

$$g(x', 0) = \frac{1}{2} \int_{-\infty}^{0} \frac{1}{\sqrt{-\xi}} \frac{d\xi}{x' - \xi}$$

$$= \frac{-1}{2} \frac{1}{\sqrt{x'}} \qquad x' > 0$$

$$= 0 \qquad x' < 0,$$

which has the same form as the shear stress τ_{xy} of the in-plane crack obtained earlier. Remarkably, the transverse component of particle velocity vanishes inside the crack. The corresponding displacement will then be constant inside the crack, and of the form $\sqrt{x'}$ ahead of the crack. In the case of step-function slip, the transverse component of displacement shows a symmetric impulsive form ($\sim \log|x'|$), as can be seen in Figure 15.3, which qualitatively agrees with the observed form for the Parkfield earthquake, as discussed in Section 14.2.1. The solution for the crack, on the other hand, does not show the symmetric

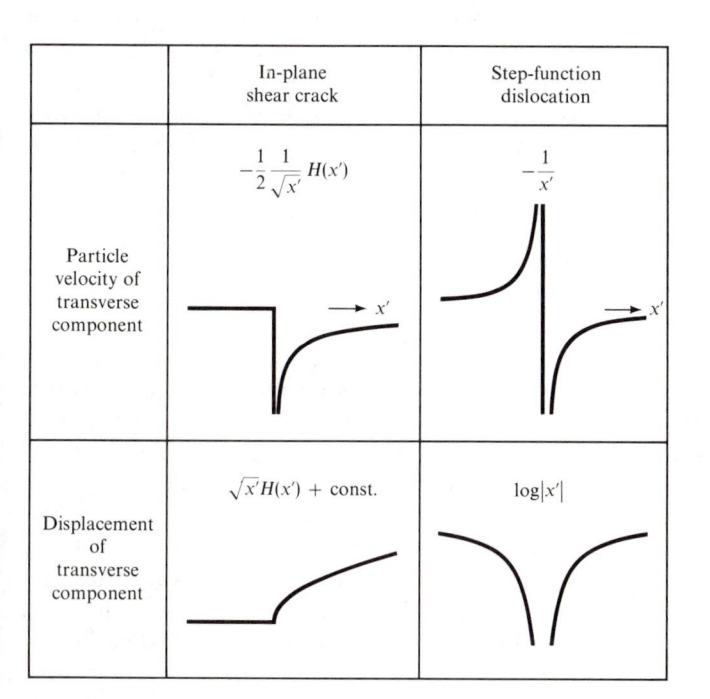

	In-plane shear crack	Step-function dislocation
Particle velocity of transverse component	$-\dfrac{1}{2}\dfrac{1}{\sqrt{x'}} H(x')$	$-\dfrac{1}{x'}$
Displacement of transverse component	$\sqrt{x'} H(x') + \text{const.}$	$\log|x'|$

FIGURE **15.3**

Particle velocity and displacement normal to the fault plane for a shear crack and for a step-function shear dislocation.

impulsive form, but an asymmetric step-like form of $\sqrt{x'}H(x')$ + const., as shown in Figure 15.3.

Equation (15.8) shows that $g(x, y)$ is the convolution of $f(x, y)$ with $\frac{1}{2}(1/\sqrt{-x})H(-x)$. For $k = \omega p > 0$, the Fourier transform of the latter function can be obtained as

$$\frac{1}{2}\int_{-\infty}^{0} \frac{1}{\sqrt{-x}} e^{-ikx}\, dx = \frac{1}{2}\int_{-i\infty}^{0} \frac{1}{\sqrt{-x}} e^{-ikx}\, dx \qquad \text{(Change the path to the}$$
$$\text{negative imaginary } x\text{-axis.)}$$

$$= \tfrac{1}{2}e^{i\pi/4}\int_{0}^{\infty} \frac{1}{\sqrt{y}} e^{-ky}\, dy \qquad \text{(putting } x = -iy)$$

$$= e^{i\pi/4}\int_{0}^{\infty} e^{-kz^2}\, dz \qquad \text{(putting } y = z^2)$$

$$= \frac{1}{2}\sqrt{\frac{\pi}{k}}\, e^{i\pi/4} = \frac{1}{2}\sqrt{\frac{\pi}{\omega p}}\, e^{i\pi/4}.$$

In the frequency domain, therefore, $g(x, y)$ (seismic motion caused by propagating semi-infinite cracks) has an amplitude spectrum proportional to $1/\sqrt{\omega}$ times the spectrum of $f(x, y)$ (seismic motion caused by propagating dislocation of step-function slip) and the phase is shifted by $\pi/4$. This phase shift in the x-coordinate corresponds to a delay of $\pi/4$ in the time axis. Because the $1/\sqrt{\omega}$ factor will attenuate higher frequencies, the motion caused by the propagating crack is smoother than the motion caused by the propagating dislocation with step-function slip.

15.1.2   Energetics at the crack tip

As the crack tip propagates, slip occurs across the fault plane. Neglecting friction, the traction on the fault plane vanishes over the part where slip is occurring. It seems, therefore, that there is no work done on a crack except for the work against friction. A closer look, however, reveals that a finite amount of work is done at the crack tip per unit distance of its propagation. Since the crack tip is moving, it is not obvious how to calculate this work. Let us first derive a general formula for two-dimensional cracks following Freund (1972). For compactness in equations, we shall use x_i ($i = 1, 2, 3$) coordinates, put the crack plane at $x_2 = 0$, and let its tip propagate toward the $+x_1$-direction with velocity v. As shown in Figure 15.4, we consider an external surface S_e fixed to the solid body, with the crack surface S_c already formed and an internal surface S_t enclosing and traveling with the crack tip.

In the volume V bounded by these three surfaces, the body obeys Hooke's law, the equation of motion, and strain-displacement relations:

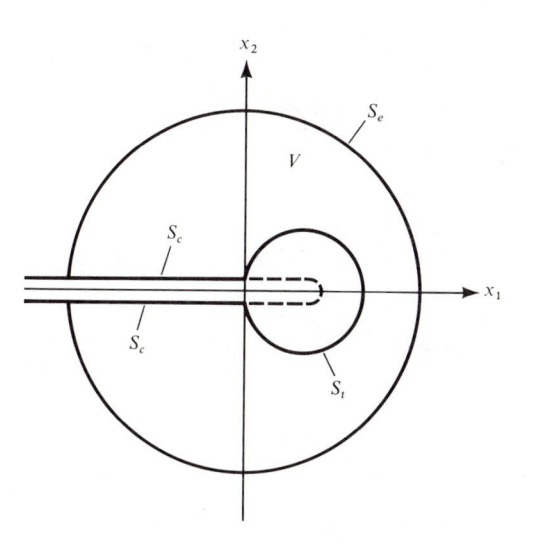

FIGURE **15.4**

$$\tau_{ij} = c_{ijkl}e_{kl},$$

$$\rho \ddot{u}_i = \tau_{ij,j}, \tag{15.11}$$

$$e_{ij} = \tfrac{1}{2}(u_{i,j} + u_{j,i}), \tag{15.12}$$

where τ_{ij}, e_{ij}, c_{ijkl}, and u_i are components of stress, strain, elastic constants, and displacement. We assume that body forces are absent.

On the surfaces S_e and S_t, traction T_i is given as

$$T_i = \tau_{ij}n_j,$$

where n_j is the normal of the surfaces pointing out of V.

The rate of work of the tractions on S_e, the rate of increase of kinetic energy, and that of strain energy in V are, respectively,

$$W = \int_{S_e} T_i \dot{u}_i \, dS,$$

$$\dot{K} = \frac{d}{dt} \int_V \tfrac{1}{2}\rho \dot{u}_i \dot{u}_i \, dV, \tag{15.13}$$

$$\dot{U} = \frac{d}{dt} \int_V \tfrac{1}{2}\tau_{ij}e_{ij} \, dV.$$

The energy flow g into the crack tip can now be obtained as a limit of the flow into the inside of S_t:

$$g = W - \lim_{S_t \to 0} [\dot{K} + \dot{U}].$$

Since S_t is moving along with the crack tip, the region V in equation (15.13) is time-dependent. Thus both $\dot{K}$ and $\dot{U}$ consist of the change in energy occurring inside V and the flux of energy through the boundary S_t. They are

$$\dot{K} = \int_V \rho \dot{u}_i \ddot{u}_i \, dV + \int_{S_t} \tfrac{1}{2} \rho \dot{u}_i \dot{u}_i v_n \, dS,$$

$$\dot{U} = \int_V \tau_{ij} \dot{u}_{i,j} \, dV + \int_{S_t} \tfrac{1}{2} \tau_{ij} u_{i,j} v_n \, dS \qquad (\text{using } \tau_{ij} = \tau_{ji}),$$

where v_n is the normal component of velocity of a point on S_t. Replacing the first integrand for $\dot{U}$ by $(\tau_{ij} \dot{u}_i)_{,j} - \tau_{ij,j} \dot{u}_i$ and applying the divergence theorem to $\int_V (\tau_{ij} \dot{u}_i)_{,j} \, dV$, we find

$$g = \int_{S_e} T_i \dot{u}_i \, dS - \lim_{S_t \to 0} \left[\int_V (\rho \dot{u}_i \ddot{u}_i - \tau_{ij,j} \dot{u}_i) \, dV + \int_{S_e + S_c} \tau_{ij} n_j \dot{u}_i \, dS \right.$$

$$\left. + \int_{S_t} (\tau_{ij} n_j \dot{u}_i + \tfrac{1}{2} \tau_{ij} u_{i,j} v_n + \tfrac{1}{2} \rho \dot{u}_i \dot{u}_i v_n) \, dS \right]$$

$$= - \lim_{S_t \to 0} \int_{S_t} (\tau_{ij} n_j \dot{u}_i + \tfrac{1}{2} \tau_{ij} \dot{u}_{i,j} v_n + \tfrac{1}{2} \rho \dot{u}_i \dot{u}_i v_n) \, dS, \qquad (15.14)$$

where we used (15.11) and (15.12). The contribution from S_c vanishes because $v_n = 0$ and because the crack surface is traction-free. (As mentioned in the beginning of this section, we chose to neglect the effect of friction here.)

Now, coming back to the coordinate frame (x', y) moving with the tip, we shall choose a rectangular surface shown in Figure 15.5 as S_t. The side lengths of the rectangle are 2δ in the x-direction and 2ε in the y-direction. If we shrink the width ε to zero, the contribution from the sides $x = \pm \delta$ vanishes. Since v_n vanishes on the sides $y = \pm \varepsilon$, equation (15.1) is simplified to

$$g = \lim_{\delta \to 0} \int_{-\delta}^{\delta} \mathbf{T}(x', 0) \cdot [\dot{\mathbf{u}}(x', +0) - \dot{\mathbf{u}}(x', -0)] \, dx'.$$

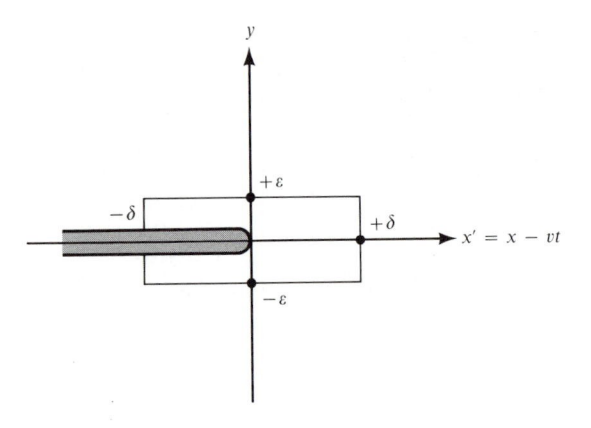

FIGURE **15.5**

Thus, by putting equation (15.6) into the above integral, the rate of work for the anti-plane case is obtained as

$$g = \frac{AvK}{2\sqrt{2\pi}} \lim_{\delta \to 0} \int_{-\delta}^{\delta} \frac{H(-x') \, H(x')}{\sqrt{-x'} \, \sqrt{x'}} \, dx'$$

and, using equation (15.7) for the in-plane case, as

$$g = \frac{A'vK'}{2\sqrt{2\pi}} \lim_{\delta \to 0} \int_{-\delta}^{\delta} \frac{H(-x')H(x') \, dx'}{\sqrt{-x'} \, \sqrt{x'}}.$$

The integrand in the above formulas vanishes except at $x' = 0$, where it is infinite because $H(0) = \frac{1}{2}$. Actually, the integrand behaves like a Dirac delta function. To show this, let us consider the following integral:

$$\int_{-\infty}^{\infty} \frac{H(x')H(x - x')}{\sqrt{x'} \, \sqrt{x - x'}} \, dx' = \int_{0}^{x} \frac{dx'}{\sqrt{x'} \, \sqrt{x - x'}} = \pi H(x).$$

(See Box 12.1.) It then follows that

$$\int_{-\infty}^{\infty} \frac{H(x') \, H(-x')}{\sqrt{x'} \, \sqrt{-x'}} \, dx' = \pi H(0) = \frac{\pi}{2}.$$

Thus, for the anti-plane crack, the rate of work spent at the crack tip is

$$g = \sqrt{\frac{\pi}{2}} \frac{AvK}{4} = \frac{vK^2}{2\mu} \left(1 - \frac{v^2}{\beta^2}\right)^{-1/2}, \qquad (15.15)$$

and for the in-plane crack

$$g = \frac{v}{8} \frac{v^2 K'^2}{\mu \beta^2} \left[\left(1 - \frac{v^2}{\alpha^2}\right)^{1/2} - \left(1 - \frac{v^2}{2\beta^2}\right)^2 \left(1 - \frac{v^2}{\beta^2}\right)^{-1/2}\right]^{-1}. \qquad (15.16)$$

The above result may be obtained without using a value of $H(x)$ at $x = 0$. From equations (15.9) and (15.12) for stress and particle velocity of the anti-plane crack, the first term of the integrand of (15.14) is given by

$$2\tau_{yz}(x', y)\dot{w}(x', y) = \frac{vK^2 y}{2\pi\mu(x'^2 + \gamma^2 y^2)},$$

where K is the stress-intensity factor defined earlier and $\gamma = \sqrt{1 - v^2/\beta^2}$. Putting $y = \pm\varepsilon$ into the above formula and integrating from $x' = -\delta$ to

$x' = +\delta$, we get

$$g = \lim_{\varepsilon \to 0} \lim_{\delta \to 0} \frac{vK^2}{\pi\mu\gamma} \tan^{-1}\left(\frac{\varepsilon}{\delta\gamma}\right) = \frac{vK^2}{2\mu\gamma}, \tag{15.17}$$

which confirms the result given in (15.15).

In the case of the anti-plane crack, the energy flow at the tip vanishes when $K = 0$, i.e., when the rupture velocity is equal to the shear velocity. In the case

---

BOX **15.2**

Fracture criteria

Since most materials fracture when stressed beyond some critical level, it is natural to describe the condition for fracture by a critical applied stress, or strength of material. It was found, however, that the fracture strength of a given material varies greatly, and the theories built around the concept of strength as a material constant were incapable of accounting for diversity in fracture behavior.

A breakthrough was made by A. A. Griffith in 1920. He assumed the existence of flaws in material in the form of cracks. Creating new crack surfaces requires an increase of the free surface energy. This energy must be supplied from the surrounding medium for the crack to extend. Griffith's fracture criterion is based on the balance of consumed surface energy and the supply of mechanical energy for an infinitesimal virtual increase in crack length. In this section, we have just calculated the rate of supply of mechanical energy to the crack tip when the crack tip moves at a constant speed ((15.15) and (15.16)). In Section 15.2.1, we shall use the Griffith concept of energy balance in deriving the equation of motion for a crack tip (15.53).

On the other hand, fracture mechanics was formulated around the concept of stress-intensity factor by G. R. Irwin and his associates in about 1950. It was found that the Griffith fracture criterion is equivalent to the existence of a critical stress-intensity factor. If the stress-intensity factor exceeds the critical value, the crack will extend. We shall call this the Irwin criterion.

In equations (15.15) and (15.16), we have shown that the energy flow into the crack tip is determined by the stress-intensity factor K, or K', and the rupture-propagation velocity v. Therefore, at the initiation of crack extension, when $v = 0$, the energy flow and stress-intensity factor are uniquely related, demonstrating the equivalence of Griffith and Irwin criteria. The equivalence relation is shown explicitly in (15.22) setting $v = 0$ there for an anti-plane crack and in (15.79) for an in-plane crack.

For a finite rupture velocity v, both the Griffith surface energy and the critical stress-intensity factor may depend on v. In Sections 15.2.2 and 15.2.3, we shall consider two fracture criteria. In the Griffith criterion, we assume that the surface energy does not depend on v; in the Irwin criterion, we assume that the critical stress intensity factor is independent of v. Figure 15.21 compares the motion of the crack tip obtained by the two criteria.

of the in-plane crack, it vanishes when the rupture velocity is equal to the Rayleigh-wave velocity. Thus, at these velocities, energy needed for creating new surfaces of the crack cannot be supplied to the crack tip. In this sense, they are the terminal velocities of crack propagation. Equation (15.16) shows that if the rupture velocity exceeds the Rayleigh-wave velocity, g becomes negative. In other words, the crack-tip becomes a source of energy flow instead of a sink. This is physically unacceptable, and the speed of the in-plane shear crack cannot exceed the Rayleigh-wave velocity. This conclusion, however, will be modified in Section 15.2.3, where we discuss the rupture propagation in a medium with finite cohesive force.

15.1.3 Cohesive force

The solutions for stress and particle velocity around the propagating crack tip obtained in Section 15.1.1 are still not quite realistic, because both become infinity at the crack tip. Any material has a finite strength and cannot withstand stress beyond some limit. The singularities can, however, be eliminated by defining the *cohesive force* introduced by Barenblatt (1959), which is distributed inside the crack near the tip and which opposes the external stress.

Let us consider the case of an anti-plane crack, and put the total traction on the ruptured surface ($x' \leq 0$) as

$$\sigma_{yz}(x', 0) = \sigma_{yz}^d + \sigma_c(x'). \tag{15.18}$$

Here σ_{yz}^d is due to dynamic friction and acts all over the crack, but the cohesive force (per unit area) $\sigma_c(x')$ is nonzero only in $-d < x' \leq 0$, where d is the length of the end region, as shown in Figure 15.6. The distribution of the cohesive force will generate a concentration of τ_{yz} ahead of the crack tip, with

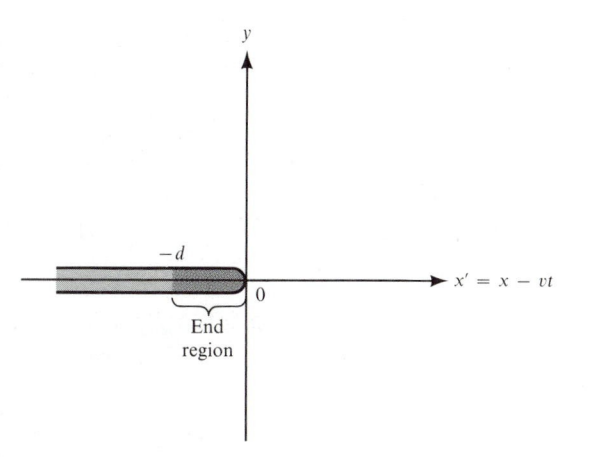

FIGURE **15.6**

the stress-intensity factor given by

$$-\sqrt{\frac{2}{\pi}} \int_{-d}^{0} \sigma_c(\xi)(-\xi)^{-1/2} \, d\xi. \tag{15.19}$$

The derivation of the above expression is given in Box 15.3. The original stress singularity due to the external stress may be eliminated if we choose the cohesive force $\sigma_c(\xi)$ that satisfies the condition

$$K = \sqrt{\frac{2}{\pi}} \int_{-d}^{0} \sigma_c(\xi)(-\xi)^{-1/2} \, d\xi. \tag{15.20}$$

With this choice of $\sigma_c(\xi)$, the stress component $\sigma_{yz}(x',0)$ will be finite and continuous at the crack tip. Since the slip velocity $\Delta \dot{w}$ is the Hilbert transform of the shear stress times a constant, as shown in equation (15.3), the singularity of $\Delta \dot{w}$ is also removed if the shear stress becomes continuous there.

If d is small, the elastic field due to the cohesive force is limited near the crack tip and does not affect the field outside the immediate vicinity of the crack tip. Then the energy flow into the crack tip through the external surface will be the same as given in (15.15) for the case of no cohesive force. This energy flow is absorbed to create a new surface of the crack. Expressing the surface energy per unit area as G, we have

$$g = 2Gv, \tag{15.21}$$

where the factor 2 accounts for both faces of the crack. From (15.15) and (15.21), we find

$$G = \frac{K^2}{4\mu} \left(1 - \frac{v^2}{\beta^2} \right)^{-1/2} \tag{15.22}$$

(A similar relation may be obtained for an in-plane shear crack using (15.16).)

In order to get a rough estimate for the highest frequencies involved in seismic motion caused by propagation of a crack, we shall assume that the cohesive force is uniformly distributed over the end region. The corresponding stress-intensity factor is

$$K = \sqrt{\frac{2}{\pi}} \int_{-d}^{0} \sigma_c(-\xi)^{-1/2} \, d\xi = \frac{2\sigma_c}{\sqrt{\pi}} \sqrt{2d}, \tag{15.23}$$

where σ_c is the cohesive force per unit area. Putting (15.23) into (15.22), we find

$$G = \frac{2\sigma_c^2 d}{\mu\pi} \left(1 - \frac{v^2}{\beta^2} \right)^{-1/2} \tag{15.24}$$

This equation gives a relation among important quantities that determine the seismic motion around the crack tip, and about which we know very little. In general, G, σ_c, and d may depend on the rupture velocity.

Since d is the measure of distance over which slip is resisted, the longer the d, the slower the slip at the initial stage. Therefore, we expect higher slip velocity and acceleration as d gets shorter. The characteristic time constant t_d may be given by d/v:

$$t_d = d/v$$
$$= \frac{\pi \mu G [1 - (v^2/\beta^2)]^{1/2}}{2\sigma_c^2 v}. \tag{15.25}$$

This is the time constant that controls the highest end of the seismic spectrum. Static experiments on rock samples in the laboratory give G on the order of 10^3 erg/cm^2 and σ_c on the order of 10^9 dyne/cm^2. For a rough estimate, we shall assume that their order of magnitude remains the same in the dynamic case, so that for $\beta = 3.5$ km/sec, $v = 3$ km/sec, and $\mu = 3 \times 10^{11}$ dyne/cm^2, we get

$$t_d = 10^{-9} \text{ sec.}$$

Thus we expect radiation of seismic waves with frequency up to a gigahertz if the laboratory values are applicable.

In the actual field situation, G may increase with crack length. The stress around the crack tip increases as the crack length increases. (As shown in Box 15.1, the static stress-intensity factor increases for larger cracks.) Consequently, the volume of the region of microcracks and plastic deformation will increase. This region will absorb energy, making the apparent value of G greater for larger earthquakes.

The highest frequency contained in the usual earthquake records is on the order of 100 Hz. Assuming that the cohesive stress σ_c in the actual fault gouge is on the order of 10^8 dyne/cm^2, the value of G corresponding to $t_d = 0.01$ sec will be around 10^8 erg/cm^2 from (15.25), which is many orders of magnitude greater than the laboratory values.

The physical meaning of cohesive force may become clearer if we write it, instead of equation (15.18), in the form of constitutive equations, such as

$$\sigma_{yz}(x', 0) = \sigma_{yz}^d + \sigma_c[\Delta w(x')] \qquad x' \leq 0$$

or   (15.26)

$$\sigma_{yz}(x', 0) = \sigma_{yz}^d + \sigma_c[\varepsilon(x')] \qquad x \leq 0,$$

where ε is the plastic strain in the fault gouge and Δw is the equivalent slip

between the fault surfaces corresponding to the plastic strain. If the width of the fault gouge is b, we may consider that $\Delta w = b\varepsilon$. The specific surface energy G can be expressed as $G = \frac{1}{2} \int_0^\infty \sigma_c(D) \, dD$, where the factor $\frac{1}{2}$ accounts for the two surfaces of the crack. Relation (15.26) may be determined by laboratory experiment on the stress-strain relation of rock samples and field study on gouge width, and it may be appropriate to allow σ_c to depend on $\Delta\dot{w}$ as well as on Δw. Once the relation is known, the slip function can be calculated by an iterative method. We start with an initial guess of the slip function $\Delta w(x')$ and obtain the corresponding cohesive force from (15.26). Then we can calculate the stress-intensity factor K from (15.20). At a distance sufficiently far from the tip compared with the scale of the end region, the cohesive force no longer governs the slip function, which is determined instead by the macroscopic crack parameters, such as the shape, length, and stress drop. Knowing σ_{yz} for $-\infty < x' < \infty$, the slip velocity can be obtained by the Hilbert transform (see equation (15.3)). The resultant slip function will be used as the second trial function for revising the cohesive force. The iteration proceeds until the slip function converges to a final solution. Ida (1972, 1973) used this method to calculate the slip function and its time derivatives for various cases of the cohesive force diagram $\sigma_c[\Delta w]$ (assuming a semi-infinite crack with constant stress drop as the macroscopic model) and discussed the maximum acceleration and velocity in terms of this material property. Andrews (1976) extended Ida's work and incorporated the cohesive force in a finite-difference calculation of crack propagation (discussed in Section 15.2.3 and Fig. 15.26), which holds promise as an approach for combining numerical analysis of rupture propagation with laboratory results on rock mechanics.

15.1.4   Near field of a growing elliptical crack

In Section 14.1.6, we studied the far-field body waves from an elliptical crack growing with constant velocity and keeping the same shape. Neglecting the stopping phase, we found that the initial rise of far-field displacement is proportional to the square of time measured from the onset. The corresponding acceleration showed a finite jump discontinuity at the onset. In this section, we shall consider seismic motion in the near field of the growing elliptical crack.

Let us assume initially a state of uniform stress $\boldsymbol{\sigma}^0$ and suppose that a plane shear crack nucleates at the origin at time $t = 0$. The fault surface $S(t)$ is defined in Cartesian coordinates by the ellipse

$$S(t) = [x_3 = 0: x_1^2/u^2 + x_2^2/v^2 \le t^2],$$

which (see Fig. 15.7) has axes growing steadily at speeds u and v, each less than (or equal to) the shear-wave speed β. The shear stresses across plane $x_3 = 0$

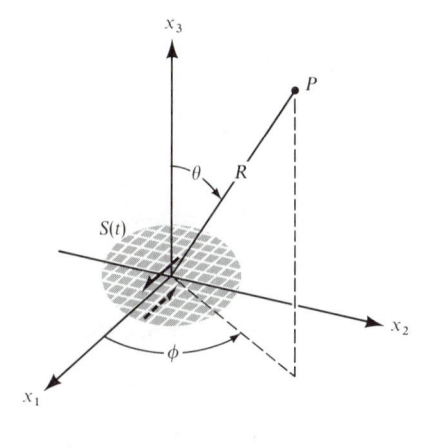

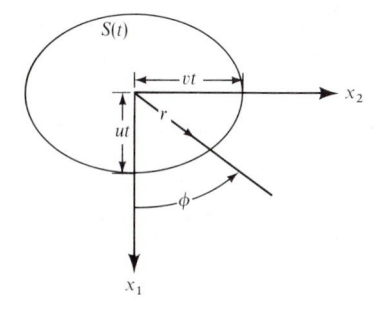

FIGURE **15.7**

are influenced by waves emanating from the point of nucleation, but after arrival of the rupture they drop to new values prescribed over $S(t)$.

To describe the problem further, let $\mathbf{u}$ be displacement from the initial (pre-stressed, static) position, with $\boldsymbol{\tau}$ as the stress tensor due to $\mathbf{u}$ (so that $\boldsymbol{\sigma}^0 + \boldsymbol{\tau}$ is the total stress). Within an infinite homogeneous medium, $\mathbf{u}$ and $\boldsymbol{\tau}$ have certain symmetric properties with respect to the crack plane $x_3 = 0$ (see Problem 14.2): from equation (14.37) or an argument similar to the one used in the in-plane problem (Section 14.2.4), when the discontinuity across the crack plane is restricted to the parallel displacement, we find that τ_{33}, u_1, and u_2 are odd functions of x_3. They must therefore vanish at $x_3 = 0$ wherever they are continuous there. Thus we have the following boundary conditions:

$$\tau_{33} = 0 \qquad \text{everywhere on } x_3 = 0 \qquad (15.27)$$

and

$$u_1 = u_2 = 0 \qquad \text{on } x_3 = 0 \text{ but off } S(t). \qquad (15.28)$$

Burridge and Willis (1969) found the following simple solution for the slip function across a growing elliptical shear crack:

$$\begin{pmatrix} u_1 \\ u_2 \end{pmatrix} = \begin{pmatrix} a \\ b \end{pmatrix} \left(t^2 - \frac{x_1^2}{u^2} - \frac{x_2^2}{v^2} \right)^{1/2} \qquad \text{on } x_3 = +0 \text{ and } S(t)$$

$$= \begin{pmatrix} 0 \\ 0 \end{pmatrix} \qquad \text{on } x_3 = +0 \text{ but off } S(t).$$

(15.29)

The elastic field generated by this slip function under the conditions (15.27) and (15.28) indeed gives the shear-stress jump (τ_{13}, τ_{23}) constant in time and space on $S(t)$. It was found that τ_{13} is proportional to a and that τ_{23} is proportional to b, where a and b are particle-velocity components at the center of the crack, as seen from (15.29). For simplicity, we shall take the x_1-axis in the direction of maximum initial shear, so that no drop occurs in the stress component τ_{23}. In this case, $b = 0$ and the slip component u_2 vanishes. On the moving part of the fault, τ_{13} is constant, and we can think of the total shear stress $\sigma_{13}^0 + \tau_{13}$ as being proportional to σ_{33}^0 via a dynamic coefficient of friction according to the Coulomb law of friction.

Following Richards (1973, 1976), we shall take the following steps for computing the elastic field radiated from the growing crack:

i) Fourier transformation for x_1 and x_2; Laplace transformation for t:

$$f(x_1, x_2, x_3, t) \rightarrow f(k_1, k_2, x_3, s),$$

where f is any dependent variable (such as a displacement component) of interest. Boundary conditions on $x_3 = 0$ are thus transformed to

$$\tau_{33} = 0, \quad u_1 = \frac{4\pi auv}{(s^2 + k_1^2 u^2 + k_2^2 v^2)^2}, \quad u_2 = 0.$$

ii) Transformation of the wave equation and use of potentials to derive algebraic expressions for $\mathbf{u}(k_1, k_2, x_3, s)$. The double Fourier inverse transform is taken, yielding the forward Laplace transform as an explicit double integral over the whole (k_1, k_2) plane. A rotation and stretch of the (k_1, k_2) plane to variables (w, q) is carried out via the de Hoop transformation

$$k_1 = (s/\alpha)(q \cos \phi - w \sin \phi),$$

$$k_2 = (s/\alpha)(q \sin \phi + w \cos \phi),$$

where α is the P-wave speed. The Laplace-transformed P-wave component of displacement at position $\mathbf{x}$ then has the form

$$\mathbf{u}^P(\mathbf{x}, s) = (1/s^2) \int_0^\infty dw \int_{-\infty}^\infty dq \mathbf{F}(q, w, \phi) e^{-st}, \qquad (15.30)$$

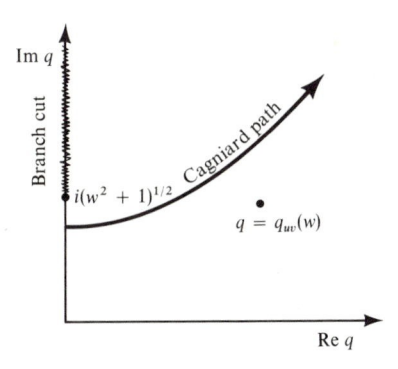

FIGURE **15.8**

There is a pole at q_{uv} near the Cagniard path for evaluating (15.30). [From Richards, 1976a.]

where $\mathbf{F}$ is known, $t = t(q, w, \theta) \equiv [-iq \sin \theta + (1 + q^2 + w^2)^{1/2} \cos \theta](R/\alpha)$, and the spherical polars (R, θ, ϕ) for $\mathbf{x}$ are shown in Figure 15.7. It can be shown that only the positive real q-axis is needed for the integration in (15.30). There is a similar expression for the S-wave component.

iii) Application of Cagniard's method, turning the q-integral into the Laplace transform of the w-integral, so that displacement in the time domain is recognized as a single integral over w. A complication arises because of singularities of the integrand $\mathbf{F}$, as shown in Figure 15.8. This is a diagram of the complex q-plane, and it shows that between the real-axis path of integration needed in (15.30) and the Cagniard path (on which the expo-nent $t(q, w, \theta)$ in (15.30) is real), the integrand has a pole. In fact, it is a second-order pole, denoted by q_{uv}, and is due to the moving nature of the source: it is necessary to pick up residues in converting to the Cagniard path, giving the form

$$\mathbf{u}^P(\mathbf{x}, s) = \frac{1}{s^2} \int_0^\infty dw \int_0^\infty dt \mathbf{F}(q(t), w, \phi) e^{-st} \frac{dq}{dt}$$

$$+ \int_0^\infty dw \mathbf{R}(q_{uv}, w, \phi, s) e^{-st}(q_{uv}, w, \theta), \qquad (15.31)$$

From the first term on the right-hand side here, one can invert to the time domain in the usual fashion (i.e., by reversing the order of integration and recognizing the result as a forward Laplace transform), obtaining a single integral over w. The second term on the right-hand side of (15.3) is already in the form suitable for recognition as the Laplace transform of a function of time, and results in an algebraic closed-form expression. This method, an algebraic expression resulting from an integral of residues, was first developed by Gakenheimer and Miklowitz (1969) for solving Lamb's problem with a moving source.

The complete seismogram can be calculated only numerically. Figure 15.9 shows theoretical record sections for x_1- and x_3-components of acceleration near a left-lateral strike-slip fault. The coordinates for the four stations are

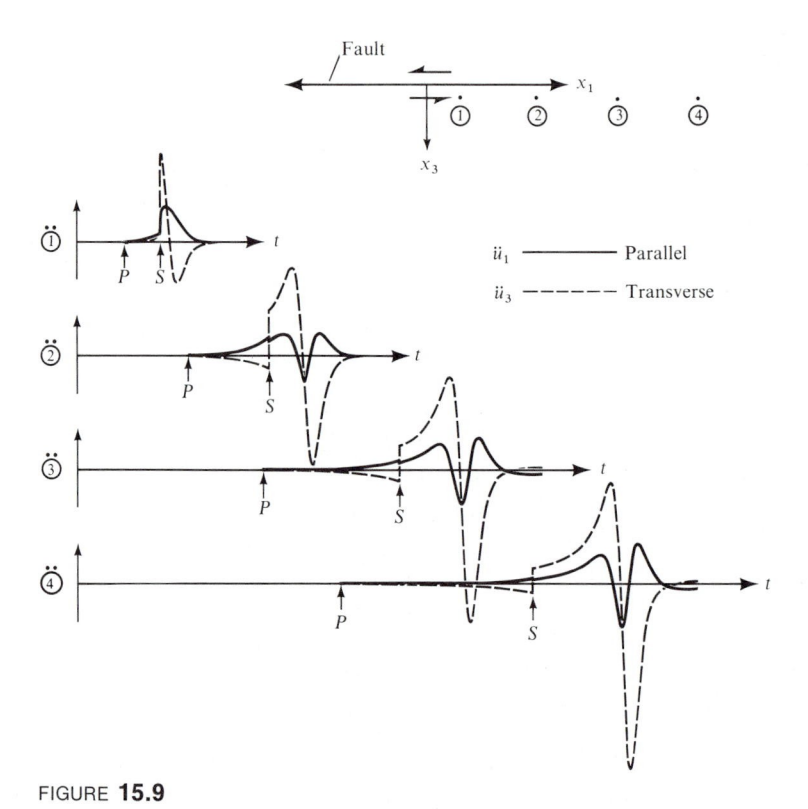

FIGURE **15.9**

Synthetic seismograms for x_1- and x_3-components of acceleration at stations shown at the top. [From Richards, 1976a.]

$(1, 1.5, 0.5)$, $(4, 1.5, 0.5)$, $(7, 1.5, 0.5)$, and $(10, 1.5, 0.5)$. The density of the medium is 2.7 grams/cm^3, the P-wave velocity is 5.2 km/sec, and the S-wave velocity is 3 km/sec. The rupture speed in the x_1-direction is 90% of the Rayleigh-wave velocity, and that in the x_2-direction is 90% of the S-wave velocity. We see, in this case, small P-waves, sharp step-like S-waves arriving from the nucleation point, and large acceleration associated with the passage of the crack tip. The amplitude of waves from the nucleation point decreases with distance, whereas the acceleration associated with passage of the crack-tip increases because the stress-intensity factor increases with increasing crack length.

The corresponding displacement records are shown in Figure 15.10. As discussed in Section 15.1.1, the transverse component shows a step-like waveform rather than a symmetric, impulsive form. The parallel components show a very slow rise beginning at the arrival of P-waves from the nucleation point, and do not show any sharp mark due to passage of the rupture front. This shows the difficulty of accurately estimating rupture velocity from displacement measurements at points off the crack plane.

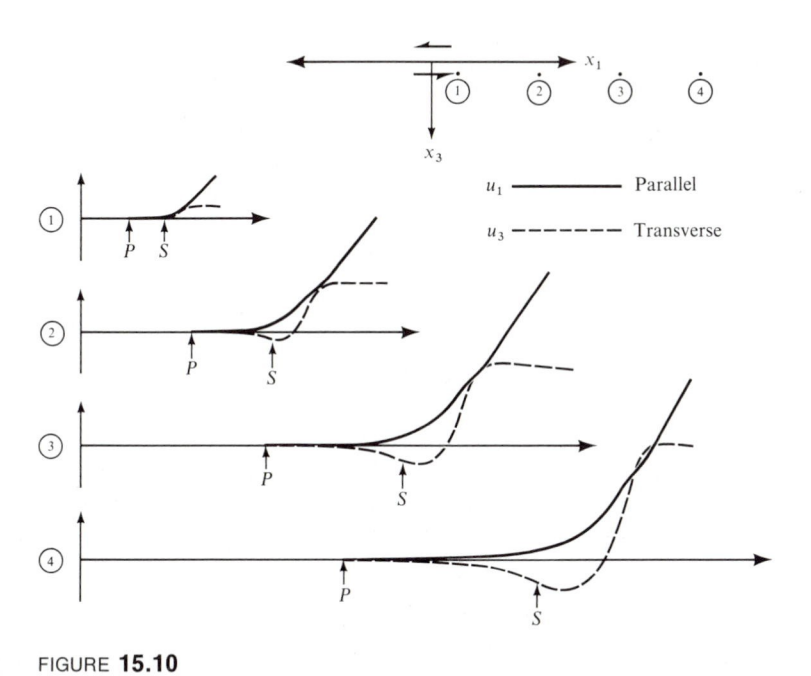

FIGURE **15.10**

Synthetic seismograms for x_1- and x_3-components of displacement at
stations shown at the top. [From Richards, 1976a.]

Compact formulas can be obtained for approximate waveforms corresponding to the arrivals of P- and S-waves from the nucleation point. At the
arrival time $t = R/\alpha$, we find that the acceleration has a jump discontinuity:

$$\ddot{\mathbf{u}}^P = \frac{4uv\beta^2 \cos\theta \sin\theta}{\alpha^3(1 - D\sin^2\theta)^2} a \cos\phi \frac{H(t - R/\alpha)}{R} \hat{\mathbf{R}},$$

where $D = (u^2 \cos^2\phi + v^2 \sin^2\phi)/\alpha^2$. (R, θ, ϕ) are the spherical coordinates
shown in Figure 15.7. The vector $\ddot{\mathbf{u}}^P$ points to the radial direction from the
nucleation point, given by the unit vector $\hat{\mathbf{R}}$. The acceleration due to shear waves
from the nucleation point shows a jump discontinuity at $t = R/\beta$:

$$[\ddot{u}_1^S, \ddot{u}_2^S, \ddot{u}_3^S] = \frac{2uv}{(1 - D\alpha^2 \sin^2\theta/\beta^2)^2} \{\cos\theta[a, 0, 0]$$

$$- a\sin\theta\cos\phi[\sin 2\theta\cos\phi, \sin 2\theta\sin\phi, \cos 2\theta]\} \frac{H(t - R/\beta)}{R}.$$

The high-frequency asymptote of the acceleration is therefore proportional to
ω^{-1}, and the corresponding displacement spectrum has a high-frequency
asymptote like ω^{-3}, in agreement with previous results (equation (14.28)). The

radiation pattern of these waves shows a double-couple symmetry modified by the factors $(1 - D \sin^2 \theta)^{-2}$ for P-waves and $(1 - D\alpha^2 \sin^2 \theta/\beta^2)^{-2}$ for S-waves.

Another compact form of approximate solution can be obtained for singularities of particle velocity and traction components near the crack tip. Let us denote the arrival time of the crack tip at $(x_1, x_2, 0)$ as t_c, so that

$$t_c = (x_1^2/u^2 + x_2^2/v^2)^{1/2}.$$

The particle velocity $\dot{u}_1$ on the plane $x_3 = 0$ is given by boundary conditions (15.28) and (15.29) as

$$\dot{u}_1 \sim a\sqrt{t_c/2}(t - t_c)^{-1/2}H(t - t_c). \tag{15.32}$$

Singularities in $\dot{u}_2$ and τ_{33} are absent on $x_3 = 0$, since these quantities are zero throughout the plane. Singularities in the remaining velocity and traction components on $x_3 = 0$ are

$$\dot{u}_3 \sim \frac{2\beta^2 Va \cos \phi}{U\alpha^2 F^{1/2}B_s} \left[\tfrac{1}{2}\alpha^2/\beta^2 + B_s^2 - B_p B_s\right] \sqrt{\frac{t_c}{2}}(t_c - t)^{-1/2}H(t_c - t),$$

$$\tau_{13} \sim \frac{4\mu\beta^2 a}{U^2\alpha^3 FB_s} [(B_p - B_s)B_s V^2 \cos^2 \phi + \tfrac{1}{4}(U^2 B_s^2 F - V^2 \cos^2 \phi)(\alpha^2/\beta^2)]$$

$$\times \sqrt{t_c/2}(t_c - t)^{-1/2}H(t_c - t), \tag{15.33}$$

$$\tau_{23} \sim \frac{4\mu\beta^2 a \cos \phi \sin \phi}{\alpha^3 FB_s} [B_s B_p - B_s^2 - \tfrac{1}{4}(\alpha^2/\beta^2)]$$

$$\times \sqrt{t_c/2}(t_c - t)^{-1/2}H(t_c - t),$$

where μ is the rigidity and all capital letter symbols are dimensionless quantities given by

$$U = u/\alpha,$$

$$V = v/\alpha,$$

$$F = U^2 \sin^2 \phi + V^2 \cos^2 \phi,$$

$$B_p^2 + 1 = B_s^2 + \frac{\alpha^2}{\beta^2} = \frac{U^4 \sin^2 \phi + V^4 \cos^2 \phi}{U^2 V^2 F}.$$

Since the singularities (15.32) and (15.33) describe local properties of the motion at points near the crack tip, it is instructive to work with a coordinate system related naturally to the local geometry. Figure 15.11 shows such a system, using directions of the normal, the tangent, and the binormal (i.e., the

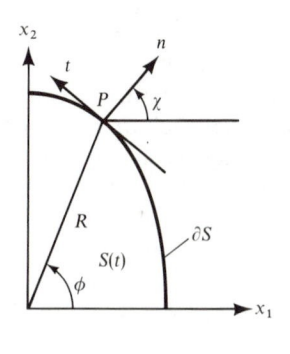

FIGURE **15.11**

Local Cartesian coordinates in the normal (n) and tangent (t) directions. [From Richards, 1976a.]

x_3-axis). Tensor components are rotated to

$$\tau_{3n} = \tau_{31} \cos \chi + \tau_{23} \sin \chi,$$

$$\tau_{t3} = -\tau_{31} \sin \chi + \tau_{23} \cos \chi.$$

Letting v_n be the velocity of rupture in direction n, we find that

$$v_n = \frac{uv}{\alpha} [F/(U^4 \sin^2 \phi + V^4 \cos^2 \phi)]^{1/2},$$

which is simply related to B_p and B_s.

We can now resolve the local motion into in-plane components (u_n and τ_{zn}) and anti-plane components (u_t and τ_{tz}). Then the singularities for the in-plane components are

$$\dot{u}_n \sim \frac{a v_n V \cos \phi}{\alpha U F^{1/2}} \sqrt{t_c/2} (t - t_c)^{-1/2} H(t - t_c),$$

$$\tau_{3n} \sim \frac{-\mu \beta^2 v_n V a R(1/v_n)}{B_s U F^{1/2}} \cos \phi \sqrt{t_c/2} (t_c - t)^{-1/2} H(t_c - t),$$

where $R(1/v_n) = [(\alpha^2/\beta^2)(v_n^2/\beta^2) - 4B_s B_p + 4B_s^2]/(\alpha^2 v_n^2)$, so that $R(1/v_n)$ is the Rayleigh function,

$$R(p) \equiv \left(\frac{1}{\beta^2} - 2p^2\right)^2 - 4p^2 \left(p^2 - \frac{1}{\alpha^2}\right)^{1/2} \left(p^2 - \frac{1}{\beta^2}\right)^{1/2}, \qquad (15.34)$$

and singularities for the anti-plane components are

$$\dot{u}_t \sim \frac{-a v_n U \sin \phi}{\alpha V F^{1/2}} \sqrt{t_c/2} (t - t_c)^{-1/2} H(t - t_c),$$

$$\tau_{t3} \sim -\frac{\mu U v_n a B_s}{V F^{1/2} \alpha^2} \sin \phi \sqrt{t_c/2} (t_c - t)^{-1/2} H(t_c - t).$$

In agreement with results obtained in Section 15.1.1, the in-plane stress
singularity will vanish wherever the rupture velocity is the Rayleigh-wave
velocity, and the anti-plane singularity will vanish wherever the rupture velocity
is the shear velocity (then, $B_s = 0$). The energy flow into the crack tip per unit
length of rupture front can be obtained in the same way as for the two-
dimensional crack. Integrating the work rate over the area enclosing the crack
tip and moving with it, we find the rate of energy flow into the crack tip as

$$g = \lim_{\substack{\delta^+ \to 0 \\ \delta^- \to 0}} \int_{-\delta^-}^{\delta^+} (\tau_{3n} \, \Delta \dot{u}_n + \tau_{t3} \, \Delta \dot{u}_t) \, dn$$

$$= \frac{\pi \mu \beta^2 v_n}{4\alpha^3 U^2 V^2 B_s F} \, a^2 t_c [B_s^2 (v_n/\beta)^2 U^4 \sin^2 \phi - R(1/v_n)\alpha^2 v_n^2 V^4 \cos^2 \phi]. \quad (15.35)$$

For v_n less than Rayleigh wave speed c_R, $R(1/v_n)$ is negative and the crack tip is
an energy sink for both anti-plane and in-plane motions. But for $v_n > c_R$,
$R(1/v_n)$ is positive and the crack tip becomes the apparent energy source for
in-plane motion. This is unrealistic for a pure in-plane crack, but may be possible
if the energy flow supplied by the anti-plane component (positive if $v_n < \beta$) can
compensate for it. The motion at $\phi = 0$ is purely in-plane, and the terminal
velocity of the crack tip will be the Rayleigh-wave velocity c_R. The motion at
$\phi = 90°$ is purely anti-plane, with its terminal velocity being the shear-wave
velocity β. For arbitrary ϕ, setting $g = 0$ in (15.35) will give the terminal velocity.
The resultant terminal crack will be approximately elliptical, with major and
minor axes growing at speeds β and c_R.

The above discussion of terminal velocity is based only on the rates of energy
balance at the crack tip. In the case of an in-plane shear crack, it is possible that
the stress associated with P- and S-waves running ahead of the crack tip can
overcome the cohesive force (if finite), and the rupture velocity will exceed the
Rayleigh-wave velocity, reaching eventually the P-wave velocity. We shall come
back to this point in Section 15.2.3.

15.1.5   The far-field spectrum for a circular crack that stops

So far we have considered only the cases in which the crack grows with a constant
velocity. The results gave us some insight into the slip function expected for
a shear crack and also some understanding of its elastic near field. To understand its far field, however, we must solve a more difficult problem, in which the
growth of the crack is stopped.

Let us consider a circular crack that is nucleated at its center at time $t = 0$,
expanded with a constant velocity v, and suddenly stopped at a radius r_c. Up
to the time of stopping, $t = r_c/v$, the problem is self-similar and the slip function
given in (15.29) for $u = v$ gives the exact solution. If we freeze the motion at this

instant, we get the kinematic model depicted in Figure 14.10, for which a compact solution for the far field is given in equation (14.25). This freezing of motion is unrealistic, because it violates causality. At the instant of stopping, the points inside the crack have not yet sensed the termination of crack growth. The slip function of another kinematic model shown in Figure 14.12 is more plausible, and in fact the ramp-function slip at the crack center is quite appropriate, although the slip function at other points should have a square-root rise.

The high-frequency asymptote of the far-field displacement spectrum was determined by the form of the slip function in space near the crack tip, as discussed in Section 14.1.6. For the step-function rise, the asymptote is expected to be $\omega^{-3/2}$, and for the square-root rise, ω^{-2}. As discussed in Section 15.1.3, the cohesive force smooths these singularities over the length of the end region. The rupture velocity divided by this length will give the upper limit of frequency to which the asymptote is applicable.

Because of the difficulty in dealing with multiple diffraction at the edges of the crack, no analytic solution is known for the elastic field of a growing crack that stops. Burridge (1969) used a numerical solution of the integral-equation representation of the problem to solve some finite in-plane and anti-plane cracks. A similar method, developed by Hamano (1974), has been used by Das and Aki (1977a). Finite-difference or finite-element methods have also been used for similar problems by Hanson et al. (1971), Dieterich (1973), and Andrews (1975). Here we shall outline the work of Madariaga (1976), who used a finite-difference method to calculate the far-field seismic spectrum from a growing circular crack that stops. As we shall see in his results, the finite mesh size and some smoothing procedures introduce an artificial end region similar to that due to cohesive force.

We shall use the same notations and coordinate system as used for an elliptical crack in the preceding section (Fig. 15.7). We shall again assume that the stress drop on the crack occurs only in the τ_{13} component. Similarly, τ_{33} vanishes on the plane $x_3 = 0$, and u_1 and u_2 vanish outside the crack on the plane $x_3 = 0$. The boundary conditions on $x_3 = 0$ are therefore

$$\left. \begin{array}{c} \tau_{13} = -p_0 \\ \tau_{23} = 0 \end{array} \right\} \quad \text{for } r < \min(vt, r_c),$$

$$u_1 = u_2 = 0 \quad \text{for } r > \min(vt, r_c),$$

and   (15.36)

$$\tau_{33} = 0 \quad \text{for all } r$$

(p_0 is the stress drop, as discussed in more detail in Section 15.2).

The circular shape of the crack, which has a final radius of r_c, suggests cylindrical coordinates (r, ϕ, z) as the most convenient system to study the problem.

We can rewrite the boundary conditions (15.36) as

$$\left.\begin{array}{l} \tau_{rz} = -p_0 \cos \phi \\ \tau_{\phi z} = p_0 \sin \phi \end{array}\right\} \quad r < \min(vt, r_c),$$

$$u_r = u_\phi = 0 \quad r > \min(vt, r_c),$$

and $\tau_{zz} = 0$ for all r.

These boundary conditions have a simple sinusoidal azimuthal dependence. Consequently, we find that the ϕ-dependence of displacement components is either $\sin \phi$ or $\cos \phi$. They can be written as

$$u_r = u(r, z, t) \cos \phi,$$

$$u_\phi = v(r, z, t) \sin \phi,$$

$$u_z = w(r, z, t) \cos \phi.$$

The corresponding stress components can also be written in the same form:

$$\tau_{rr} = \Sigma_{rr}(r, z, t) \cos \phi,$$

$$\tau_{\phi\phi} = \Sigma_{\phi\phi}(r, z, t) \cos \phi,$$

$$\tau_{zz} = \Sigma_{zz}(r, z, t) \cos \phi,$$

$$\tau_{rz} = \Sigma_{rz}(r, z, t) \cos \phi,$$

$$\tau_{r\phi} = \Sigma_{r\phi}(r, z, t) \sin \phi,$$

$$\tau_{z\phi} = \Sigma_{z\phi}(r, z, t) \sin \phi.$$

Three components of particle velocity, $\dot{u}$, $\dot{v}$, $\dot{w}$, and six stress components make up nine unknowns, for which we have a system of nine first-order differential equations: three equations of motion and six equations for Hooke's law. Denoting the partial derivative by a comma followed by the variable in which the derivative is taken, the equations can be written as

$$\rho \dot{u}_{,t} = \frac{1}{r} (r\Sigma_{rr})_{,r} + \frac{1}{r} (\Sigma_{r\phi} - \Sigma_{\phi\phi}) + \Sigma_{rz,z},$$

$$\rho \dot{v}_{,t} = \frac{1}{r} (r\Sigma_{r\phi})_{,r} + \frac{1}{r} (\Sigma_{r\phi} - \Sigma_{\phi\phi}) + \Sigma_{z\phi,z},$$

$$\rho \dot{w}_{,t} = \frac{1}{r} (r\Sigma_{r\phi})_{,r} + \frac{1}{r} \Sigma_{z\phi} + \Sigma_{zz,z},$$

$$\Sigma_{rr,t} = (\lambda + 2\mu)\dot{u}_{,r} + \frac{\lambda}{r}(\dot{u} + \dot{v}) + \lambda\dot{w}_{,z},$$

$$\Sigma_{zz,t} = \lambda\dot{u}_{,r} + \frac{\lambda}{r}(\dot{u} + \dot{v}) + (\lambda + 2\mu)\dot{w}_{,z},$$

$$\Sigma_{\phi\phi,t} = \lambda\dot{u}_{,r} + (\lambda + 2\mu)(\dot{u} + \dot{v})/r + \lambda\dot{w}_{,z},$$

$$\Sigma_{r\phi,t} = \mu\dot{v}_{,r} - \mu(\dot{u} + \dot{v})/r,$$

$$\Sigma_{z\phi,t} = \mu\dot{v}_{,z} - \mu\dot{w}/r,$$

$$\Sigma_{zr,t} = \mu\dot{w}_{,r} + \mu\dot{u}_{,z},$$

where λ, μ are the Lamé constants and ρ is the density. We have to solve this equation subject to the following boundary conditions on $z = 0$:

$$\Sigma_{rz} = -\Sigma_{\phi z} = -p_0 \qquad r < \min(vt, r_c),$$

$$\dot{u} = \dot{v} = 0 \qquad r > \min(vt, r_c),$$

and

$$\Sigma_{zz} = 0 \qquad \text{for all } r.$$

The slip components Δu_1 and Δu_2 in the original coordinates can be written in terms of u and v at $z = 0$:

$$\Delta u_1 = 2u \cos^2 \phi - 2v \sin^2 \phi,$$

$$\Delta u_2 = (u + v) \sin 2\phi.$$

In the case of self-similar cracks studied in the preceding section, Δu_2 vanishes. In the present case, Δu_2 does not necessarily vanish but is found to be practically negligible; i.e., $u \sim -v$, so that

$$\Delta u_1 = 2u = -2v. \tag{15.37}$$

Interestingly, Δu_1 is independent of ϕ.

Madariaga (1976) solved the above problem by the finite-difference method using a staggered grid (Section 13.6.1) in which the velocities are defined at times $k \Delta t$ and the stresses at times $(k + \frac{1}{2}) \Delta t$, where Δt is the time-grid interval. The grid-point assignment for each of the nine stress-particle velocity components is shown in Figure 15.12.

Figure 15.13 shows the slip function $\Delta u(r, t) = u(r, +0, t) - u(r, -0, t)$ at several points on the crack. The rupture starts at $t = 0$ and expands with

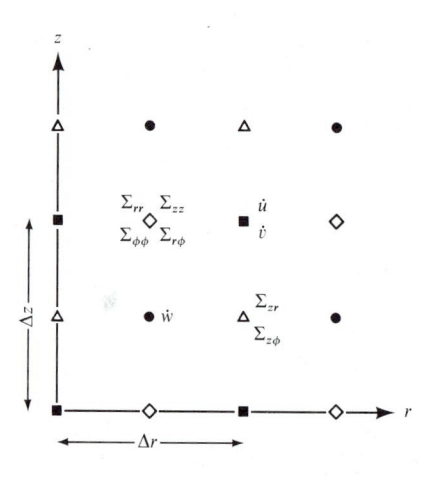

FIGURE **15.12**

Grid-point assignment for each of nine stress/particle-velocity components.

velocity 0.9β, where β is the shear velocity. The slip is measured with $p_0 r_c/\mu$ as the unit. The time t and radial distance r' are normalized to r_c/α and r_c, respectively, where α is the P-wave velocity. The slip function in time is shown at the center ($r = 0$) and at four other points at intervals of $0.2 r_c$. The arrow for each curve indicates the arrival of P-waves, at the point where the slip is shown, originating from the terminal periphery of the crack at the instant of rupture stop. The figure also shows the static slip expected at the center $r = 0$, indicating a significant overshoot of dynamic slip. When the slip reaches the maximum and its velocity becomes zero, the slip is held fixed. (This would actually occur if static friction were large enough.) A closed circle indicates the time of slip arrest at each point. The broken curve at the initial rise shows the square-root function expected for the analytic solution. The numerical solution shows a

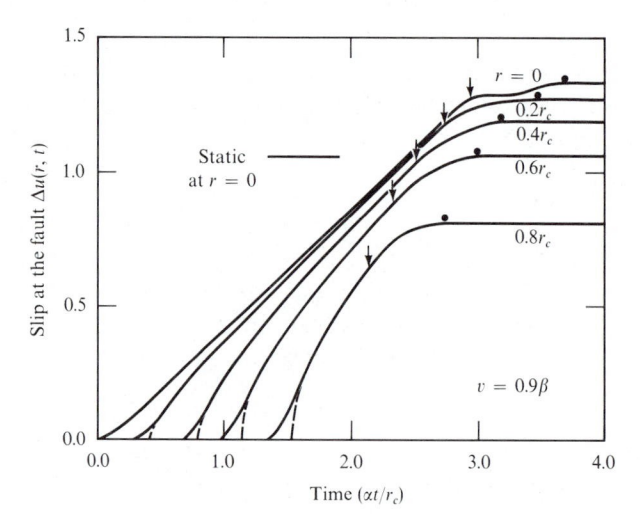

FIGURE **15.13**

Slip function at several distances from the center on the circular crack plotted against time. See text for explanation of symbols. [From Madariaga, 1976.]

less sharp rise because of the smoothing. This is an example of the effect of artificially introduced end region discussed earlier.

The far-field displacement waveform corresponding to the slip function Δu_1 was given by equation (14.13) as

$$\Omega(\mathbf{x}, t) = \int_{\Sigma} \Delta \dot{u}_1 \left(\boldsymbol{\xi}, t - \frac{R - (\boldsymbol{\xi} \cdot \boldsymbol{\gamma})}{c} \right) d\Sigma,$$

where $\boldsymbol{\gamma}$ is the unit vector pointing to the receiver, $\boldsymbol{\xi}$ is the position vector of $d\Sigma$, and c is the velocity of P- or S-waves. Writing the Fourier transform of $\Delta \dot{u}_1(\boldsymbol{\xi}, t)$ as $\Delta \dot{u}_1(\boldsymbol{\xi}, \omega)$, the far-field displacement spectrum can be obtained from equation (14.14) as

$$\Omega(\mathbf{x}, \omega) = e^{i\omega R/c} \int_{\Sigma} \Delta \dot{u}_1(\boldsymbol{\xi}, \omega) \exp[-i\omega(\boldsymbol{\xi} \cdot \boldsymbol{\gamma})/c] \, d\Sigma.$$

In our case, since $d\Sigma = r \, dr \, d\phi$ and Δu_1 is independent of ϕ, as shown in (15.37), we get

$$\Omega(\mathbf{x}, \omega) = e^{i\omega R/c} \int_0^{r_c} r \, dr \, \Delta \dot{u}_1(r, \omega) \int_{-\pi}^{\pi} \exp\left[i \frac{\omega r}{c} \sin \theta \cos(\phi - \phi_0) \right] d\phi,$$

where we have used

$$\boldsymbol{\xi} \cdot \boldsymbol{\gamma} = r \sin \theta \cos(\phi - \phi_0).$$

Using the property of a Bessel function given prior to (6.7), we find

$$\Omega(\mathbf{x}, \omega) = e^{i\omega R/c} 2\pi \int_0^{r_c} r \, dr \, \Delta \dot{u}_1(r, \omega) J_0 \left(r \frac{\omega}{c} \sin \theta \right). \tag{15.38}$$

This equation shows that the far-field displacement spectrum is a Hankel transform of $\Delta \dot{u}_1(r, \omega)$ in r. As discussed for a more general case in Section 14.1.3, the far-field spectrum can recover the slip function only for the wavenumber less than ω/c, because $|\sin \theta| \leq 1$ for real θ.

The numerical solutions for $\Delta \dot{u}_1(r, t)$ are Fourier-transformed in t and Hankel-transformed in r to find the far-field spectrum $\Omega(\mathbf{x}, \omega)$ by (15.38). Figure 15.14 shows the resultant $|\Omega(\mathbf{x}, \omega)|$ for P- and S-waves at three receiver directions ($\theta = 0$ corresponds to the normal to the crack plane) from a circular crack with rupture velocity $v = 0.9\beta$. The spectra are flat at low frequencies and decay roughly as ω^{-2}. If the nucleation phase determines the high-frequency asymptote, we should have obtained ω^{-3}. We must, therefore, conclude that the stopping phase dominates the high-frequency spectrum, and the power of asymptotic decay is more like 2 rather than 3 in the case of a circular crack that suddenly stops.

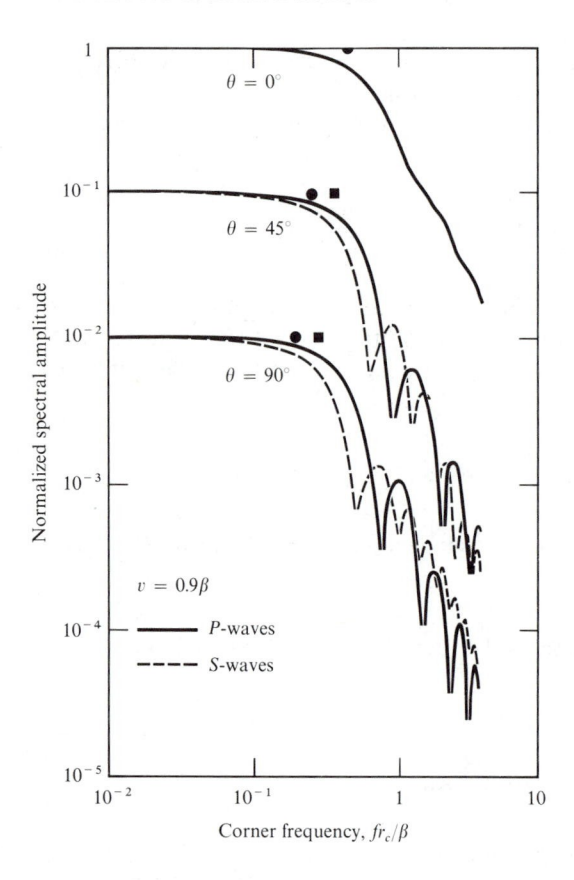

FIGURE **15.14**

Far-field spectra $|\Omega(\mathbf{x}, \omega)|$ for
P- and S-waves in the case of
rupture velocity $v = 0.9\beta$. [From
Madariaga, 1976.]

The corner frequencies of the spectra were determined by the intersection
of the low-frequency level and the high-frequency asymptote. They are indicated in Figure 15.14 by a closed circle for S-waves and a square for P-waves.
The results for corner frequencies for various directions θ and rupture velocities
are summarized in Figure 15.15. The corner frequencies are given in units of
β/r_c for rupture velocities 0.6β, 0.7β, and 0.9β. Although the corner frequency
increases with the rupture velocity, the variation is not very strong for the
range of rupture velocity considered here. The average values of the corner
frequencies over all directions, for the case of $v = 0.9\beta$, are given by

$$f_c^P \text{ (in Hz)} = 0.32\beta/r_c$$

$$f_c^S \text{ (in Hz)} = 0.21\beta/r_c$$

for P- and S-waves, respectively. The above equations predict considerably
lower corner frequencies (by about a factor of 2) than Brune's (1970) formula,
which was derived from a simple kinematic approach and which has been

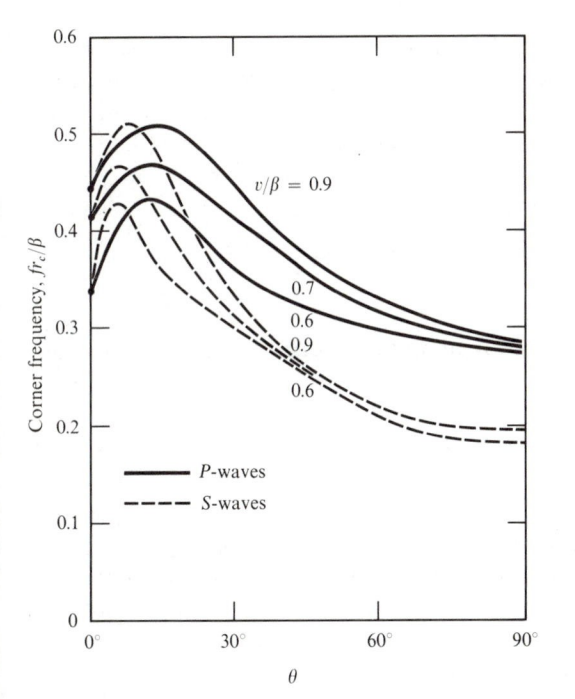

FIGURE **15.15**

Corner frequencies of *P*- and *S*-waves plotted against radiation direction for various rupture velocities. [From Madariaga, 1976.]

widely used in the interpretation of observed seismic spectra. The corner frequency for *P*-waves is higher than for *S*-waves, as expected from the earlier result of Molnar et al. (1973) for kinematic models with a similar slip function (see Figs. 14.12 and Section 14.1.6).

15.2 Dynamics of Spontaneous Rupture Propagation

One of the most challenging problems in seismology is to predict the occurrence of an earthquake and the resultant seismic motion from the study of physical properties of rocks in the epicentral region and the tectonic stress existing in the region. In order to approach this problem, we must go beyond the treatment of rupture propagation in the preceding section, in which the nucleation, propagation, and stopping of a rupture front are arbitrarily prescribed. There are three important lines of work to be done before we solve this problem. First, we must study the properties of fault-gouge material, such as the specific surface energy, length of the end region, static and dynamic friction and their distribution in space. Second, we must determine the tectonic stress acting on the fault zone. Third, we must be able to predict the entire rupture phenomena from beginning to end solely on the basis of the stress condition and material properties of the fault zone. In this section, we shall consider the last aspect of

the problem for an idealized case. We shall concentrate on how the stress distribution and the fracture criterion determine the movement of a crack tip, and (consequently) the slip function. For simplicity, we shall start with the case of an anti-plane crack, following the pioneering work of Kostrov (1966).

15.2.1   Spontaneous propagation of an anti-plane shear crack: General theory

Using the (x, y, z) coordinates shown in Figure 14.20, we shall define the crack as

$$x_1 < x < x_2, \quad -\infty < z < \infty, \quad \text{and } y = 0.$$

For an anti-plane case (Section 14.2.3), only the z-component of displacement $w(x, y, t)$ exists, and the stress tensor has only two nonvanishing elements: $\tau_{zx} = \mu(\partial w/\partial x)$ and $\tau_{yz} = \mu(\partial w/\partial y)$. The problem is two-dimensional, nothing depending on z. The equation of motion in this case reduces to the wave equation

$$\frac{1}{\beta^2} \frac{\partial^2 w}{\partial t^2} = \frac{\partial^2 w}{\partial x^2} + \frac{\partial^2 w}{\partial y^2}, \tag{15.39}$$

where $\beta = \sqrt{\mu/\rho}$ is the shear velocity.

Suppose that initially the crack is absent and the body is in equilibrium with an initial state of stress σ^0. We shall take this initial state as the reference state and measure the displacement relative to this state. The total stress is taken as $\sigma = \sigma^0 + \tau$, where τ (the *incremental* stress) is derived from $\mathbf{u}$ by Hooke's law. Initial conditions are that w and $\partial w/\partial t$ are zero for $t = 0$. When the crack is formed, the traction on the crack drops to the dynamic frictional stress. The only changing component of traction on the crack ($y = 0$) is σ_{yz}, and it changes from, say, σ_{yz}^0 to σ_{yz}^d. We shall equate the stress drop $\sigma_{yz}^0(x, 0) - \sigma_{yz}^d(x, 0, t)$ to $p(x, t)$. The appropriate boundary condition for traction on the crack for the above choice of reference state is then given by

$$\tau_{yz} = -p(x, t) \qquad \text{for } x_1 < x < x_2, y = 0. \tag{15.40}$$

In order to find the boundary condition outside of (x_1, x_2) on $y = 0$, we shall write the solution of equation (15.39) in the following form:

$$w = \iint w^+(\omega, k) \exp(-i\omega t + ikx - vy) \, d\omega \, dk \qquad y > 0$$

$$= \iint w^-(\omega, k) \exp(-i\omega t + ikx + vy) \, d\omega \, dk \qquad y < 0,$$

where $v = \sqrt{k^2 - \omega^2/\beta^2}$ and $\mathrm{Re}\, v \geq 0$ because of the radiation condition. Then the continuity of traction τ_{yz} across $y = 0$ gives

$$w^+(\omega, k) = -w^-(\omega, k),$$

and hence $w(x, y, t)$ must be an odd function of y. An odd function of y must vanish at $y = 0$ if it is continuous there. Since w is continuous at $y = 0$ outside the crack, we have

$$w(x, y, t) = 0 \qquad x < x_1, x_2 < x, y = 0. \tag{15.41}$$

Equations (15.40) and (15.41) together give the (mixed boundary condition at $y = 0$. Because of the symmetry, it is sufficient to obtain a solution only in the half-space $y < 0$.

To solve this boundary-value problem, let us start with the representation theorem (2.43), using the Green function that satisfies the stress-free condition at $y = 0$. Such a Green function can be obtained by first finding the Green function G for a full space corresponding to a line body-force impulse located at $(x, 0, t)$:

$$\rho \frac{\partial^2 G}{\partial t_0^2} - \mu \frac{\partial^2 G}{\partial x_0^2} - \mu \frac{\partial^2 G}{\partial y_0^2} = \delta(x_0 - x)\, \delta(y_0)\, \delta(t_0 - t).$$

Apart from differences in notation, this is the equation (6.41) that is solved by (6.42), and for our present purposes we obtain the solution as

$$G(x_0, t_0; x, t) = \frac{H\{(t_0 - t) - [(x_0 - x)^2 + y_0^2]^{1/2}/\beta\}}{2\pi\mu R},$$

where $R^2 = (t_0 - t)^2 - \{(x_0 - x)^2 + y_0^2\}/\beta^2$ and $H\{\ \}$ is the unit step function. Since the above Green function happened to satisfy the stress-free condition at $y_0 = 0$, the Green function on the surface $y_0 = 0$ is merely $2G(x_0, t_0; x, t)$, where the effect of reflection is taken care of by doubling the amplitude.

Suppose we know the traction τ_{yz} everywhere on $y = 0$,

$$\tau_{yz}(x, 0, t) = \tau(x, t).$$

We can write the solution by putting $2G(x_0, t_0; x, t)$ into the representation theorem (2.43) as

$$w(x_0, y_0, t_0) = \frac{1}{\pi\mu} \iint_S \frac{\tau(x, t)}{R}\, dx\, dt \qquad y_0 < 0, \tag{15.42}$$

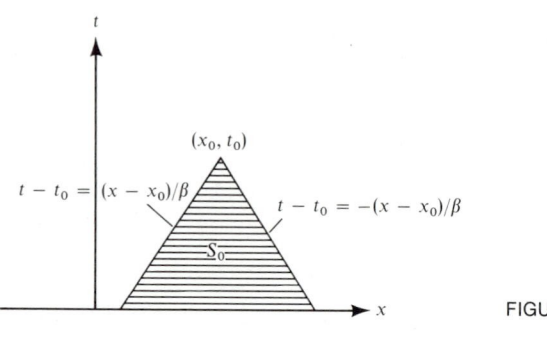

FIGURE **15.16**

where S is that part of the xt-plane which lies inside the cone

$$\beta^2(t_0 - t)^2 - (x_0 - x)^2 - y_0^2 \geq 0 \qquad 0 \leq t \leq t_0.$$

For $y_0 = 0^-$, we obtain

$$w(x_0, 0^-, t_0) = \frac{1}{\pi\mu} \iint_{S_0} \frac{\tau(x, t) \, dx \, dt}{\sqrt{(t_0 - t)^2 - (x_0 - x)^2/\beta^2}}, \qquad (15.43)$$

where S_0 is the triangle

$$\beta^2(t_0 - t)^2 - (x_0 - x)^2 \geq 0 \qquad 0 \leq t \leq t_0,$$

as shown by the shaded area in Figure 15.16.

Since we do not know $\tau(x, t)$ for the whole area of S_0, equation (15.43) does not immediately give the solution. The boundary condition (15.41), however, gives the following equation for $x_0 < x_1$ and $x_0 > x_2$:

$$\iint_{S_0} \frac{\tau(x, t) \, dx \, dt}{\sqrt{(t_0 - t)^2 - (x_0 - x)^2/\beta^2}} = 0.$$

$\tau(x, t)$ is known in some parts of the above integration region S_0, shown in Figure 15.17, where the loci of crack tips are indicated by $x_1(t)$ and $x_2(t)$. The subregion S_1 lies inside the crack, and $\tau(x, t)$ is known there from (15.40). On the other hand, $\tau(x, t)$ vanishes in the subregion $S_0 - S_1 - S_2$, for which $x > x_2(0) + \beta t$, because any disturbances from the crack have not yet reached this subregion. Finally, the value of $\tau(x, t)$ in subregion S_2 is unknown. Thus, as long as S_0 does not intersect $x_1(t)$ (when the disturbances from the left crack tip have not yet reached the observation point), for $x_0 < x_1$ and $x_0 > x_2$:

$$\iint_{S_2} \frac{\tau(x, t) \, dx \, dt}{\sqrt{(t_0 - t)^2 - (x_0 - x)^2/\beta^2}} = \iint_{S_1} \frac{p(x, t) \, dx \, dt}{\sqrt{(t_0 - t)^2 - (x_0 - x)^2/\beta^2}}. \qquad (15.44)$$

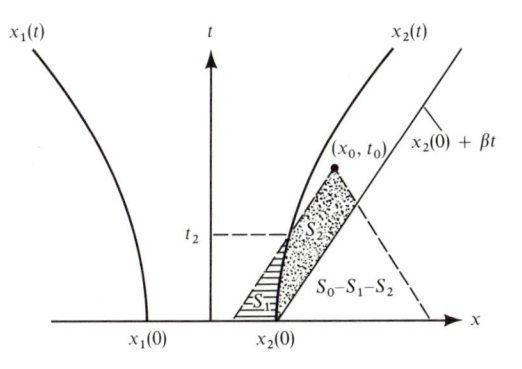

FIGURE **15.17**

$\tau(x, t)$ is known in S_1, but unknown in S_2. It vanishes in S_0-S_1-S_2.

To solve the above integral equation for $\tau(x, t)$ in S_2, we shall first make the following transformation of variables:

$$\xi = (\beta t - x)/\sqrt{2},$$
$$\eta = (\beta t + x)/\sqrt{2}. \tag{15.45}$$

Then (15.44) can be rewritten as

$$\int_{-x_2(0)/\sqrt{2}}^{\xi_0} \frac{d\xi}{\sqrt{\xi_0 - \xi}} \int_{\eta_2(\xi)}^{\eta_0} \frac{\tau(\xi, \eta)\, d\eta}{\sqrt{\eta_0 - \eta}} = \int_{-x_2(0)/\sqrt{2}}^{\xi_0} \frac{d\xi}{\sqrt{\xi_0 - \xi}} \int_{-\xi}^{\eta_2(\xi)} \frac{p(\xi, \eta)\, d\eta}{\sqrt{\eta_0 - \eta}}, \tag{15.46}$$

where $\eta_2(\xi)$ is the solution of

$$\eta_2 - \xi = \sqrt{2} x_2 \left(\frac{\eta_2 + \xi}{\sqrt{2}\beta} \right),$$

which defines the position of the right crack tip in ξ and η. The integration limits for ξ and η can be easily found from Figure 15.18. Clearly, (15.46) will be satisfied if

$$\int_{\eta_2(\xi)}^{\eta_0} \frac{\tau(\xi, \eta)\, d\eta}{\sqrt{\eta_0 - \eta}} = \int_{-\xi}^{\eta_2(\xi)} p(\xi, \eta) \frac{d\eta}{\sqrt{\eta_0 - \eta}}. \tag{15.47}$$

The above equation is in the form of Abel's integral equation for $\tau(\xi, \eta)$. The solution is described in Box 12.1, and in our case we find

$$\tau(\xi_0, \eta_0) = \frac{1}{\pi} \frac{d}{d\eta_0} \int_{\eta_2(\xi_0)}^{\eta_0} \frac{d\eta_1}{\sqrt{\eta_0 - \eta_1}} \int_{-\xi_0}^{\eta_2(\xi_0)} p(\xi_0, \eta) \frac{d\eta}{\sqrt{\eta_1 - \eta}}.$$

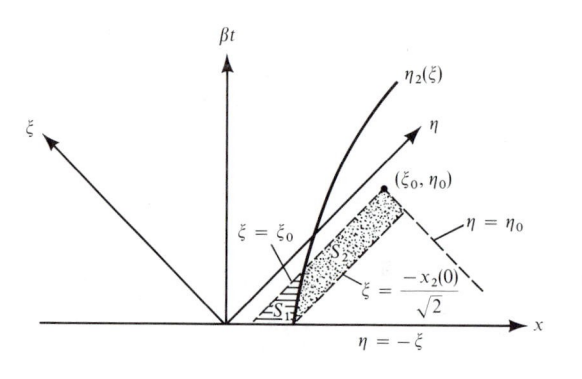

FIGURE **15.18**

Change of integral variables from (x, t) to (ξ, η).

Since

$$\int_{\eta_2}^{\eta_0} \frac{d\eta_1}{\sqrt{\eta_0 - \eta_1}\sqrt{\eta_1 - \eta}} = \sin^{-1}\left(1 + \frac{2(\eta_0 - \eta_1)}{\eta - \eta_0}\right)\Bigg|_{\eta_1 = \eta_2}^{\eta_1 = \eta_0},$$

the above equation reduces to

$$\tau(\xi_0, \eta_0) = \frac{1}{\pi\sqrt{\eta_0 - \eta_2(\xi_0)}} \int_{-\xi_0}^{\eta_2(\xi_0)} p(\xi_0, \eta) \frac{\sqrt{\eta_2(\xi_0) - \eta}}{\eta_0 - \eta} \, d\eta. \quad (15.48)$$

The path of integration is along $\xi = \xi_0$, which corresponds to path $t - t_0 = (x - x_0)/\beta$ in the xt-coordinates. Transforming back to xt-coordinates, and referring to Figures 15.17 and 15.18, we obtain

$$\tau(x_0, t_0) = \frac{1}{\pi\sqrt{x_0 - x_2(t_2)}} \int_{x_0 - \beta t_0}^{x_2(t_2)} p[x, t_0 + (x - x_0)/\beta] \frac{\sqrt{x_2(t_2) - x}}{x_0 - x} \, dx$$

$$(15.49)$$

for $x_0 > x_2(t_0)$, where t_2 is the solution of

$$\beta t_0 - x_0 = \beta t_2 - x_2(t_2).$$

In other words, t_2 is the time at which the crack-tip locus $x_2(t)$ intersects the integration path. The above expression is valid for the time interval $0 < t_0 < [x_0 - x_1(0)]/\beta$. A similar result may be obtained for the region $x_0 < x_1$ for the time interval $0 < t_0 < [x_2(0) - x_0]/\beta$. To determine $\tau(x_0, t_0)$ for later periods, additional subregions of S_0 with unknown $\tau(x, t)$ appear, corresponding to repeated diffraction of the waves at the crack boundary.

Equation (15.49) shows that the stress $\tau(x_0, t_0)$ becomes infinity when the crack tip arrives at the receiver, so that $x_0 = x_2(t_0)$. At a given time t_0 prior to

arrival, the distance between the crack tip and the receiver is $x_0 - x_2(t_0)$. Using the stress-intensity factor K defined in Section 15.1.1, $\tau(x_0, t_0)$ near and ahead of the crack tip can be written as

$$\tau(x_0, t_0) \sim \frac{K}{\sqrt{2\pi[x_0 - x_2(t_0)]}}. \tag{15.50}$$

On the other hand, as can be seen from Figure 15.19,

$$x_0 - x_2(t_2) = \beta(t_0 - t_2),$$

$$x_2(t_0) - x_2(t_2) \sim \dot{x}_2(t_0)(t_0 - t_2),$$

and therefore

$$\frac{x_0 - x_2(t_0)}{x_0 - x_2(t_2)} \sim 1 - \dot{x}_2(t_0)/\beta. \tag{15.51}$$

Comparing (15.49), (15.50), and (15.51), we find that

$$K = \frac{\sqrt{1 - \dot{x}_2(t_0)/\beta}}{\sqrt{\pi/2}} \int_{x_2(t_0) - \beta t_0}^{x_2(t_0)} p\{x, t_0 - [x_2(t_0) - x]/\beta\} \frac{dx}{\sqrt{x_2(t_0) - x}} \tag{15.52}$$

near the crack tip, where $x_0 \sim x_2(t_2) \sim x_2(t_0)$. The integration path is a straight line, shown connecting $[t_0, x_2(t_0)]$ and $[0, x_2(t_0) - \beta t_0]$ in Figure 15.19.

In Section 15.1.2, we showed that the tip of an anti-plane crack moving with a subsonic velocity v absorbs energy at a rate given by

$$g = \frac{vK^2}{2\mu}(1 - v^2/\beta^2)^{-1/2}. \tag{15.15 again}$$

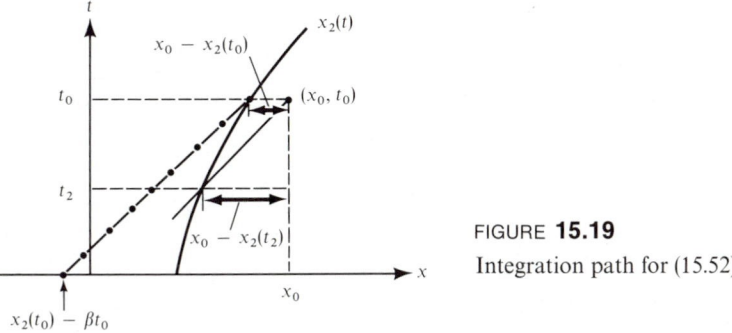

FIGURE **15.19**

Integration path for (15.52).

Expressing the surface energy required to create a new unit area as G, we have

$$G = \frac{g}{2v} = \frac{K^2}{4\mu}(1 - v^2/\beta^2)^{-1/2}.$$

BOX **15.3**

The stress-intensity factor associated with cohesive force alone

Here we shall show that equation (15.52) can be used to derive (15.19) for the case of a crack tip moving at constant velocity.

Let the coordinate in the x-direction in a frame moving with the crack tip at a velocity v be x'. Then $x' = x - x_2(t)$, where $x_2(t) = \text{const.} + vt$.

We have previously defined p as the stress drop $\sigma_{yz}^0 - \sigma_{yz}^d$. But if a cohesive force is considered, as in (15.18), the stress on the fault plane becomes $\sigma_{yz}^d + \sigma_c$, so that the stress drop is $\sigma_{yz}^0 - \sigma_{yz}^d - \sigma_c$, i.e., it is augmented by an amount $-\sigma_c$. The effect of the cohesive force is therefore to add a stress concentration, with the stress-intensity factor derived from (15.52) by replacing p with $-\sigma_c$. The integration is limited to the region $-d \leq x' \leq 0$ in which $\sigma_c \neq 0$, and all this range is included in the integration limits of (15.52).

As can easily be seen from the figure,

$$x_2(t_0) - x = \frac{\beta}{\beta - v}(-x') \quad \text{and} \quad dx = \frac{\beta}{\beta - v}dx'.$$

Therefore, equation 15.52 is transformed to

$$K = -\sqrt{\frac{2}{\pi}} \int_{-d}^{0} \sigma_c(x') \frac{dx'}{\sqrt{-x'}},$$

which is the result used earlier in (15.19). Note that K is independent of v in this case.

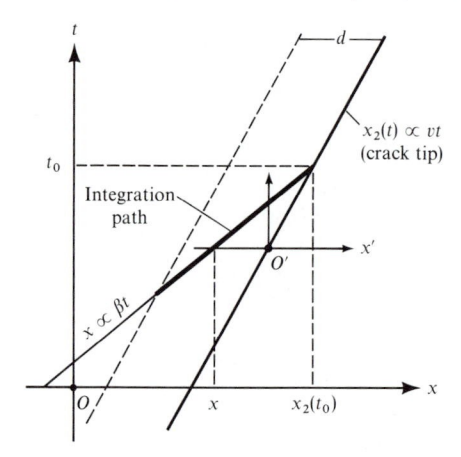

Combining this equation with (15.52) (dropping the subscript 0 from t_0), we obtain

$$K = 2(\mu G)^{1/2}(1 - v^2/\beta^2)^{1/4}$$

$$= \sqrt{\frac{2}{\pi}}(1 - v/\beta)^{1/2} \int_{x_2 - \beta t}^{x_2} p[x, t - (x_2 - x)/\beta] \frac{dx}{\sqrt{x_2 - x}}$$

or

$$\int_{x_2 - \beta t}^{x_2} p[x, t - (x_2 - x)/\beta] \frac{dx}{\sqrt{x_2 - x}} = (2\mu\pi G)^{1/2} \left(\frac{1 + v/\beta}{1 - v/\beta}\right)^{1/4}, \quad (15.53)$$

where $v = \dot{x}_2(t)$. This equation, first derived by Kostrov (1966), gives the velocity of the crack tip for given $p(x, t)$ and G. The above equation holds only when

$$\int_{x_2 - \beta t}^{x_2} p[x, t - (x_2 - x)/\beta] \frac{dx}{\sqrt{x_2 - x}} \geq (2\mu\pi G)^{1/2}.$$

Otherwise, the crack tip does not move.

Once the locus $x_2(t)$ of the crack tip is determined, $\tau(x_0, t_0)$ can be calculated by equation (15.48). Then we can use (15.42) to determine the displacement at any point. In fact, the displacement inside the crack can be determined using only the stress drop $p(x, t)$ inside the crack. Transforming the variables (x, t) to (ξ, η) by equation (15.45), we can rewrite (15.43) as

$$w(\xi_0, \eta_0) = \frac{1}{\pi\mu} \iint \frac{\tau(\xi, \eta) \, d\xi \, d\eta}{\sqrt{2} \sqrt{\xi_0 - \xi} \sqrt{\eta_0 - \eta}} \qquad \text{on } y = 0^-.$$

Let us divide the area of integration into four parts, as shown in Figure 15.20. In S_1 and S_3, $-\tau(\xi, \eta)$ is given as the stress drop $p(\xi, \eta)$. In S_2, $\tau(\xi, \eta)$ is unknown, but determined by equation (15.48) using $p(\xi, \eta)$. In the remaining parts of S_0, $\tau(\xi, \eta)$ vanishes.

From equation (15.47), for a point (ξ_1, η_1) close to the crack-tip locus but outside of the crack, we have

$$\int_{\eta_2(\xi)}^{\eta_1} \frac{\tau(\xi_1, \eta) \, d\eta}{\sqrt{\eta_1 - \eta}} - \int_{-\xi_1}^{\eta_2(\xi)} \frac{p(\xi_1, \eta) \, d\eta}{\sqrt{\eta_1 - \eta}} = 0.$$

Our integral with respect to η for the areas S_1 and S_2 is exactly of the above form, with $\eta_0 = \eta_1$. Thus the contributions from S_1 and S_2 vanish. The only

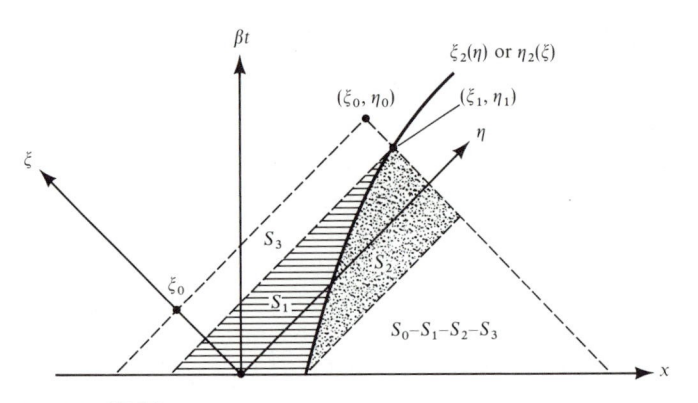

FIGURE **15.20**

contribution comes from S_3, so that the displacement on $y = 0^-$ is

$$
w(\xi_0, \eta_0) = \frac{-1}{\sqrt{2\pi\mu}} \int_{\xi_2(\eta_0)}^{\xi_0} \frac{d\xi}{\sqrt{\xi_0 - \xi}} \int_{-\xi}^{\eta_0} \frac{p(\xi, \eta)\, d\eta}{\sqrt{\eta_0 - \eta}}, \tag{15.54}
$$

where $\xi = \xi_2(\eta)$ is the locus of the crack tip in the (ξ, η) plane. The above equation gives the fault slip ($2w$) as a function of stress drop $p(\xi, \eta)$ and crack-tip location $\xi_2(\eta)$. The above equation was used by Ida (1973) in a study of spontaneous rupture propagation that is described later.

15.2.2   Examples of spontaneous anti-plane crack propagation

Let us find how the equation of motion (15.53) for a crack tip derived in the preceding section works using simple examples.

A SEMI-INFINITE CRACK

Consider an unbounded body under a uniform shear stress σ_{yz}^0. A crack appears instantly at $t = 0$ over a half-plane $y = 0$, $x < 0$. Assuming that the dynamic friction is zero, a stress drop of σ_{yz}^0 occurs instantaneously for $y = 0$, $x < 0$. We shall find the position $x_2(t)$ of the crack tip for $t > 0$ using the equation of crack-tip motion. Since

$$
p(x, t) = \sigma_{yz}^0 \qquad \text{for } x < x_2(t),
$$

we have from equation (15.53)

$$
\int_{x_2 - \beta t}^{x_2} \frac{\sigma_{yz}^0\, dx}{\sqrt{x_2 - x}} = (2\mu\pi G)^{1/2} \left(\frac{1 + \dot{x}_2/\beta}{1 - \dot{x}_2/\beta} \right)^{1/4}. \tag{15.55}
$$

The left-hand side of the above equation is equal to $2\sigma^0_{yz}\sqrt{\beta t}$. The above equation cannot be satisfied for t smaller than t_c given by

$$2\sigma^0_{yz}\sqrt{\beta t_c} = (2\mu\pi G)^{1/2}. \tag{15.56}$$

In other words, the crack tip does not propagate until time t_c. Once this time is passed, the crack-tip motion is governed by equation (15.55), i.e.,

$$\sqrt{\frac{t}{t_c}} = \left(\frac{1 + \dot{x}_2/\beta}{1 - \dot{x}_2/\beta}\right)^{1/4}.$$

Solving for $\dot{x}_2$ and integrating with respect to t from t_c to t, we find

$$x_2(t) = \beta t - \beta t_c\left[1 + 2\tan^{-1}\left(\frac{t}{t_c}\right) - \frac{\pi}{2}\right].$$

The crack tip starts moving at $t = t_c$ with zero initial velocity, rapidly reaching a terminal velocity β. Figure 15.21 shows motion of the crack tip

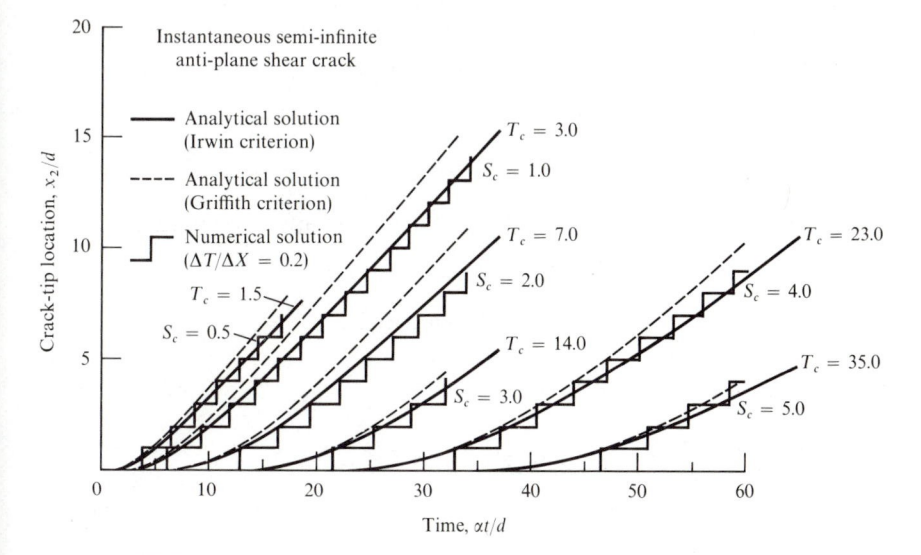

FIGURE **15.21**

The crack-tip location x_2 as a function of time for various values of T_c, where $T_c = \alpha t_c/d$, α is the compressional wave velocity, t_c is the rupture starting time defined in (15.56), and d is the grid length used in the numerical solution described in Section 15.2.3. S is the parameter of a fracture criterion used in the numerical solution, and $1 + S = S_c/(\sigma^0_{xy} - \sigma^d_{xy})$, where S_c is the critical stress difference defined in (15.75). Broken curves correspond to the criterion of constant surface energy, and solid curves to the criterion of constant critical stress-intensity factor. [From Das and Aki, 1977a.]

for different t_c. The solid lines correspond to the Irwin criterion in which the critical stress-intensity factor is assumed to be a material constant, independent of rupture velocity (see Problem 15.2). The step-like curves are obtained by a numerical method that is discussed later.

A SEMI-INFINITE CRACK THAT STOPS

The above classic example given by Kostrov (1966) was extended by Husseini et al. (1975) to include the stopping of crack-tip motion. The crack-tip motion can be stopped by placing a barrier of high surface energy along the fault plane or by limiting the prestressed region to a finite size. In either case, the following condition is imposed on the stress drop $p(x, t)$ over the initial semi-infinite crack:

$$p(x, t) = 0 \qquad \text{for } x < -a, t > 0,$$
$$= p_0 \qquad \text{for } -a < x < x_2(t), t > 0,$$

where p_0 is a constant. This is intended to simulate a finite crack without in-troducing complex multiple diffractions at crack edges. For a given p_0 and specific surface energy G_0 at $x = 0$, a must be greater than βt_c so that the rup-ture can be initiated. From (15.56), the condition is

$$a > \frac{\mu \pi G_0}{2 p_0^2}.$$

The rupture can be stopped by making G increase with x. For example, con-sider a linearly increasing surface energy

$$G(x) = (1 + mx)G_0.$$

From equation (15.53), we get

$$\dot{x}_2(t) = \beta \frac{t^2 - t_c^2(1 + mx_2)^2}{t^2 + t_c^2(1 + mx_2)^2} \qquad x_2 - \beta t > -a$$

$$= \beta \frac{(x_2 + a)^2/\beta^2 - t_c^2(1 + mx_2)^2}{(x_2 + a)^2/\beta^2 + t_c^2(1 + mx_2)^2} \qquad x_2 - \beta t < -a. \tag{15.57}$$

The stopping position of the crack tip, x_s, may be obtained from the second equation in (15.57) by setting $\dot{x}_2(t) = 0$. Then

$$x_s = \frac{a - \beta t_c}{m \beta t_c - 1}.$$

Since $x_s > 0$ for a real stopping, the rate m of increase in specific energy must be greater than $(\beta t_c)^{-1}$ for the crack tip to stop. The motion of the crack tip can be obtained by solving the differential equation (15.57).

Another simple case of a barrier is a step-like increase in G:

$$G = G_0 \qquad\qquad 0 \leq x < b,$$

$$G = G_0 + \Delta G \qquad b < x.$$

In this case, for $x_2(t) > b$, we have

$$\dot{x}_2 = \frac{(x_2 + a)^2/\beta^2 - t_c^2[1 + (\Delta G/G_0)]^2}{(x_2 + a)^2/\beta^2 + t_c^2[1 + (\Delta G/G_0)]^2}, \tag{15.58}$$

and setting $\dot{x}_2 = 0$ we can solve for the stopping position of the tip,

$$x_s = \beta t_c[1 + (\Delta G/G_0)] - a.$$

Since $x_s \geq b$, an inequality has to be satisfied for the stopping to occur:

$$\beta t_c\left(1 + \frac{\Delta G}{G_0}\right) = \frac{\mu \pi (G_0 + \Delta G)}{2 p_0^2} \geq (a + b). \tag{15.59}$$

If we put this condition into equation (15.58), we find $\dot{x}_2$ to be zero or negative. Since $\dot{x}_2$ cannot be negative physically, $\dot{x}_2$ must vanish and the equality holds in (15.59). The equality means that $x_s = b$, or that the crack tip stops immediately at b if condition (15.59) holds. If not, the tip will propagate indefinitely beyond b. For example, if $G_0 = 10^4$ erg/cm^2, $(a + b) = 1$ km, $p_0 = 10$ bars, and $\mu = 3 \times 10^{11}$ dyne/cm^2, then ΔG must be about 10^7 erg/cm^2 or greater for the rupture to stop. Furthermore, the larger the length or the larger the stress drop, the greater ΔG must be to stop the rupture.

An alternative way of stopping a rupture is to limit the size of the prestressed region. For example, consider the case in which, for $t > 0$,

$$p(x, t) = 0 \qquad \text{for } x < -a$$

$$= p_0 \qquad -a < x < x_2(t) < b$$

$$= 0 \qquad b < x.$$

The equation of motion (15.53) gives the crack-tip velocity as

$$\dot{x}_2 = \frac{[f(x_2, t)]^4 - \beta^2 t_c^2}{[f(x_2, t)]^4 + \beta^2 t_c^2},$$

where

$$
\begin{aligned}
f(x_2, t) &= (\beta t)^{1/2} & x_2 < b,\, x_2 - \beta t > -a \\
&= (x_2 + a)^{1/2} & x_2 - \beta t < -a \\
&= (\beta t)^{1/2} - (x_2 - b)^{1/2} & x_2 > b,\, x_2 - \beta t > -a \\
&= (x_2 + a)^{1/2} - (x_2 - b)^{1/2} & x_2 - \beta t < -a.
\end{aligned}
$$

From the final equation, the stopping position may be obtained by setting $\dot{x}_2 = 0$. The result is

$$
x_s = \frac{(a + b)^2}{4\beta t_c} + \frac{b - a}{2} + \frac{\beta t_c}{4}.
$$

For example, if $b \sim a \sim \beta t_c$, $x_s \sim b(1 + \frac{1}{4})$, but if $b \gg a \sim \beta t_c$, $x_s \sim b(1 + b/4a)$. Thus, if the length b of the prestressed region is much greater than $\beta t_c = \mu \pi G_0/2p_0^2$, there will be a considerable overshoot of crack extension into the initially unstressed region. For the surface energy G measured in laboratories ($\sim 10^4$ ergs/cm^2) and $\tau_0 = 10$ bars, βt_c is only 50 cm. If this value applies to natural earthquakes, they will certainly overshoot. However, as mentioned in Section 15.1.3, the real value of G for earthquakes may be around 10^8 ergs/cm^2, which corresponds to βt_c of 5 km. Since G is expected to increase with earthquake magnitude because of increase in the zone of microcrack formation and plastic deformation, the overshoot may not play a very important role.

SLIP-RATE-DEPENDENT BOUNDARY CONDITION AT THE FAULT

If there should be any constitutive relation between the stress and slip or stress and slip rate, we can incorporate it in our equation of rupture propagation. For example, Ida (1973) assumed that the stress σ_{yz} on the fault is related to the slip rate $\Delta\dot{w}$ by the following equation (see Fig. 15.22):

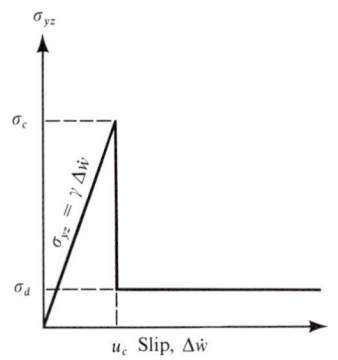

FIGURE **15.22**

Constitutive relation between the stress across the fault plane and slip rate assumed by Ida. [From Ida, 1973; copyrighted by the American Geophysical Union.

$$\sigma_{yz} = \gamma \, \Delta\dot{w} \qquad \text{for } \Delta\dot{w} \le v_c$$
$$= \sigma_{yz}^d \qquad \text{for } \Delta\dot{w} > v_c, \tag{15.60}$$

where σ_{yz} is *total* stress acting on the fault plane, i.e., the sum of the initial stress σ_{yz}^0 and the stress increment τ_{yz} due to crack formation. The slip rate $\Delta\dot{w}$ is equal to $-2\dot{w}$ for $\dot{w}$ evaluated on $y = 0^-$ (which is the side of the fault for which we have studied displacement; see, e.g., (15.42)). Although the above constitutive relation is not very realistic, it does display a transition from ductile to brittle behavior. This may grossly simulate the behavior of an earthquake fault on which creep and dynamic failure are both occurring. Suppose we start with an initially unstressed fault. As the tectonic stress increases, slow creep may occur across the fault, and the slip rate may increase in proportion to the stress. When the slip rate reaches a certain yield limit v_c, the stress may suddenly drop to the dynamic friction level σ_{yz}^d, creating an earthquake.

To incorporate the above constitutive relation into the equation of rupture propagation, we make use of

$$w(\xi_0, \eta_0) = \frac{-1}{\sqrt{2\pi}\mu} \int_{\xi_2(\eta_0)}^{\xi_0} \frac{d\xi}{\sqrt{\xi_0 - \xi}} \int_{-\xi}^{\eta_0} \frac{p(\xi, \eta) \, d\eta}{\sqrt{\eta_0 - \eta}}, \tag{15.54 again}$$

from which we can find slip $\Delta w = -2w$ inside the crack in terms of stress drop p inside the crack.

Let us assume, as before, that a semi-infinite crack suddenly appears for $x < 0$ at $t = 0$. Since, for $t < -x/\beta$ or $\eta < 0$, we expect no disturbance from the crack tip, the slip will be uniform under a uniform initial stress σ_{yz}^0. The slip for $\eta < 0$ can be expressed by equation (15.54) in terms of a uniform stress drop p_0, which is to be determined by the constitutive relation (15.60). Since the integration region is bounded by $t = 0$, $\xi_2(\eta) = -\eta$ for $\eta < 0$. Then equation (15.54) gives an easily integrable result:

$$w(\xi_0, \eta_0) = \frac{-1}{\sqrt{2\pi}\mu} \int_{-\eta_0}^{\xi_0} \frac{d\xi}{\sqrt{\xi_0 - \xi}} \int_{-\xi}^{\eta_0} \frac{p_0 \, d\eta}{\sqrt{\eta_0 - \eta}}$$

$$= -\frac{\beta p_0 t}{\mu} \qquad \text{on } y = 0^-,$$

where (15.45) was used. From equation (15.60), we have

$$\sigma_{yz} = \sigma_{yz}^0 - p_0 = \gamma \, \Delta\dot{w} = -2\gamma\dot{w} = 2\gamma\beta p_0/\mu \qquad \text{for } \beta t < -x. \tag{15.61}$$

This equation determines the stress drop p_0 occurring for $\beta t < -x$ in terms of the initial stress and the material constants, i.e.,

$$p_0 = \frac{\sigma_{yz}^0}{1 + 2\gamma\beta/\mu}.$$

For $\beta t > -x$ or $\eta > 0$, we can determine the stress drop $p(\xi, \eta)$ basically in the same way as above by solving (15.54) and (15.60) simultaneously. We must, however, use some numerical methods to solve the integral equation (15.54). Since the integration range is limited to S_3, shown in Figure 15.20, the discretized integral equation can be solved in steps, in each of which unknowns are $p(\xi_n, \eta_m)$ and $\dot{w}(\xi_n, \eta_m)$ at one discretized point (n, m). Since $p(\xi_n, \eta_m)$ and $\Delta\dot{w} = -2\dot{w}(\xi_n, \eta_m)$ must also be related by equation (15.60), the two equations can determine both p and $\Delta\dot{w}$ at the point.

In solving (15.54), the crack-tip location $\xi_2(\eta)$ must be known. Recognizing that $\xi = [\beta t - x_2(t)]/\sqrt{2}$ and $\eta = [\beta t + x_2(t)]/\sqrt{2}$ on the crack-tip locus, we have

$$\frac{d\xi_2}{d\eta} = \frac{d\xi_2/dt}{d\eta/dt} = \frac{\beta - \dot{x}_2(t)}{\beta + \dot{x}_2(t)}.$$

Then the equation (15.53) for the motion of the crack tip can be rewritten as

$$\frac{d\xi_2}{d\eta} = \frac{(2\pi\mu G)^2}{\left\{\int_{x_2 - \beta t}^{x_2} p[x, t - (x_2 - x)/\beta] \dfrac{dx}{\sqrt{x_2 - x}}\right\}^4}. \tag{15.62}$$

The above equation is valid only when

$$\int_{x_2 - \beta t}^{x_2} p[x, t - (x_2 - x)/\beta] \frac{dx}{\sqrt{x_2 - x}} \geq (2\mu\pi G)^{1/2}. \tag{15.63}$$

Otherwise, the crack tip does not move and $x_2(t) = 0$. In that case,

$$\frac{d\xi_2}{d\eta} = 1. \tag{15.64}$$

The condition (15.63) can be checked by a numerical integration of discretized $p(\xi_n, \eta_m)$. Then, either (15.62) or (15.64) is used to determine the locus of the crack tip by

$$\xi_2(\eta_{m+1}) = \xi_2(\eta_m) + \frac{d\xi_2}{d\eta} \Delta\eta,$$

where $\Delta\eta$ is the grid spacing in η.

Ida (1973) made numerical calculations for various choices of parameters v_c, γ, and σ_{yz}^d, and found two distinctly different types of rupture propagation, depending on the parameters. One type is a smooth rupture propagation in which, once the rupture starts, the crack tip is accelerated smoothly, approaching the shear velocity. An example of the smooth propagation is shown

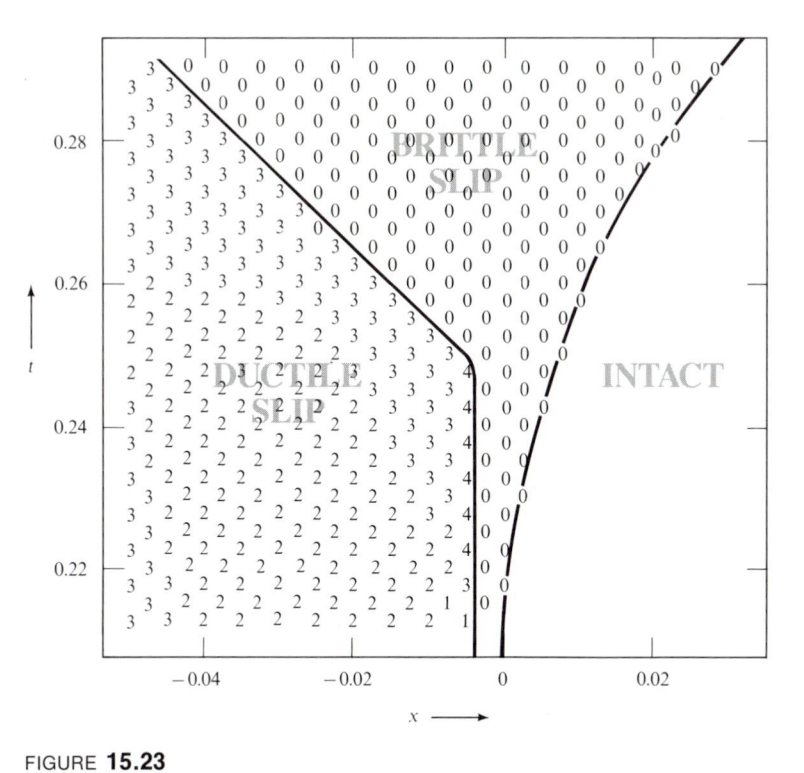

FIGURE **15.23**

Plot of $\sigma_{yz}(x, t)$ in units of p_0. A smooth rupture propagation occurs in this case. [From Ida, 1973; copyrighted by the American Geophysical Union.]

in Figure 15.23. Here the time t is measured in units of $t_c = \pi\mu G/2\beta p_0^2$. This is the delay time given in (15.56), which corresponds to the stress drop given in (15.61). We discussed the magnitude of t_c in earthquakes in the example of a semi-infinite crack that stops (p. 896). The distance x is measured in units of βt_c, and the numbers in Figure 15.23 represent $\sigma_{yz}(x, t)$ in units of p_0. The parameters are chosen as $\gamma = 2\mu/\beta$, $v_c = 2.1 \times (\beta p_0/\mu)$, and $\sigma_{yz}^d = 0$. In the case of smooth rupture propagation, the cracked region ($\sigma_{yz} = 0$ in this case of $\sigma_{yz}^d = 0$) extends in both directions.

For a slightly different choice of parameters, the mode of rupture propagation can be quite different. The result is shown in Figure 15.24 for $\gamma = 2\mu/\beta$, $v_c = 2.6 \times (\beta p_0/\mu)$, and $\sigma_{yz}^d = 0$. The rupture propagation is quite irregular; the crack tip moves for a short distance, then stops, restarts, then stops, and repeats the process. The fault plane, once cracked, can be quickly healed, because the slip rate drops below v_c. Thus the healing front follows the crack tip with a similar speed, making the effective crack length always roughly constant.

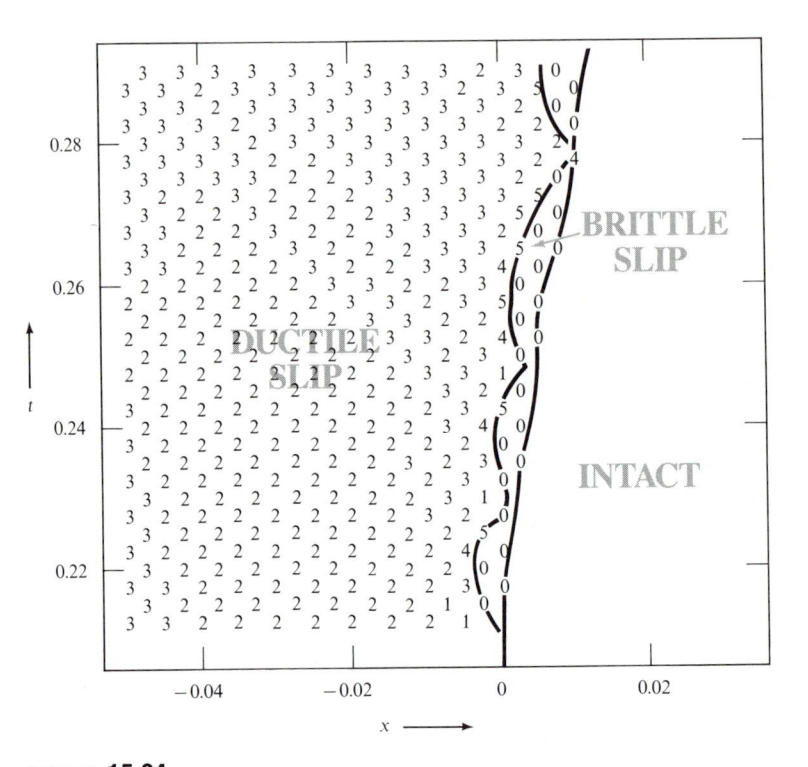

FIGURE **15.24**

Plot of $\sigma_{yz}(x, t)$ in units of p_0. A case of irregular propagation. [From Ida, 1973; copyrighted by the American Geophysical Union.]

As discussed in Chapter 14, the high-frequency spectrum of seismic waves in the far field consists primarily of contributions from rupture nucleation and stopping points. We therefore anticipate a long duration of complex high-frequency waves from an irregular rupture process, as shown in Figure 15.24. On the other hand, a smooth earthquake like the one shown in Figure 15.23 will generate large, long-period waves with distinct short-period phases associated with the initial start and the final stopping points.

Ida's result indicates that the smooth rupture occurs when μ/β (impedance associated with plane shear waves (Box 5.4)) becomes lower as compared to p_0/v_c. Thus, roughly speaking, the smooth propagation occurs when impedance in the creep region is higher than in the elastic region. For a given v_c, the smooth rupture occurs at lower frictional stress σ_{yz}^d.

COHESIONLESS CRACK

Burridge and Halliday (1971) considered an anti-plane crack nucleated along a line at a constant depth in a homogeneous half-space. The crack propagates

vertically both upward and downward. Their fracture criterion is a special case of (15.53), in which the specific surface energy G was set to zero. From (15.53), the condition that $G = 0$ can be met either by

$$v = \beta$$

or by

$$\int_{x_2 - \beta t}^{x_2} p[x, t - (x_2 - x)/\beta] \frac{dx}{\sqrt{x_2 - x}} = 0. \qquad (15.65)$$

Equation (15.65) is possible only if $p(x, t)$ changes sign along the above integration path. If (15.65) is not met, the crack tip must propagate with the shear velocity β. To slow down a propagating crack, therefore, one must postulate a negative stress drop. Taking the x-axis vertically downward, Burridge and Halliday considered the stress drop given by

$$p(x, t) = p_0(1 - x^2/b^2), \qquad (15.66)$$

where p_0 is a constant. The crack tip propagates downward and past the depth b with the velocity β until the contribution from negative p matches that from positive p to satisfy equation (15.65). Then the crack tip moves deeper with a velocity determined by equation (15.65). Thus the stress along the fault plane above the depth b drops to a lower value, but that below b jumps to a higher value. We shall come back to the cohesionless fracture in the discussion of in-plane cracks in the next section.

15.2.3 Spontaneous propagation of an in-plane shear crack

Let us now consider spontaneous propagation of an in-plane shear crack. As in the anti-plane case, the crack lies on the plane $y = 0$, extending to infinity in the z-direction but confined in the x-direction between $x_1(t)$ and $x_2(t)$ at time t. For the in-plane case, the nonvanishing displacement components are $u(x, y, t)$ and $v(x, y, t)$ (see Section 14.2.4). As in the anti-plane case studied in Section 15.2.1, we assume that initially the crack is absent and that the body is in equilibrium with an initial state of stress σ^0. We shall take this initial state as the reference state and measure the displacement relative to it. The initial conditions are then given by

$$u = v = 0 \quad \text{and} \quad \frac{\partial u}{\partial t} = \frac{\partial v}{\partial t} = 0 \quad t < 0.$$

The total stress is then $\sigma = \sigma^0 + \tau$. When the crack is formed, σ_{xy} on the crack drops from σ_{xy}^0 to the dynamic frictional stress σ_{xy}^d. Putting the stress

drop $\sigma_{xy}^0 - \sigma_{xy}^d = p(x, t)$, the boundary condition for incremental stress on the crack appropriate for the above choice of the reference state is given by

$$\tau_{xy} = -p(x, t) \qquad x_1(t) < x < x_2(t), y = 0. \qquad (15.67)$$

As shown in Section 14.2.4, the continuity of $v(x, y, t)$ and $\tau_{xy}(x, y, t)$ across $y = 0$ leads to the symmetry relation that $u(x, y, t)$ and $\tau_{yy}(x, y, t)$ are odd functions of y and that $v(x, y, t)$ and $\tau_{xy}(x, y, t)$ are even functions of y. Since τ_{yy} is continuous across $y = 0$, τ_{yy} must vanish at $y = 0$.

$$\tau_{yy} = 0 \qquad y = 0. \qquad (15.68)$$

We need another condition at $y = 0$ outside the crack. This is given by the continuity of u. Since a continuous odd function of y must vanish at $y = 0$, we have

$$u = 0 \qquad x < x_1(t), x_2(t) < x, y = 0. \qquad (15.69)$$

Equations (15.67) through (15.69) give the boundary condition at $y = 0$. Because of the symmetry, it is sufficient to obtain a solution only in the half-space $y < 0$.

We shall define the two-dimensional Green functions $g_{x\xi}(x, y, t; \xi, y, \tau)$ and $g_{y\xi}(x, y, t; \xi, y, \tau)$ for a homogeneous half-space $y \leq 0$ with free surface at $y = 0$ as the displacement components u and v observed at (x, y, t) due to a line-impulsive force applied at (ξ, y, τ) in the ξ-direction. Then, from the representation theorem (2.43), since our $g_{x\xi}$, $g_{y\xi}$ satisfy the stress-free boundary condition, and since $\tau_{yy} = 0$ on $y = 0$, we have

$$u(x, y, t) = \iint_S \tau_{xy}(\xi, \tau) g_{x\xi}(x, y, t; \xi, 0, \tau) \, d\xi \, d\tau,$$

$$v(x, y, t) = \iint_S \tau_{xy}(\xi, \tau) g_{y\xi}(x, y, t; \xi, 0, \tau) \, d\xi \, d\tau.$$

The region of integration S is, from causality, that region of the (ξ, τ) plane for which

$$\alpha^2(t - \tau)^2 - (x - \xi)^2 - y^2 \geq 0 \qquad t \geq \tau \geq 0,$$

where α is the P-wave velocity. The above representation is valid also for displacements on the crack plane $y = 0^-$, in which we are particularly interested. In this case, the region of integration is a triangle S_0 in the (ξ, τ) plane given by

$$\alpha^2(t - \tau)^2 - (x - \xi)^2 \geq 0 \qquad t \geq \tau \geq 0,$$

and we write

$$u(x, 0, t) = \iint\limits_{S_0} \tau_{xy}(\xi, \tau) g_{x\xi}(x - \xi, 0, t - \tau) \, d\xi \, d\tau,$$

$$v(x, 0, t) = \iint\limits_{S_0} \tau_{xy}(\xi, \tau) g_{y\xi}(x - \xi, 0, t - \tau) \, d\xi \, d\tau,$$

$$(15.70)$$

This notation for the Green function is appropriate for a homogeneous half space when source (ξ, η, τ) and receiver (x, y, t) are both on the free surface. Explicit formulas for $g_{x\xi}$ and $g_{y\xi}$ are easily derived by Cagniard's method (Section 6.4), and they are particularly simple when $y = \eta = 0$. The result in this case is

$$g_{x\xi}(x, 0, t) = \frac{4\sigma^2}{\pi\mu\beta^2 x} \frac{(\sigma^2 - \beta^{-2})(\sigma^2 - \alpha^{-2})^{1/2}}{R(\sigma)R^*(\sigma)} \qquad \frac{1}{\alpha} < \sigma < \frac{1}{\beta}$$

$$= \frac{1}{\pi\mu\beta^2 x} \frac{(\sigma^2 - \beta^{-2})^{1/2}}{R(\sigma)} \qquad \frac{1}{\beta} < \sigma,$$

$$g_{y\xi}(x, 0, t) = \frac{K_l}{\mu} \delta\left(t - \frac{x}{c_R}\right) + \frac{2\sigma}{\pi\mu\beta^2 x} \frac{(2\sigma^2 - \beta^{-2})(\sigma^2 - \alpha^{-2})^{1/2}(\beta^{-2} - \sigma^2)^{1/2}}{R(\sigma)R^*(\sigma)},$$

$$(15.71)$$

where $\sigma = t/x$, β is the velocity of shear waves, R is the Rayleigh function

$$R(\sigma) = (2\sigma^2 - \beta^{-2})^2 - 4\sigma^2(\sigma^2 - \alpha^{-2})^{1/2}(\sigma^2 - \beta^{-2})^{1/2},$$

and R* is defined by

$$R^*(\sigma) = (2\sigma^2 - \beta^{-2})^2 + 4\sigma^2(\sigma^2 - \alpha^{-2})^{1/2}(\sigma^2 - \beta^{-2})^{1/2},$$

$$K_l = \frac{(2\beta^2/c_R^2 - 1)^3}{\frac{16\beta^2}{c_R^2}\left[1 - \left(6 - 4\frac{\beta^2}{\alpha^2}\right)\frac{\beta^2}{c_R^2} + 6\left(1 - \frac{\beta^2}{\alpha^2}\right)\frac{\beta^4}{c_R^4}\right]},$$

and c_R is the velocity of Rayleigh waves $(R(c_R^{-1}) = 0)$. Equation (15.71) was first derived by Lamb (1904).

If $\tau_{xy}(\xi, t)$ were known on the whole x-axis, equation (15.70) would give the solution of the problem. From the boundary condition (15.67), however, the stress component is known only on the crack surface. Outside the crack, the boundary condition (15.69) is given for displacement component u. Separating the region of integration S_0 in (15.70) into a part S_1 inside the crack $[x_1(t) < x < x_2(t)]$, for which τ_{xy} is known by (15.67), and a part S_2, we can rewrite

the condition (15.69) as

$$\iint_{S_1} p(\xi, \tau) g_{x\xi}(x - \xi, 0, t - \tau) \, d\xi \, d\tau$$

$$= \iint_{S_2} \tau_{xy}(\xi, \tau) g_{x\xi}(x - \xi, 0, t - \tau) \, d\xi \, d\tau \quad \begin{array}{l} \text{for } x < x_1(t) \\ \text{or } x_2(t) < x. \end{array} \quad (15.72)$$

Kostrov (1975) obtained an analytic solution of the above equation. The result, however, is much more involved than in the case of an anti-plane crack. For example, the stress-intensity factor given by only one integration in the anti-plane case (equation (15.52)) now requires five integrations and one differentiation. Besides, the result is valid only for a crack-tip velocity less than the Rayleigh-wave velocity. It appears that a numerical approach may be more satisfactory at present.

A sophisticated method of discretizing the integral equations (15.70) and (15.72) was described by Burridge (1969). However, a more conventional method, such as the one used by Hamano (1974), seems to reproduce his result quite closely. In Hamano's method, the x-axis is divided into segments of equal interval d, and each segment is presumed to take the average value of stress and displacements over the segment. Then it is natural to replace the point-to-point Green function $g(x - \xi, 0, t - \tau)$ by a segment-to-segment Green function $\bar{\bar{g}}(x_i - \xi_j, 0, t - \tau)$, which is the averaged displacement over the ith segment due to the force distributed over the jth segment:

$$\bar{\bar{g}}(x_i - \xi_i, 0, t - \tau) = \frac{1}{d^2} \int_{x_i - (d/2)}^{x_i + (d/2)} dx \int_{\xi_j - (d/2)}^{\xi_j + (d/2)} g(x - \xi, 0, t - \tau) \, d\xi. \quad (15.73)$$

For $g_{x\xi}(x, 0, t)$ given in (15.71), $\bar{\bar{g}}_{x\xi}(x, 0, t)$ can be obtained in a compact form, as given in Das and Aki (1977a). Using the averaged Green function, the integral equation (15.72) can be discretized as

$$\sum_{\substack{j \ l \\ in \ S_1}} p(\xi_j, \tau_l) \bar{\bar{g}}_{x\xi}(x_i - \xi_j, t_k - \tau_l)$$

$$= \sum_{\substack{j \ l \\ in \ S_2}} \tau_{xy}(\xi_j, \tau_l) \bar{\bar{g}}_{x\xi}(x_i - \xi_j, t_k - \tau_l) \quad \begin{array}{l} \text{for } x_i < x_1(t_k) \\ \text{or } x_2(t_k) < x_i. \end{array} \quad (15.74)$$

The order of solving the above set of equations can be arranged so that only one unknown, $\tau_{xy}(\xi_j, \tau_l)$, occurs when each equation is solved. Once τ_{xy} is determined for the whole region, the displacement can be calculated by the discretized equation (15.70).

So far we have pretended that the locations of crack tips $x_1(t)$ and $x_2(t)$ were known. We need some fracture criterion to determine the crack-tip

locations. The simplest fracture criterion that can be easily incorporated in the discretized formulation (15.74) is to monitor the stress difference between the neighboring grid points between which the crack tip lies. The total stress at the point inside the crack is known to be σ_{yx}^d, and that outside is determined as $\sigma_{xy}^0 + \tau_{xy}$ by solving (15.74) for the incremental stress τ_{xy}. Thus the excess of stress outside the crack over that inside is $\tau_{xy} + \sigma_{xy}^0 - \sigma_{xy}^d = \tau_{xy} + p$. As soon as this stress difference exceeds a certain limit S_c, i.e.,

$$\tau_{xy} - \sigma_{xy}^d + \sigma_{xy}^0 \geq S_c, \tag{15.75}$$

we presume that rupture takes place. The crack tip is advanced to beyond the point at which the stress difference had been exceeded, and the stress at the point is set to σ_{xy}^d. The stress difference across the crack tip may be considered as a smeared-out stress concentration. We know from equation (15.7) that the stress concentration has the form of

$$\sigma = \frac{K'}{\sqrt{2\pi x'}} H(x'),$$

where $x' = x - vt$ is the distance measured from the crack tip and K' is the stress-intensity factor for in-plane cracks. Suppose that the crack tip lies in the middle of the two grid points as shown in Figure 15.25. Then the average stress over the grid immediately outside the tip will be

$$\bar{\sigma} = \frac{1}{d} \int_0^d \frac{K'}{\sqrt{2\pi x'}} dx' = 2 \frac{K'}{\sqrt{2\pi d}}.$$

In Box 15.2, we introduced Irwin's fracture criterion, which is based on the critical intensity factor K_c. The critical average stress $\bar{\sigma}$ over the grid immediately outside the tip corresponding to K_c may be written as

$$S_c = 2 \frac{K_c}{\sqrt{2\pi d}}. \tag{15.76}$$

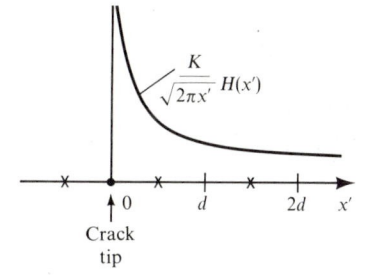

Crack
tip

FIGURE **15.25**

Grid points are shown by crosses.

Das and Aki (1977a) compared a numerical solution based on the criterion for S_c and an analytic solution based on the criterion for K_c for the case of a semi-infinite, anti-plane crack. Figure 15.21 shows the analytic solution as solid curves and the numerical solution as step-like curves. The symbol T_c attached to the analytic solution is the rupture starting time t_c defined in (15.56) normalized to α/d as $T_c = \alpha t_c/d$. For an anti-plane crack, the critical stress intensity factor K_c may be obtained by setting $v = 0$ in (15.22), i.e., $K_c^2 = 4\mu G$. Then, from (15.56), we have $\beta t_c = (\pi/2)K_c^2/(2\sigma_{yz}^0)^2$. On the other hand, the parameter S attached to the numerical solution is related to S_c by $1 + S = S_c/(\sigma_{yz}^0 - \sigma_{yz}^d)$. If our assumed relation (15.76) is correct, we should find the relation between T_c and S as $T_c = (\pi/4)^2(\alpha/\beta)(1 + S)^2 \sim 1.07(1 + S)^2$, where we take into account the assumption $\sigma_{yz}^d = 0$ made in deriving (15.56). Figure 15.21 shows that equation (15.76) gives a good approximation to the actual value for large S. For small S, the constant factor in (15.76) must be slightly larger than 2. For the range of S from 0.5 to 5, the appropriate value of the constant varies from 2.10 to 2.53. For a given S_c, S can be increased by making the grid length smaller.

Thus the fracture criterion for the critical stress difference S_c may be approximately the same as the fracture criterion for the critical intensity factor, which we called Irwin's criterion in Box 15.2. As discussed in the box, the Irwin and Griffith criteria are equivalent as far as the initiation of crack extension is concerned. However, for a finite rupture velocity, the two criteria are different, and the fracture criterion by S_c is not exactly the same as the Griffith criterion resulting in different crack-tip motions, as shown in Figure 15.21.

Andrews (1976) used Ida's description of cohesive force to introduce the Griffith criterion (Section 15.1.3) into a finite-difference calculation of the in-plane shear-crack propagation. He assumed that traction across the fault plane is related to the slip Δu by the following formulas (see Fig. 15.26):

$$\sigma(\Delta u) = \sigma_s - (\sigma_s - \sigma_d)\,\Delta u/D \qquad \Delta u < D,$$

$$\sigma(\Delta u) = \sigma_d \qquad\qquad\qquad\quad \Delta u \geq D,$$

(15.77)

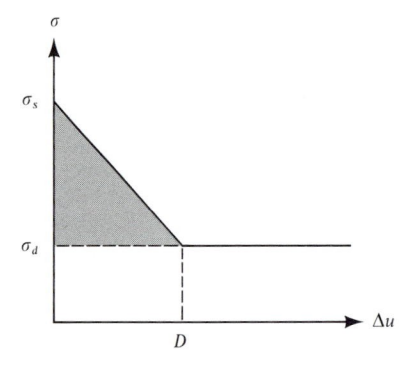

FIGURE **15.26**

Relation between cohesive force and slip used by Ida (1972) and Andrews (1976). σ_s is the upper limit of static friction at this level, instability begins and weakening occurs. For slip greater than D, the stress drops to the dynamic friction σ_d. The shaded area corresponds to surface energy, i.e., the work done against the cohesive force (see Section 15.1.3). [After Andrews, 1976; copyrighted by the American Geophysical Union.]

where σ_s is the static friction, σ_d is the dynamic friction, and D is the slip required for stress to drop to σ_d. The inelastic work done at the rupture front in excess of the work done against the dynamic frictional stress σ_d is identified as the specific surface energy (for each unit surface of newly created crack),

$$G = \tfrac{1}{4}(\sigma_s - \sigma_d)D. \tag{15.78}$$

The boundary condition on the fault plane must next be described: when the crack is not slipping,

$$|\sigma_{xy}^0 + \tau_{xy}| \le \sigma(\Delta u) \qquad \text{if } \frac{\partial \, \Delta u}{\partial t} = 0,$$

and during slip,

$$\sigma_{xy}^0 + \tau_{xy} = \sigma(\Delta u) \, \text{sign}\!\left(\frac{\partial \, \Delta u}{\partial t}\right) \qquad \text{if } \frac{\partial \, \Delta u}{\partial t} \ne 0.$$

With this boundary condition, the propagation of a crack expands symmetrically in both the $+x$ and $-x$ directions. The results are discussed in terms of two nondimensional numbers: L_c/L and $(\sigma_s - \sigma_0)/(\sigma_0 - \sigma_d)$, where σ_0 is the initial stress σ_{xy}^0 and L_c is the critical half-length of an in-plane Griffith crack, which can be obtained from equation (15.16). Taking the limit as $v \to 0$, the critical stress intensity factor K_c will satisfy the following equation:

$$G = \frac{g}{2v}$$

$$= \frac{1}{16} \frac{K_c^2}{\mu \beta^2} \lim_{v \to 0} \frac{v^2}{\left(1 - \dfrac{v^2}{\alpha^2}\right)^{1/2} - \left(1 - \dfrac{v^2}{2\beta^2}\right)^2 \left(1 - \dfrac{v^2}{\beta^2}\right)^{-1/2}}$$

$$= \frac{K_c^2}{8\mu} \frac{\lambda + 2\mu}{\lambda + \mu}. \tag{15.79}$$

From equation (9) of Box 15.1, the stress-intensity factor K' is related to the crack half-length L by $K' = (\sigma_0 - \sigma_d)\sqrt{\pi L}$. Therefore, the critical half-length L_c is given by

$$L_c = \frac{8\mu(\lambda + \mu)G}{\pi(\lambda + 2\mu)(\sigma_0 - \sigma_d)^2}. \tag{15.80}$$

Das and Aki (1977a) solved the same problem using Hamano's method, with the fracture criterion based on S_c, discussed earlier. In their case, L_c can

be calculated by putting the value of K_c obtained from (15.76) into $K = (\sigma_0 - \sigma_d)\sqrt{\pi L}$ to find

$$L_c = d\,\frac{S_c^2}{2(\sigma_0 - \sigma_d)^2}. \tag{15.81}$$

The other parameter, $(\sigma_s - \sigma_0)/(\sigma_0 - \sigma_d)$, is nothing but the parameter S used in the discussion of Figure 15.21:

$$S = \frac{S_c}{\sigma_0 - \sigma_d} - 1 = \frac{\sigma_s - \sigma_d}{\sigma_0 - \sigma_d} - 1 = \frac{\sigma_s - \sigma_0}{\sigma_0 - \sigma_d}. \tag{15.82}$$

The results of calculation by the two methods agree in general, and only Andrew's result is reproduced in Figure 15.27. There are two distinct styles of rupture propagation. If the parameter S is greater than about 1.63, the velocity of rupture propagation is always less than the Rayleigh-wave velocity c_R, and the velocity approaches c_R as the crack length increases. On the other hand, if S is less than 1.63, the rupture starts with sub-Rayleigh velocity, but as the crack length exceeds a certain limit (which depends on S), the rupture velocity exceeds the shear velocity and approaches P-wave velocity as the crack length increases. The critical value of $S = 1.63$ was obtained by Burridge (1973), using the cohesionless fracture criterion discussed in the example of anti-plane crack propagation (p. 901). The cohesionless crack cannot propagate at velocities lower than the Rayleigh-wave velocity because of its inability to sustain any stress singularity. It can propagate with the Rayleigh velocity, at which the stress-intensity factor vanishes. Burridge, however, showed that even at the Rayleigh velocity, the stress ahead of the crack at the S-wave front may exceed the static friction if S is less than 1.63. In that case, the admissible speed of the crack tip is the P-wave velocity. In Section 15.1.2, we concluded from the study

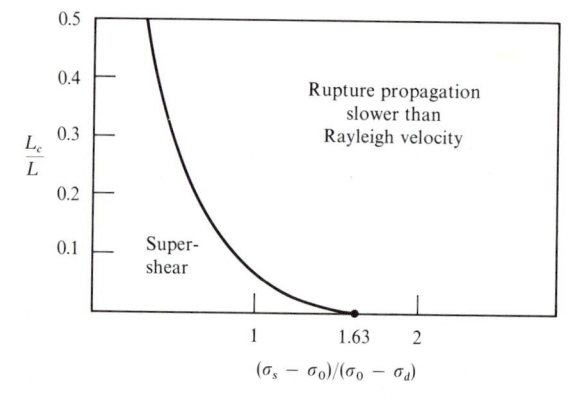

$\dfrac{L_c}{L}$

0.5
0.4
0.3
0.2
0.1

Rupture propagation
slower than
Rayleigh velocity

Super-
shear

1   1.63   2

$(\sigma_s - \sigma_0)/(\sigma_0 - \sigma_d)$

FIGURE **15.27**

Regions in which rupture propagates sub-Rayleigh velocity and supershear velocity. [After Andrews, 1976; copyrighted by the American Geophysical Union.]

of energetics at the crack tip that the speed of an in-plane shear crack cannot exceed the Rayleigh-wave velocity. For a cohesionless crack, the stress-intensity factor is always zero and there is no energy flow through the crack tip. Therefore, the conclusion from the energetics does not apply to a cohesionless crack. A more difficult question is why the numerical solutions, which apparently involve a finite energy flow through the crack-tip (as demonstrated by the agreement with analytic solutions for such cases), show rupture-propagation velocities exceeding that of Rayleigh waves. The answer lies in their fracture criterion, in which the initiation of fault slip does not require an infinite stress but only a finite stress. Thus the stress associated with P- and S-waves propagating ahead of a crack tip can cause the fault slip. Hence we may say that the super-Rayleigh-wave velocity propagation is a consequence of finite cohesive forces.

The finite cohesive force has another important consequence on what happens when the rupture propagates along a fault plane with obstacles or barriers. These barriers may be expressed by a localized high value of S_c defined in (15.75). Das and Aki (1977b) found that three different situations can occur when a crack tip passes such a barrier, depending on the relative magnitude of barrier strength to initial stress.

  i) If the initial stress is relatively high, the barrier is broken immediately.
 ii) If the initial stress is relatively low, the crack tip proceeds beyond the barrier, leaving behind an unbroken barrier.
iii) If the initial stress is intermediate, the barrier is not broken at the initial passage of the crack tip, but eventually breaks due to later increase in stress.

If the barrier encounter of type (i) occurs throughout the fault plane, rupture propagation is smooth, generates a simple impulsive seismic signal, and results in a high average stress drop. On the other hand, if the type (ii) encounter occurs at many barriers, rupture propagation becomes rough, generates a long sequence of high-frequency waves, and ends up with a low average stress drop. Type (iii) propagation generates seismograms with ripples superimposed on long-period motion. The seismic radiation becomes less dependent on direction of rupture propagation than others, because the slip on the central part of the crack occurs more or less simultaneously, resulting in an effectively symmetric source. Similar results were obtained by Mikumo and Miyatake (1978), who studied rupture propagation over a fault plane with 2-dimensional nonuniform distribution of static friction.

Thus, a variety of rupture processes can be generated by distributing a number of barriers with different strength on the fault plane to serve as models of complex actual earthquakes. The fault model with barriers is particularly important for the study of short-period motions from a large earthquake, namely, the most hazardous of earthquake waves to mankind.

SUGGESTIONS FOR FURTHER READING

Achenbach, J. D. On dynamic effects in brittle fracture. *In* S. Nemat-Nasser (editor), *Mechanics Today* (Vol. 1). New York: Pergamon Press, 1974.

Andrews, D. J. Rupture velocity of plane-strain shear cracks. *Journal of Geophysical Research*, **81**, 5679–5687, 1976.

Freund, L. B. Crack propagation in an elastic solid subject to general loading. II. Non-uniform rate of extension. *Journal of Mechanics and Physics of Solids*, **20**, 141–152, 1972.

Freund, L. B. The analysis of elastodynamic crack tip stress fields. *In* S. Nemat-Nasser (editor), *Mechanics Todays* (Vol. 3). New York: Pergamon Press, 1976.

Freund, L. B. Dynamic crack propagation. *In* F. Erdogan (editor), *The Mechanics of Fracture* (Vol. 19). American Society of Mechanical Engineers (Applied Mechanics Division), 1976.

Liebowitz, H. (editor). *A Treatise on Fracture* (Vol. 2). New York: Academic Press, 1968.

Madariaga, R. Dynamics of an expanding circular fault. *Bulletin of the Seismological Society of America*, **65**, 163–182, 1976.

Madariaga, R. High frequency radiation from crack (stress drop) models of earthquake faulting. *Geophysical Journal of the Royal Astronomical Society*. **51**, 625–651, 1977.

Orowan, E. Mechanism of seismic faulting. *In* D. Griggs and J. Handin (editors), *Rock Deformation* (Chapter 12). Geological Society of America, 1960.

Palmer, A. C. and J. R. Rice, The growth of slip surfaces in the progressive failure of over-consolidated clay, *Proceedings of the Royal Society of London*, **A332**, 527–548, 1973.

Rice, J. R. A path independent integral and the approximate analysis of strain concentration by cracks and notches. *Journal of Applied Mechanics*, **35**, 379–386, 1968.

PROBLEMS

15.1 a) If the slip across a fault surface Σ is known as a function of position and time, is this enough to determine completely the motions throughout the medium within which the fault is situated (assuming no other source is active)? If your answer is "yes," then explain why this result is only of limited use in earthquake source theory. If your answer is "no," then describe what else must be known about the source in order to determine the motions that it radiates.

b) Suppose that, instead of the slip, we know the *traction* at all times on the part of the fault surface that is undergoing slip (i.e., on $\Sigma(t)$). Is this enough to determine the motions radiated away from the fault? Comment on your answer here (yes or no) in the same fashion requested for (a) above.

15.2 For the semi-finite crack described as the example in Section 15.2.2, the stress-intensity factor K is given by

$$K = \frac{(1 - \dot{x}_2/\beta)^{1/2}}{\sqrt{\pi/2}} \int_{x_2 - \beta t}^{x_2} p(x, t - (x_2 - x)/\beta) \frac{dx}{\sqrt{x_2 - x}}$$

$$= \frac{2(1 - \dot{x}_2/\beta)^{1/2} \tau_0}{\sqrt{\pi/2}} \sqrt{\beta t}.$$

In that example, we derived the crack-tip motion assuming that the surface energy G is independent of rupture velocity. Show that, if the critical-intensity factor is constant instead of G, the crack-tip motion is described by

$$x_2(t) = \beta(t - t_c) - \beta t_c \ln t/t_c.$$

This curve is shown in Figure 15.21 together with the curves corresponding to constant G.

15.3 For an in-plane tensile crack, the rupture propagation always has velocities lower than the Rayleigh velocity, even in the case of finite cohesive force. Confirm this conclusion by investigating the sense of stress associated with the P- and S-wave part of the Green function appropriate for a tensile crack.

Bibliography

Abe, K.
    1975   Static and dynamic fault parameters of the Saitama earthquake of July 1, 1968. *Tectonophysics* 27:223–238.

Abramowitz, M., and I. A. Stegun
    1965   *Handbook of Mathematical Functions* (4th printing). U.S. Government Printing Office.

Aki, K.
    1960   Study of earthquake mechanism by a method of phase equalization applied to Rayleigh and Love waves. *Journal of Geophysical Research* 65:729–740.
    1961   Crustal structure in Japan from the phase velocity of Rayleigh waves. *Bulletin of the Earthquake Research Institute, Tokyo Univ.* 39:255–283.
    1967   Scaling law of seismic spectrum. *Journal of Geophysical Research* 72:1217–1231.
    1968   Seismic displacements near a fault. *Journal of Geophysical Research* 73:5359–5376.
    1973   Scattering of P waves under the Montana Lasa. *Journal of Geophysical Research* 78:1334–1346.
    1976   Signal to noise ratio in seismic measurements. *In* H. Aoki and S. Iizuka (editors), *Volcanoes and Tectonosphere*. Tokyo: Tokai University Press.
    1977   Three dimensional seismic velocity anomalies in the lithosphere. Method and summary of results. *Journal of Geophysics* 43:235–242.

Aki, K., A. Christoffersson, and E. S. Husebye
  1976   Determination of the three-dimensional seismic structures of the lithosphere. *Journal of Geophysical Research* 82:277–296.

Aki, K., and K. L. Larner
  1970   Surface motion of a layered medium having an irregular interface due to incident plane *SH* waves. *Journal of Geophysical Research* 75:933–954.

Aki, K., and W. H. K. Lee
  1976   Determination of three-dimensional velocity anomalies under a seismic array using first *P* arrival times from local earthquakes. 1. A homogeneous initial model. *Journal of Geophysical Research* 81:4381–4399.

Alford, R. M., K. R. Kelly, and D. M. Boore
  1974   Accuracy of finite-difference modeling of the acoustic wave equation. *Geophysics* 39:834–842.

Alsop, L. E.
  1966   Transmission and reflection of Love waves at a vertical discontinuity. *Journal of Geophysical Research* 71:3969–3984.

Alsop, L. E., G. H. Sutton, and M. Ewing
  1961a   Free oscillations of the Earth observed on strain and pendulum seismographs. *Journal of Geophysical Research* 66:631–641.

Alterman, Z. S., and F. C. Karal, Jr.
  1968   Propagation of elastic waves in layered media by finite difference methods. *Bulletin of the Seismological Society of America* 58:367–398.

Alterman, Z. S., and A. Rotenberg
  1969   Seismic waves in a quarter plane. *Bulletin of the Seismological Society of America* 59:347–368.

Ambuter, B. P., and S. C. Solomon
  1974   An event-recording system for monitoring small earthquakes. *Bulletin of the Seismological Society of America* 64:1181–1188.

Andersen, N. O.
  1974   On the calculation of filter coefficients for maximum entropy spectral analysis. *Geophysics* 39:69–72.

Anderson, J. G.
  1976   Motions near a shallow rupturing fault: Evaluation of effects due to the free surface. *Geophysical Journal of the Royal Astronomical Society* 46:575–593.

Anderson, J. G., and P. G. Richards
  1975   Comparison of strong ground motion from several dislocation models. *Geophysical Journal of the Royal Astronomical Society* 42:347–373.

Andrews, D. J.
  1975   From anti-moment to moment: Plane strain models of earthquakes that stop. *Bulletin of the Seismological Society of America* 65:163–182.
  1976   Rupture velocity of plane-strain shear cracks. *Journal of Geophysical Research* 81:5679–5687.

Ang, D. D., and M. L. Williams
   1959   The dynamic stress field due to an extensional dislocation. *Proceedings of the Fourth Midwestern Conference on Solid Mechanics.* Austin: Univ. Texas Press.

Archambeau, C. B.
   1968   General theory of elasto-dynamic source fields. *Review of Geophysics* 16:241–288.

Asada, T.
   1957   Observations of nearby microearthquakes with ultrasensitive seismometers. *Journal of the Physics of the Earth* 5:83–113.

Backus, G., and F. Gilbert
   1967   Numerical applications of a formalism for geophysical inverse problems. *Geophysical Journal of the Royal Astronomical Society* 13:247–276.
   1968   The resolving power of gross earth data. *Geophysical Journal of the Royal Astronomical Society* 16:169–205.
   1970   Uniqueness in the inversion of inaccurate gross Earth data. *Philosophical Transactions of the Royal Society of London* A266:123–192.

Barenblatt, G. I.
   1959   The formation of equilibrium cracks during brittle fracture. General ideas and hypotheses. *Journal of Applied Mathematics and Mechanics* 23:622–636.

Benioff, H., F. Press, and S. W. Smith
   1961   Excitation of the free oscillations of the earth by earthquakes. *Journal of Geophysical Research* 66:605–620.

Ben-Menahem, A.
   1961   Radiation of seismic surface waves from finite moving sources. *Bulletin of the Seismological Society of America* 51:401–435.

Ben-Menahem, A., and M. N. Toksöz
   1963   Source mechanism from spectra of long period surface waves. *Journal of Geophysical Research* 68:5207–5222.

Berckhemer, H.
   1962   Die ausdehnung der Bruchfläche im Erdbeben herd und ihr Einfluss auf das seismische Wellen spektrum. *Gerlands. Beitr. Geophys.* 71:5–26.

Bessonova, E. N., V. M. Fishman, V. Z. Ryaboyi, and G. A. Sitnikova
   1974   The tau method for inversion of travel times. I. Deep seismic sounding data. *Geophysical Journal of the Royal Astronomical Society* 36:377–398.

Blackman, R. B., and J. W. Tukey
   1958   *The Measurement of Power Spectra.* New York: Dover Publications Inc., p. 190.

Boatwright, J., and D. M. Boore
   1975   A simplification in the calculation of motions near a propagating dislocation. *Bulletin of the Seismological Society of America* 65:133–138.

Bôcher, M.
   1909   *An Introduction to the study of Integral Equations.* Cambridge University Press, p. 9.

Bogdanov, V. I., and V. M. Graizer
  1976   Determination of remaining ground displacement from seismograms. *Dokladi, Acad. Nauk, S.S.S.R.* 229:59–62

Boore, D. M.
  1970   Love waves in a non-uniform wave-guide: Finite difference calculation. *Journal of Geophysical Research* 75:1512–1527.
  1972   Finite difference methods for seismic wave propagation in heterogeneous materials. *In* B. A. Bolt (editor), *Seismology: Surface Waves and Earth Oscillations* (Methods in Computational Physics, Vol. 11). New York: Academic Press.

Boore, D. M., K. Larner, and K. Aki
  1971   Comparison of two independent methods for the solution of wave scattering problems: Response of a sedimentary basin to vertically incident SH waves. *Journal of Geophysical Research* 76:558–569.

Boore, D. M., and Zoback, M. D.
  1974   Near-field motions from kinematic models of propagating faults. *Bulletin of the Seismological Society of America* 64:321–342.

Bouchon, M.
  1973   Effect of topography on surface motion. *Bulletin of the Seismological Society of America* 63:615–632.
  1979   Predictability of ground displacement and velocity at proximity of an earthquake. An example: The Parkfield earthquake of 1966. *Journal of Geophysical Research*. In press.

Bouchon, M., and K. Aki
  1977a   Near-field of a seismic source in a layered medium with irregular interfaces. *Geophysical Journal of the Royal Astronomical Society* 50:669–684.
  1977b   Discrete wave-number representation of seismic-source wave fields. *Bulletin of the Seismological Society of America* 67:259–277.

Brune, J. N.
  1960   Radiation patterns of Rayleigh waves from the southeast Alaska earthquake of July 10, 1958. *Publications of the Dominion Observatory of Ottowa* 24(10): 373–383.
  1970   Tectonic stress and the spectra of seismic shear waves from earthquakes. *Journal of Geophysical Research* 75:4997–5009.

Brune, J. N., and J. Dorman
  1963   Seismic waves and Earth structure in the Canadian Shield. *Bulletin of the Seismological Society of America* 53:167–210.

Budiansky, B., and R. J. O'Connell
  1976   Elastic moduli of a cracked solid. *International Journal of Solid Structures* 12:81–97.

Buland, R.
  1976   The mechanics of locating earthquakes. *Bulletin of the Seismological Society of America* 66:173–187.

Burg, J. P.
    1964   Three-dimensional filtering with an array of seismometers. *Geophysics* 29:693–713.
    1967   Maximum entropy spectral analysis. Paper presented at the 37*th Annual International SEG Meeting*, Oklahoma City, Oklahoma, October 31, 1967.

Burridge, R.
    1969   The numerical solution of certain integral equations with non-integrable kernels arising in the theory of crack propagation and elastic wave diffraction. *Philosophical Transactions of the Royal Society of London* A265:353–381.
    1973   Admissible speeds for plane-strain self-similar shear cracks with friction but lacking cohesion. *Geophysical Journal of the Royal Astronomical Society* 35:439–456.

Burridge, R., and G. S. Halliday
    1971   Dynamic shear cracks with friction as models for shallow focus earthquakes. *Geophysical Journal of the Royal Astronomical Society* 25:261–283.

Burridge, R., and J. Willis
    1969   The self-similar problem of the expanding elliptical crack in an anisotropic solid. *Proceedings of the Cambridge Philosophical Society* 66:443–468.

Capon, J.
    1969   High-resolution frequency-wavenumber spectrum analysis. *Proceedings of the Institute of Electrical and Electronic Engineers* 57:1408–1418.
    1974   Characterization of crust and upper mantle structure under Lasa as a random medium. *Bulletin of the Seismological Society of America* 64:235–266.

Capon, J., R. J. Greenfield, R. J. Koller, and R. T. LaCoss
    1968   Short-period signal processing results for the large aperture seismic array. *Geophysics* 33:452–472.

Capon, J., R. J. Greenfield, and R. T. LaCoss
    1969   Long-period signal processing results for the large aperture seismic array. *Geophysics* 34:305–329.

Carotta, R., and D. Michon
    1967   Continuous analysis of the velocity function and of the move out corrections. *Geophysical Prospecting* 15:584–597.

Chernov, L. A.
    1960   *Wave Propagation in a Random Medium.* New York: McGraw-Hill.

Claerbout, J. F.
    1964   Detection of *P*-waves from weak sources at great distances. *Geophysics* 29:197–211.
    1968   Synthesis of a layered medium from its acoustic transmission response. *Geophysics* 33:264–269.
    1970   Coarse grid calculations of waves in inhomogeneous media with application to delineation of complicated seismic structure. *Geophysics* 35:407–418.
    1976   *Fundamentals of Geophysical Data Processing.* New York: McGraw-Hill.

Claerbout, J. F., and F. Muir
    1973   Robust modeling with erratic data. *Geophysics* 38:826–844.

Clowes, R. M., E. R. Kanasewich, and G. L. Cumming
    1968   Deep crustal seismic reflections at near-vertical incidence. *Geophysics* 33: 441–451.

Cooley, J. W., P. A. W. Lewis, and P. D. Welch
    1969   The fast Fourier transform and its applications. *Institute of Electrical and Electronic Engineers Transactions, Education* 12:27–34.

Cooley, J. W., and J. W. Tukey
    1965   An algorithm for the machine computation of complex Fourier series. *Math. Comput.* 19:297–301.

Crank, J., and P. Nicolson
    1947   A practical method for numerical evaluation of solutions of partial differential equations of the heat conduction type. *Proceedings of the Cambridge Philosophical Society* 43:50–67.

Crosson, R. S.
    1976a   Crustal structure modeling of earthquake data. 1. Simultaneous least square estimation of hypocenter and velocity parameters. *Journal of Geophysical Research* 81:3036–3046.
    1976b   Crustal structure modeling of earthquake data. 2. Velocity structure of the Puget Sound region, Washington. *Journal of Geophysical Research* 81: 3047–3054.

Dahlen, F. A.
    1969   The normal modes of a rotating, elliptical Earth. II. Near-resonance multiplet coupling. *Geophysical Journal of the Royal Astronomical Society* 18:397–436.
    1974   On the ratio of P-wave to S-wave corner frequencies for shallow earthquake sources. *Bulletin of the Seismological Society of America* 64:1159–1180.

Dainty, A. M., M. N. Toksöz, K. R. Anderson, P. J. Pines, Y. Nakamura, and G. Latham
    1974   Seismic scattering and shallow structure of the moon in Oceanus Procellarum. *The Moon* 9:11–29.

Dantzig G. B.
    1963   *Linear Programming and Extensions*. Princeton University Press.

Das, S., and K. Aki
    1977a   A numerical study of two-dimensional spontaneous rupture propagation. *Geophysical Journal of the Royal Astronomical Society* 50:643–668.
    1977b   Fault plane with barriers: A versatile earthquake model. *Journal of Geophysical Research* 82:5658–5670.

DeGolyer, E.
    1935   Notes on the early history of applied geophysics in the petroleum industry. *Journal of the Society of Petroleum Geophysicists* 6:1–10.

Der, Z., R. Massé, and M. Landisman
    1970   Effects of observational errors on the resolution of surface waves at intermediate distances. *Journal of Geophysical Research* 75:3399–3409.

Dewey, J. W.
1972   Seismicity and tectonics of Western Venezuela. *Bulletin of the Seismological Society of America* 62:1711–1751.

Dewey, J., and P. Byerly
1969   The early history of seismometry (to 1900). *Bulletin of the Seismological Society of America* 59:183–277.

Dieterich, J. H.
1973   A deterministic near-field source model. *Proceedings of the 5th World Conference of Earthquake Engineers*, Rome.

Dorman, J., M. Ewing, and J. Oliver
1960   Study of shear-velocity distribution in the upper mantle by mantle Rayleigh waves. *Bulletin of the Seismological Society of America* 50:87–115.

Dorman, J., and M. Ewing
1962   Numerical inversion of seismic surface wave dispersion data and crust-mantle structure in the New York–Pennsylvania area. *Journal of Geophysical Research* 67:5227–5241.

Douglas, A.
1967   Joint epicentre determination. *Nature* 215:47–48.

Dziewonski, A. M., and F. Gilbert
1976   The effect of small aspherical perturbations on travel-times and a re-examination of the corrections for ellipticity. *Geophysical Journal of the Royal Astronomical Society* 44:7–17.

Dziewonski, A. M., and A. L. Hales
1972   Numerical analysis of dispersed seismic waves. *In* B. A. Bolt (editor), *Seismology: Surface Waves and Earth Oscillations* (Methods in Computational Physics, Vol. 11). New York: Academic Press.

Edmonds, A. R.
1960   *Angular Momentum In Quantum Mechanics*. Princeton University Press.

Ellsworth, W. L.
1977   Three-dimensional structure of the crust and mantle beneath the Island of Hawaii. Ph.D. thesis, Massachusetts Institute of Technology, Cambridge.

Engdahl, E. R.
1973   Relocation of intermediate depth earthquakes in the central Aleutians by seismic ray tracing. *Nature, Physical Science* 245:23–25.

Engdahl, E. R., E. A. Flinn, and C. F. Romney
1970   Seismic waves reflected from the Earth's inner core. *Nature* 228:852–853.

Eshelby, J. D.
1949   Uniformly moving dislocations. *Proceedings of the Physical Society* A62:307–314.
1957   The determination of the elastic field of an ellipsoidal inclusion, and related problems. *Proceedings of the Royal Society of London* A241:376–396.

Evernden, J. F.
1967   Magnitude determination at regional and near-regional distances in the United States. *Bulletin of the Seismological Society of America* 57:591–639.

1969   Precision of epicenters obtained by small numbers of worldwide stations. *Bulletin of the Seismological Society of America* 59:1365–1398.

Ewing, J. A.
1881   The earthquake of March 8, 1881. *Transactions of the Seismological Society of Japan* 3:121–128.

Ewing, M., A. P. Crary, and H. N. Rutherford
1937   Geophysical investigations in the emerged and submerged Atlantic coastal plain, methods and results. *Bulletin of the Geological Society of America* 48:753–802.

Ewing, M., W. Jardetzky, and F. Press
1957   *Elastic Waves in Layered Media.* New York: McGraw-Hill, p. 380.

Ewing, M., J. L. Worzel, J. B. Hersey, F. Press, and G. R. Hamilton
1950   Seismic refraction measurements in the Atlantic Ocean Basin. *Bulletin of the Seismological Society of America* 40:233–242.

Faddeev, L. D.
1967   Properties of the *S*-matrix of the one-dimensional Schrödinger equation. *American Mathematical Society Translations Series* 2 65:139–166.

Fessenden, R.
1914   Patent application on sonic sounder, filed on April 2, 1914.
1917   Patent application on methods and apparatus for locating ore bodies, filed on January 15, 1917.

Filson, J., and T. V. McEvilly
1967   Love wave spectra and the mechanism of the 1966 Parkfield sequence. *Bulletin of the Seismological Society of America* 57:1245–1259.

Flinn, E. A.
1965   Confidence regions and error determinations for seismic event location. *Review of Geophysics* 3:157–185.

Frank, F. C.
1949   On the equations of motion of crystal dislocations. *Proceedings of the Physical Society* A62:131–134.

Franklin, J. N.
1970   Well-posed stochastic extensions of ill-posed linear problems. *Journal of Mathematical Analysis and Applications* 31:682–716.

Frasier, C. W.
1970   Discrete time solution of plane *P-SV* waves in a plane layered medium. *Geophysics* 35:197–219.

Freedman, H. W.
1967   A statistical discussion of *P* residuals from explosions. Part II. *Bulletin of the Seismological Society of America* 57:545–561.

Freund, L. B.
1972   Energy flux into the tip of an extending crack in an elastic solid. *Journal of Elasticity* 2:341–348.

Furuya, I.
1969   Predominant period and magnitude. *Journal of Physics of the Earth* 17:119–126.

Gakenheimer, D. C., and J. Miklowitz
    1969   Transient excitation of an elastic half-space by a point load travelling on
            the surface. *Journal of Applied Mechanics (Trans. ASME, Ser. E)* 36:
            505–515.

Geller, R. J.
    1976   Scaling relations for earthquake source parameters and magnitudes. *Bulletin
            of the Seismological Society of America* 66:1501–1523.

Gerver, M. L.
    1970   Inverse problem for the one-dimensional wave equation. *Geophysical
            Journal of the Royal Astronomical Society* 21:337–357.

Gerver, M. L., and V. Markushevitch
    1966   Determination of a seismic wave velocity from the travel time curve. *Geo-
            physical Journal of the Royal Astronomical Society* 11:165–173.

Gilbert, F.
    1971a   The diagonal sum rule and averaged eigenfrequencies. *Geophysical Journal
            of the Royal Astronomical Society* 23:119–123.
    1971b   Ranking and winnowing gross Earth data for inversion and resolution.
            Geophysical Journal of the Royal Astronomical Society 23:125–128.

Gilbert, F., and A. M. Dziewonski
    1975   An application of normal mode theory to the retrieval of structural param-
            eters and source mechanisms from seismic spectra. *Philosophical Transac-
            tions of the Royal Society of London* A278:187–269.

Gold, B., and C. M. Rader
    1969   *Digital Processing of Signals.* New York: McGraw-Hill.

Goupillaud, P. L.
    1961   An approach to inverse filtering of near-surface layer effects from seismic
            records. *Geophysics* 26:754–760.

Griffith, A. A.
    1920   The phenomena of rupture and flow in solids. *Philosophical Transactions
            of the Royal Society of London* 221:163.

Griggs, D. T., and F. Press
    1961   Probing the Earth with nuclear explosions. *Journal of Geophysical Research*
            66:237–258.

Gutenberg, B.
    1913   Über die Konstitution des Erdinnern, erschlossen aus Erdbebenbeobach-
            tungen. *Zeitschr. für Geophys.* 14:1217–1218.

Gutenberg, B., and C. F. Richter
    1954   *Seismicity of the Earth* (2nd ed.). Princeton University Press.
    1956   Magnitude and energy of earthquakes. *Annali di Geofisica* 9:1–15.

Haddon, R. A. W., and J. R. Cleary
    1974   Evidence for scattering of seismic PKP waves near the mantle-core bound-
            ary. *Physics of the Earth and Planetary Interiors* 8:211–234.

Hales, A. L.
    1972   The travel times of *P* seismic waves and their relevance to the upper mantle
            velocity distribution. *Tectonophysics* 13:447–482.

Hamano, Y.
1974 Dependence of rupture-time history on the heterogeneous distribution of stress and strength on the fault plane (abstract). *EOS, Transactions of the American Geophysical Union* 55:362.

Hanson, M. E., A. R. Sanford, and R. J. Shaffer
1971 A source function for a dynamic bilateral brittle shear failure. *Journal of Geophysical Research* 76:3375–3383.

Harkrider, D. G.
1964 Surface waves in multilayered elastic media 1. Rayleigh and Love waves from buried sources in a multilayered elastic half space. *Bulletin of the Seismological Society of America* 54:627–679.

Hartley, H. O.
1961 The modified Gauss-Newton method for fitting of non-linear regression functions by least squares. *Technometrics* 3:269–280.

Haskell, N. A.
1964 Radiation pattern of surface waves from point sources in a multi-layered medium. *Bulletin of the Seismological Society of America* 54:377–394.
1966 Total energy and energy spectral density of elastic wave radiation from propagating faults. II. *Bulletin of the Seismological Society of America* 56:125–140.
1969 Elastic displacements in the near-field of a propagating fault. *Bulletin of the Seismological Society of America* 59:865–908.

Healy, J. H.
1963 Crustal structure along the coast of California from seismic-refraction measurements. *Journal of Geophysical Research* 68:5777–5787.

Heaton, T. H. and D. V. Helmberger
1977 A study of the strong ground motion of the Borrego Mt., California, earthquake. *Bulletin of the Seismological Society of America* 67:315–330.

Helmberger, D. V.
1968 The crust-mantle transition in the Bering Sea. *Bulletin of the Seismological Society of America* 58:179–214.

Herglotz, G.
1907 Über das Benndorfsche Problem der Fortpflanzungsgeschwindigkeit der Erdbebenstrahlen. *Zeitschr. für Geophys.* 8:145–147.

Herrin, E.
1968 Seismological tables for *P* phases. *Bulletin of the Seismological Society of America* 58:1193–1242.
1969 Regional variation of *P*-wave velocity in the upper mantle beneath North America. *In* P. J. Hart (editor), *The Earth's Crust and Upper Mantle* American Geophysical Union Monograph (Vol. 13).

Herrin, E., and J. Taggart
1962 Regional variations in the P_n velocity and their effect on the location of epicenters. *Bulletin of the Seismological Society of America* 52:1037–1046.

Higuchi, S.
1932 On Love dispersion in a complicated surface layer (in Japanese). *Zisin*, Series 1, 4:271–276.

Hill, M. N.
    1963   Single-ship seismic refraction shooting in the sea. Ideas and observations on progress in the study of the seas. *The Earth Beneath the Sea* (Vol. 3). New York: Interscience.

Hirasawa, T., and W. Stauder
    1965   On the seismic body waves from a finite moving source. *Bulletin of the Seismological Society of America* 55:1811–1842.

Hodgson, J. H.
    1957   The null vector as a guide to regional tectonic patterns. *Publications of the Dominion Observatory of Ottawa* 20:369–384.

Honda, H.
    1962   Earthquake mechanism and seismic waves. *Journal of Physics of the Earth* 10(2):1–97.

Hong, T. -L., and D. V. Helmberger
    1977   Generalized ray theory for a dipping structure. *Bulletin of the Seismological Society of America* 67:995–1008

Hong, T. -L., and D. V. Helmberger
    1978   Glorified Optics and wave propagation in nonplanar structure. *Bulletin of the Seismological Society of America* 68:1313–1358.

Housner, G. W., and M. D. Trifunac
    1967   Analysis of accelerograms—Parkfield earthquake. *Bulletin of the Seismological Society of America* 57:1193–1220.

Husebye, E. S., A. Christoffersson, K. Aki, and C. Powell
    1976   Preliminary results on the 3-dimensional seismic structure of the lithosphere under the USGS Central California Seismic Array. *Geophysical Journal of the Royal Astronomical Society* 46:319–340.

Husseini, M. I., D. B. Jovanovich, M. J. Randall, and L. B. Freund
    1975   The fracture energy of earthquakes. *Geophysical Journal of the Royal Astronomical Society* 43:367–385.

Ida, Y.
    1972   Cohesive force across the tip of a longitudinal shear crack and Griffith's specific surface energy. *Journal of Geophysical Research* 77:3796–3805.

    1973   Stress concentration and unsteady propagation of longitudinal shear cracks. *Journal of Geophysical Research* 78:3418–3429.

Iyer, H. M., L. C. Pakiser, D. J. Stuart, and D. H. Warren
    1969   Project Early Rise: Seismic probing of the upper mantle. *Journal of Geophysical Research* 74:4409–4441.

Jackson, D. D.
    1972   Interpretation of inaccurate, insufficient, and inconsistent data. *Geophysical Journal of the Royal Astronomical Society* 28:97–109.

Jeffreys, H.
    1939   *Theory of Probability.* Oxford: Clarendon Press, p. 160.

Jeffreys, H., and K. E. Bullen
    1940   *Seismological tables.* British Association, Gray-Milne Trust.

Johnson, C.
1972   Regionalized Earth models from linear programming methods. Master's thesis, Massachusetts Institute of Technology, Cambridge.

Johnson, L. E., and F. Gilbert
1972   Inversion and inference for teleseimic ray data. *In* B. A. Bolt (editor), *Seismology: Body Waves and Sources* (Methods in Computational Physics, Vol. 12). New York: Academic Press.

Julian, B. R.
1970   Ray tracing in arbitrarily heterogeneous media. Technical Note 1970–45, Lincoln Laboratory, Massachusetts Institute of Technology, Cambridge.

Jordan, T. H.
1972   Estimation of the radial variation of seismic velocities and density in the Earth. Ph.D. thesis, California Institute of Technology, Pasadena.

Julian, B. R., and D. Gubbins
1977   Three-dimensional seismic ray tracing. *Journal of Geophysics* 43:95–113.

Kanai, K., and T. Tanaka
1961   On microtremors. VIII. *Bulletin of the Earthquake Research Institue*, Tokyo Univ. 39:97–114.

Kanamori, H.
1973   Mode of strain release associated with major earthquakes in Japan. *Annual Review, Earth and Planetary Sciences* 1:213–239.

Kanamori, H., and D. L. Anderson
1975   Theoretical basis of some empirical relations in seismology. *Bulletin of the Seismological Society of America* 65:1073–1095.

Kanasewich, E. R., and G. L. Cumming
1965   Near-vertical-incidence seismic reflections from the 'Conrad' discontinuity. *Journal of Geophysical Research* 70:3441–3446.

Kawasaki, I., Y. Suzuki, and R. Sato
1975   Seismic waves due to a shear fault in a semi-infinite medium. II. Moving source. *Journal of Physics of the Earth* 23:43–61.

Keilis-Borok, V. I., and T. B. Yanovskaya
1967   Inverse seismic problems (structural review). *Geophysical Journal of the Royal Astronomical Society* 13:223–233.

Kennett, B. L. N.
1972a   Seismic waves in laterally inhomogeneous media. *Geophysical Journal of the Royal Astronomical Society* 27:301–325.
1972b   Seismic wave scattering by obstacles on interfaces. *Geophysical Journal of the Royal Astronomical Society* 28:249–266.

Knopoff, L.
1958   Energy release in earthquakes. *Geophysical Journal of the Royal Astronomical Society* 1:44–52.
1972   Observation and inversion of surface wave dispersion. *Tectonophysics* 13:497–520.

Knopoff, L., M. J. Berry, and F. A. Schwab
  1967   Tripartite phase velocity observations in laterally heterogeneous regions. *Journal of Geophysical Research* 72:2595–2601.

Koehler, H., and M. T. Taner
  1977   The direct and inverse problems relating reflection coefficients and reflection response for horizontally layered media. *Geophysics* 42:1199–1206.

Kosminskaya, I. P., and Y. V. Riznichenko
  1964   Seismic studies of the Earth's crust in Eurasia. *In* H. Odishaw (editor), *Solid Earth and Interface Phenomena* (Research in Geophysics, Vol. 2) Cambridge: M.I.T. Press.

Kostrov, B. V.
  1966   Unsteady propagation of longitudinal shear cracks, *Journal of Applied Mathematics and Mechanics* 30:1241–1248.

  1975   On the crack propagation with variable velocity. *International Journal of Fracture* 11:47–56.

LaCoss, R. T.
  1971   Data adaptive spectral analysis methods. *Geophysics* 36:661–675.

LaCoss, R. T., E. J. Kelly, and M. N. Toksöz
  1969   Estimation of seismic noise structure using arrays. *Geophysics* 34:21–38.

Lamb, H.
  1904   On the propagation of tremors over the surface of an elastic solid. *Philosophical Transactions of the Royal Society of London* A203:1–42.

Lanczos, C.
  1961   *Linear Differential Operators*. London: Van Nostrand.

Landers, T., and J. F. Claerbout
  1972   Numerical calculations of elastic waves in laterally inhomogeneous media. *Journal of Geophysical Research* 77:1476–1482.

Landisman, M., A. Dziewonski, and Y. Satô
  1969   Recent improvements in the analysis of surface wave observations. *Geophysical Journal of the Royal Astronomical Society* 17:369–403.

Langston, C. A.
  1976   Body wave synthesis for shallow earthquake sources: Inversion for source and Earth structure parameters. Ph.D. thesis, California Institute of Technology, Pasadena.

  1977   The effect of planar dipping structure on source and receiver responses for constant ray parameter. *Bulletin of the Seismological Society of America* 67:1029–1050.

Larner, K. L.
  1970   Near-receiver scattering of teleseismic body waves in layered crust-mantle models having irregular interfaces. Ph.D. thesis, Massachusetts Institute of Technology, Cambridge.

Laster, S. J., and A. F. Linville
  1966   Application of multichannel filtering for the separation of dispersive modes of propagation. *Journal of Geophysical Research* 71:1669–1701.

Lee, W. H. K., and J. C. Lahr
    1972   HYPO 71: A computer program for determining hypocenter, magnitude, and first motion pattern of local earthquakes. *U.S. Geological Survey*, Open File Report.

Lentini, M., and V. Pereyra
    1977   An adaptive finite difference solver for nonlinear two-point boundary problems with mild boundary layers. *Journal of Numerical Analysis* 14:91–111.

Leontovich, M., and V. Fok
    1946   Solution of the problem of propagation of electromagnetic waves along the Earth's surface by the parabolic equation method. *Zh. Eksp. Teor. Fiz.* 16:557. (See also Chapter 11 of V. A. Fock, *Electromagnetic Diffraction and Propagation Problems*, New York: Pergamon Press, 1965.)

Levenberg, K.
    1944   A method for the solution of certain non-linear problems in least squares. *Quarterly of Applied Mathematics* 2:164–168.

Levinson, N.
    1949   The Wiener RMS error criterion in filter design and prediction. *Extrapolation, Interpolation, and Smoothing of Stationary Time Series with Engineering Applications* (Appendix B of N. Wiener). New York: John Wiley, pp. 129–148.

Liebfried, G., and H. D. Dietze
    1949   Zur theorie der Schraufenversetzung. *Zeitschr. für Geophys.* 126:790–808.

Lippman, B. A.
    1953   Note on the theory of gratings. *Journal of the Optical Society of America* 43:408.

Lowes, F. J.
    1971   A comment on statistical estimates of amplitude and phase corrections. *Geophysical Journal* 22:227–228.

Luh, P. C.
    1973   Free oscillations of the laterally inhomogenous Earth: Quasi-degenerate multiplet coupling. *Geophysical Journal of the Royal Astronomical Society* 32:203–218.

Luh, P. C., and A. M. Dziewonski
    1975   Theoretical seismograms for the Colombian earthquake of 1970 July 31. *Geophysical Journal of the Royal Astronomical Society* 43:679–695.

Lysmer, J., and L. A. Drake
    1972   A finite element method for seismology. *In* B. A. Bolt (editor), *Seismology: Surface Waves and Earth Oscillations* (Methods in Computational Physics, Vol. 11). New York: Academic Press.

McConnell, R. K., Jr., R. N. Gupta, and J. T. Wilson
    1966   Compilation of deep crustal structure seismic refraction profiles. *Review of Geophysics* 4:41–100.

Madariaga, R.
    1972   Toroidal free oscillations of the laterally heterogeneous Earth. *Geophysical Journal of the Royal Astronomical Society* 27:81–100.

1976   Dynamics of an expanding circular fault. *Bulletin of the Seismological Society of America* 66:639–666.

1978   The dynamic field of Haskell's rectangular dislocation fault model. *Bulletin of the Seismological Society of America* 68:869–888.

Madariaga, R., and K. Aki
1972   Spectral splitting of toroidal free oscillations due to lateral heterogeneity of the Earth's structure. *Journal of Geophysical Research* 77:4421–4431.

Mallet, R.
1848   On the dynamics of earthquakes. *Irish Academic Transactions* 21:50–106.

Marquardt, D. W.
1963   An algorithm for least squares estimation of non-linear parameters. *Journal of the Society of Industrial and Applied Mathematics* 11:431–441.

Mendiguren, J.
1972   Source mechanism of a deep earthquake from analysis of worldwide observations of free oscillations. Ph.D. thesis, Massachusetts Institute of Technology, Cambridge.

1973a   High resolution spectroscopy of the Earth's free oscillations, knowing the earthquake source mechanism. *Science* 179:179–180.

1973b   Identification of free oscillation spectral peaks for 1970 July 31, Colombian deep shock using the excitation criterion. *Geophysical Journal of the Royal Astronomical Society* 33:281–321.

Mikumo, T., and T. Miyatake
1978   Dynamical rupture process on a three-dimensional fault with non-uniform frictions and near-field seismic waves. *Geophysical Journal of the Royal Astronomical Society* 54:417–438.

Milne, J.
1881   Notes on the horizontal and vertical motion of the earthquake of March 8, 1881. *Transactions of the Seismological Society of Japan* 3:129–136.

Mitra, M.
1966   Surface displacement produced by an underground fracture. *Geophysics* 31:204–213.

Molnar, P., K. H. Jacob, and K. McCamy
1973   Implications of Archambeau's earthquake source theory for slip on faults. *Bulletin of the Seismological Society of America* 63:101–104.

Molnar, P., B. E. Tucker, and J. N. Brune
1973   Corner frequencies of P and S waves and models of earthquake sources. *Bulletin of the Seismological Society of America* 63:2091–2104.

Müller, G.
1973   Amplitude studies of core phases. *Journal of Geophysical Research* 78:3469–3490.

Ness, N. F., J. C. Harrison, and L. B. Slichter
1961   Observations of the free oscillations of the Earth. *Journal of Geophysical Research* 66:621–629.

Niazy, A.
  1973   Elastic displacements caused by a propagating crack in an infinite medium;
          an exact solution. *Bulletin of the Seismological Society of America* 63:
          357–379.
  1975   An exact solution for a finite, two-dimensional moving dislocation in an
          elastic half-space with application to the San Fernando earthquake of 1971.
          Bulletin of the Seismological Society of America 65:1797–1826.

Nolet, G.
  1975   Higher Rayleigh modes in Western Europe. *Geophysical Research Letters*
          2:60.
  1976   Higher modes and the determination of upper mantle structure. Thesis,
          Vening Meinesz Laboratory, Geophysics Department, State University of
          Utrecht, Utrecht, the Netherlands.

Odegard, M. E., and G. H. Sutton
  1972   The Cannikin airborne seismic experiment [abstract] Cordilleran Section
          Program. *Geological Society of America* p. 213.

Oldham, R. D.
  1900   On the propagation of earthquake motion to great distances. *Philosophical
          Transactions of the Royal Society of London* A194:135–174.

Oliver, J., and L. Murphy
  1971   WWNSS: seismology's global network for observing stations. *Science* 174:
          258.

Oliver, J., M. Dobrin, S. Kaufman, R. Meyer, and R. Phinney
  1976   Continuous seismic reflection profiling of the deep basement, Hardeman
          County, Texas. *Bulletin of the Geological Society of America* 87:1537–1546.

Ottaviani, M.
  1971   Elastic-wave propagation in two evenly-welded quarter-spaces. *Bulletin of
          the Seismological Society of America* 61:1119–1152.

Pakiser, L. C., and J. S. Steinhart
  1964   Explosion seismology in the western hemisphere. *In* H. Odishaw (editor),
          Research in Geophysics. Cambridge: M.I.T. Press.

Papoulis, A.
  1965   *Probability, Random Variables, and Stochastic Processes.* New York:
          McGraw-Hill.

Patton, H.
  1977   Source and propagation effects of Rayleigh waves from central Asian earth-
          quakes. Ph.D. thesis, Massachusetts Institute of Technology, Cambridge.

Pekeris, C. L.
  1947   Note on the scattering of radiation in an inhomogeneous medium. *Physical
          Review* 71:268.
  1948   Theory of propagation of explosive sound in shallow water. *Geological
          Society of America Memoirs* No. 27.

Phinney, R. A.
  1964   Structure of the Earth's crust from spectral behavior of long-period body
          waves. *Journal of Geophysical Research* 69:2997–3017.

Pisarenko, V. F.
    1970   Statistical estimates of amplitude and phase corrections. *Geophysical Journal* 20:89–98.

Press, F.
    1956   Determination of crustal structure from phase velocity of Rayleigh waves. Part 1. Southern California. *Bulletin of the Geological Society of America* 67:1647–1658.
    1968   Earth models obtained by Monte Carlo inversion. *Journal of Geophysical Research* 73:5223–5234.
    1970   Earth models consistent with geophysical data. *Physics of the Earth and Planetary Interiors* 3:3–22.

Raitt, R. W., G. G. Shor, Jr., T. J. G. Francis, and G. B. Morris
    1969   Anisotropy of the Pacific upper mantle. *Journal of Geophysical Research* 74:3095–3109.

Randall, M. J.
    1973   Spectral peaks and earthquake source dimensions. *Journal of Geophysical Research* 78:2609–2611.

Rautian, T. C., and V. I. Khalturin
    1978   The use of coda for determination of the earthquake spectrum. *Bulletin of the Seismological Society of America* 68:923–948.

Rebeur-Paschwitz, E. von
    1889   The earthquake of Tokio, April 18, 1889. *Nature* 40:294–295.

Richards, P. G.
    1973   The dynamic field of a growing plane elliptical shear crack. *International Journal of Solids and Structures* 9:843–861.
    1976   Dynamic motions near an earthquake fault: A three-dimensional solution. *Bulletin of the Seismological Society of America* 66:1–31.

Richter, C. F.
    1958   *Elementary Seismology.* San Francisco: W. H. Freeman and Company.

Robinson, E. A.
    1957   Predictive decomposition of seismic traces. *Geophysics* 22:767–778.
    1963   Mathematical development of discrete filters for the detection of nuclear explosions. *Journal of Geophysical Research* 68:5559–5568.
    1966   Multichannel z-transforms and minimum delay. *Geophysics* 31:482–500.

Roller, J. C., and W. H. Jackson
    1966   Seismic wave propagation in the upper mantle: Lake Superior, Wisconsin to Central Arizona. *Journal of Geophysical Research* 71:5933–5941.

Sabatier, P. C.
    1976   On geophysical inverse problems and constraints. *Journal of Geophysics* 43:115–138.

Sato, R.
    1975   Fast computation of theoretical seismograms for an infinite medium. Part I. Rectangular fault. *Journal of Physics of the Earth* 23:323–331.

Sato, T., and T. Hirasawa
    1973   Body wave spectra from propagating shear cracks. *Journal of Physics of the Earth* 21:415–431.

Satô, Y.
   1955   Analysis of dispersed surface waves. *Bulletin of the Earthquake Research Institute*, Tokyo Univ. 33:33–48.
   1958   Attenuation, dispersion and the waveguide of the G waves. *Bulletin of the Seismological Society of America* 48:231–251.

Savage, J. C.
   1966   Radiation from a realistic model of faulting. *Bulletin of the Seismological Society of America* 56:577–592.
   1972   Relation of corner frequency to fault dimensions. *Journal of Geophysical Research* 27:3788–3795.

Scheidegger, A. E., and P. L. Willmore
   1957   The use of a least squares method for the interpretation of data from seismic surveys. *Geophysics* 22:9–22.

Schneider, W. A.
   1971   Developments in seismic data processing and analysis (1968–1970). *Geophysics* 36:1043–1073.

Schneider, W. A., and M. M. Backus
   1968   Dynamic correlation analysis. *Geophysics* 33:105–126.

Schneider, W. A., K. L. Larner, J. P. Burg, and M. M. Backus
   1964   A new data-processing technique for the elimination of ghost arrivals on reflection seismograms. *Geophysics* 29:783–805.

Shannon, C. E.
   1949   *The Mathematical Theory of Communication.* University of Illinois Press.

Sheriff, R. E.
   1968   Glossary of terms used in geophysical exploration. *Geophysics* 33:183–228.
   1969   Addendum to "Glossary of terms used in geophysical exploration." *Geophysics* 34:255–270.

Shor, G. G., Jr., and R. W. Raitt
   1969   Explosion seismic refraction studies of the crust and upper mantle in the Pacific and Indian Ocean. *In* P. J. Hart (editor), *The Earth's Crust and Upper Mantle.* American Geophysical Union Monograph (Vol. 13).

Slichter, L. B.
   1932   The theory of the interpretation of seismic travel time curves in horizontal structures. *Physics* 3:273–295.

Smith, E. G. C.
   1976   Scaling the equations of condition to improve conditioning. *Bulletin of the Seismological Society of America* 66:2075–2081.
   1978   The statistical properties of least-squares hypocentre estimates—a reappraisal. Manuscript in preparation.

Smith, W. D.
   1974   A non-reflecting plane boundary for wave propagation problems. *Journal of Computational Physics* 15:492–503.

Solomon, S. C., and B. R. Julian
   1974   Seismic constraints on ocean-ridge mantle structure: Anomalous fault plane solutions from first motions. *Geophysical Journal of the Royal Astronomical Society* 38:265–285.

Springer, D. L., and R. L. Kinnaman
    1971   Seismic source summary for U.S. underground nuclear explosions, 1961–1970. *Bulletin of the Seismological Society of America* 61:1073–1098.

Starr, A. T.
    1928   Slip on a crystal and rupture in a solid due to shear. *Proceedings of the Cambridge Philosophical Society* 24:489–500.

Steinhart, J. S.
    1964   Lake Superior seismic experiment: Shots and travel times. *Journal of Geophysical Research* 69:5335–5352.

Swanger, H. J., and D. M. Boore
    1978   Simulation of strong-motion displacements using surface-wave modal superposition. *Bulletin of the Seismological Society of America* 68:907–922.

Sykes, L. R.
    1967   Mechanism of earthquakes and nature of faulting on the mid-oceanic ridges. *Journal of Geophysical Research* 72:2131–2153.

Takahashi, T.
    1955   Analysis of dispersion curves of Love-waves. *Bulletin of the Earthquake Research Institute*, Tokyo Univ. 33:287–296.

Takahasi, R., and K. Hirano
    1941   Seismic vibrations of soft ground. *Bulletin of the Earthquake Research Institute*, Tokyo Univ. 19:534–543.

Toksöz M. N., J. W. Minear, and B. R. Julian
    1971   Temperature field and geophysical effects of a downgoing slab. *Journal of Geophysical Research* 76:1113–1138.

Tsai, Y. B., and K. Aki
    1971   Amplitude spectra of surface waves from small earthquakes and underground nuclear explosions. *Journal of Geophysical Research* 76:3440–3452.

Tukey, J. W.
    1959   Equalization and pulse shaping techniques applied to the determination of initial sense of Rayleigh waves. *Report on a Panel of Seismic Improvement*, Appendix 9 (L. V. Berkner, Chairman).
    1965   Data analysis and the frontiers of geophysics. *Science* 148:1283–1289.

Turner, H. H.
    1922   On the arrival of earthquake waves at the antipodes, and the measurement of the focal depth of an earthquake. *Monthly Notice of the Royal Astronomical Society, Geophysical Supplement* 1:1–13.

Wadati, K.
    1928   Shallow and deep earthquakes. *Geophysical Magazine* (Tokyo) 1:162–202.

Walsh, J. B.
    1965   The effect of cracks on the compressibility of rock. *Journal of Geophysical Research* 70:381–389.

Ware, J. A., and K. Aki
    1969   Continuous and discrete inverse scattering problems in a stratified elastic medium. 1. Plane waves at normal incidence. *Journal of the Acoustical Society of America* 45:911–921.

Weidner, D. J.
    1972   Rayleigh waves from mid-ocean ridge earthquakes: Source and path effects. Ph.D. thesis, Massachusetts Institute of Technology, Cambridge.
    1974   Rayleigh wave phase velocities in the Atlantic Ocean. *Geophysical Journal of the Royal Astronomical Society* 36:105–139.

Weidner, D. J., and K. Aki
    1973   Focal depth and mechanism of mid-ocean ridge earthquakes. *Journal of Geophysical Research* 78:1818–1831.

Wesson, R. L.
    1971   Travel time inversion for laterally inhomogeneous crustal velocity models. *Bulletin of the Seismological Society of America* 61:729–746.

Wiechert, E.
    1910   Bestimmung des weges der Erdbebenwellen im Erdinnern. I. Theoretisches. *Phys. Z.* 11:294–304.

Wiener, N.
    1949   *Extrapolation, Interpolation, and Smoothing of Stationary Time Series with Engineering Applications.* New York: John Wiley.

Wiggins, R. A.
    1972   The general linear inverse problem: Implication of surface waves and free oscillations for Earth structure. *Review of Geophysics and Space Physics* 10:251–285.

Wiggins, R. A., and E. A. Robinson
    1965   Recursive solution to the multichannel filtering problem. *Journal of Geophysical Research* 70:1885–1891.

Wold, H.
    1938   *Stationary Time Series.* Stockholm: Almquist and Wiksell.

Yang, J. P., and W. H. K. Lee
    1976   Preliminary investigations of computational methods for solving the two-point seismic ray-tracing problem in a heterogeneous and isotropic medium. *U.S. Geological Survey, Open-file Report,* 76–707.

Zienkiewicz, O. C., and Y. K. Cheung
    1968   *The Finite Element Method in Structural and Continuum Mechanics.* New York: McGraw-Hill.

Index